HUSSERL OR FREGE?

Husserl or Frege?

Meaning, Objectivity, and Mathematics

Claire Ortiz Hill

and

Guillermo E. Rosado Haddock

OPEN COURT
Chicago and La Salle, Illinois

To order books from Open Court, call 1-800-815-2280.

Open Court Publishing Company is a division of Carus Publishing Company.

© 2000 by Carus Publishing Company

First printing 2000

Printed and bound in the United States of America.

Library of Congress Cataloging-in-Publication Data

Hill, Claire Ortiz.
    Husserl or Frege? : meaning, objectivity, and mathematics / Claire Ortiz Hill and Guillermo E. Rosado Haddock.
        p. cm.
    Includes bibliographical references (p.) and index.
    ISBN 0-8126-9417-1 (cloth : alk. paper)
        1. Husserl, Edmund, 1859-1938. 2. Frege, Gottlob, 1848-1925.
    3. Husserl, Edmund, 1859-1938—Contributions in mathematics. I. Rosado Haddock, Guillermo E., 1945-II. Title.

B3279.H94 H53 2000
193—dc21
                                                                        00-020279

Rosado Haddock dedica este libro a su madre Asia y a la memoria de su tía y segunda madre América. Ortiz Hill dedica este libro a Harriet Blume y a la memoria de Frank Blume, su esposo y compañero inseparable.

TABLE OF CONTENTS

ACKNOWLEDGMENTS

First of all, Professor Rosado Haddock and I must thank Tinna Stoyanova, his wife, for all her help in preparing the final manuscript. Professor Rosado Haddock also thanks Mr. Joel Donato, Director of the LABCAD of the University of Puerto Rico at Río Piedras for some technical help and advice. The sculpture on the cover of the book is the work of Paris artist Jacqueline Wegmann. We are grateful to her for allowing us to enhance our work with hers. The photograph is courtesy of Jean-René Teilhac. Many thanks are also due to the booksellers of Chemin des philosophes, 1 rue des Feuillantines, Paris 75005, which specializes in rare and out-of-print books. They are very erudite and have helped in many ways.

In addition, we wish to thank the following editors and publishers for granting us the permission to republish the various essays. The editor of *Kant-Studien* has granted us permission to republish Professor Rosado Haddock's paper entitled "Remarks on Sense and Reference in Frege and Husserl," which first appeared in December of 1982 in volume 73, 4, pp. 425–39 of that journal. Professor Rosado Haddock's article on "Identity Statements in the Semantics of Sense and Reference" originally appeared in 1982 in volume 25, 100, pp. 399–411 of *Logique et Analyse* and has been republished here with the kind permission of Jean Paul Van Bendegem. Taylor & Francis Group has granted us permission to reproduce "On Frege's Two Notions of Sense" by Professor Rosado Haddock, which was originally published in 1986 in volume 7, 1, pp. 31–41, of *History and Philosophy of Logic*. Yale University Press has kindly granted us the permission to reprint material taken from several chapters of my book *Rethinking Identity and Metaphysics*, which they published in 1997 and for which they hold the copyright. This particular material was subsequently published as "The Varied Sorrows of Logical Abstraction" in volume 8, nos. 1–3, pp. 53–82 of the journal *Axiomathes*, dated 1997. Roberto Poli, the editor of *Axiomathes*, has granted us permission to republish it in that form. Barry Smith, the editor of *The Monist*, has kindly granted us permission to republish "Frege's Attack on Husserl and Cantor," which appeared in July 1994, volume 77, no. 3, pp. 345–57 of that journal. Fred van der Zee of Editions Rodopi has given us permission to reprint my "Abstraction and Idealization in Edmund Husserl and Georg Cantor Prior to 1895" written for *Poznan Studies in the Philosophy of the Sciences and the Humanities* for a volume on abstraction and idealization. The journal *Diálogos* has kindly granted us permission to reprint Professor Rosado Haddock's "Interderivability of Seemingly Unrelated Mathematical Statements and the Philosophy of Mathematics," which appeared in 1992 in volume 59, pp. 121–34 of that journal, and his "On Antiplatonism and its Dogmas," which appeared there in 1996 in volume 67, pp. 7–38. My "Husserl and Frege on Substitutivity" (originally published in 1994 in *Mind, Meaning and Mathematics*, L. Haaparanta, ed., pp. 113–40), "Husserl and Hilbert on Completeness" (originally published in 1995 in *From Dedekind to Gödel*, J. Hintikka, ed., pp. 143–63) , and "Did Georg Cantor Influence Edmund Husserl?" (originally published in 1997 in volume 113 of *Synthese*, pp. 145–70) and Professor Rosado Haddock's "On Husserl's Distinction Between State of Affairs (*Sacherverhalt*) and Situation of Affairs (*Sachlage*)" (originally published in 1991 in *Phenomenology and the Formal Sciences*, T. Seebohm et al., eds., pp. 35–48) and "Husserl's Epistemology of Mathematics and the Foundation of Platonism in Mathematics" (originally published in 1987 in volume 4, 2 of *Husserl Studies*, pp. 81–102) have been reprinted here with kind permission from Kluwer Academic Publishers. "To be a Fregean or to be a Husserlian: That is the Question for Platonists" was published just recently in 1999 by the American Mathematical Society in *Advances in Contemporary Logic and Computer Science*, edited by Walter Carnielli and Itala D'Ottaviano.

Claire Ortiz Hill

Claire Ortiz Hill

INTRODUCTION

Philosophy of logic and mathematics has been the bailiwick of the analytic school of philosophy, many of whose basic tenets are antithetical to those of the phenomenological school founded by Edmund Husserl. This fact, along with unfortunately many other factors, has conspired to keep Husserl's ideas on meaning, objectivity, logic, and mathematics from penetrating very deeply, or very well, mainstream discussions of the subject in the United States, Britain, and other English-speaking countries, where analytic philosophy especially developed and flourished during the twentieth century.

Yet Husserl was the student and assistant (1878–1881, 1883) of the great mathematician Karl Weierstrass and for fifteen years (1886–1901) he was the colleague and close friend of Georg Cantor, the creator of set theory. Husserl then spent as many years (1901–1916) in David Hilbert's circle in Göttingen. Thus, Husserl was in close professional contact over extended periods of time (almost continuously between the ages of 19 and 57!) with the greatest and most influential German mathematicians of his time, men with whom Bertrand Russell, Gottlob Frege, and their followers rarely, if ever, came into contact. Husserl tangled long and hard with the very ideas that went into the making of analytic philosophy, and the problems he struggled to solve are still under discussion now.

The principal goal of this collection of papers is to work to integrate Husserl's thought into philosophical discussions in which it rightfully belongs by establishing the legitimate ties between his ideas and those of philosophers and mathematicians who have been more readily accepted into the pantheon reserved for those deemed to have made significant contributions to the field.

Before tackling this, however, it is important to name some of the many frustrating obstacles that have stood in the way of any accurate assessment of Husserl's contribution to the field. It is also important to review the history of the attempts that have been made to integrate Husserl's ideas into mainstream philosophy of logic and mathematics.

Two devastating wars on the European continent that ultimately left Europe (Germany and the former Austro-Hungarian Empire in particular) divided into two opposing ideological camps obviously played no small role in clouding the picture. Intellectual ties were severed and obscured. Many voices were stifled or altogether silenced. Valuable documents were lost or destroyed.

The very same political events also uprooted many German-speaking intellectuals and transplanted them to English-speaking countries. This grafting of continental ideas onto American and British philosophy fostered the dissemination and rich cross-fertilization of ideas, but it also cut those ideas off from their original roots and context.

In conjunction with this, key terms of the German language common to so many of the thinkers were translated in disparate and often misleading ways. In particular, language that Husserl used in common with those of his German-speaking contemporaries who actually had a hand in shaping mainstream twentieth-century philosophy of logic and mathematics made its way into English in a sometimes unrecognizable and frequently indigestible form. This often distorted the issues and made it exceedingly difficult to make sense of what Husserl wanted to say, thus doing its share to break down the transmission of his ideas to the English-speaking world. So, when interested, analytic philosophers, who have generally been loath to learn foreign languages, have only too often found themselves facing a daunting number of linguistic barriers.

In addition, phenomenologists themselves have been loath to study Husserl's ideas about mathematics and logic, which they have considered to be the product of an early, immature, pre-phenomenological period of his philosophical career. To do so they would have had to acquire the expertise necessary to manipulate ideas that they have not usually found interesting to them in order to communicate with philosophers whose general orientation they do not share. One consequence of this lack of priority Husserlians have been willing to accord to Husserl's work in this field has been that texts indispensable to piecing together his views (his unpublished writings and lectures for example) were for a long time only available in archives in the Austrian shorthand that Husserl used. Needless to say, they are for the most part still not available in English. So analytic philosophers looking into Husserl's writings still mainly find a mass of writings on precisely the kind of views they oppose, expressed in language that they could only find repulsive.

Besides, Husserl himself was unfortunately not always as explicit as one would like when it came to spelling out the precise nature of the links binding his ideas to those of his contemporaries. He rarely named names and when he did he seemed to believe that connections were obvious that are all but invisible to us nowadays. For example, in a note to *Ideas* §72, he wrote that the close relation between his own concept of definiteness and the axiom of completeness introduced by Hilbert in his foundations of arithmetic would be apparent to every mathematician without further remark, a remark that I have tried to elucidate in chapter 10.

So many such factors conspired to obscure Husserl's contribution to twentieth-century philosophy of logic and mathematics that the necessary facts may have never come to the fore but for interest in the subject aroused by Dagfinn Føllesdal, a disciple of the pre-eminent analytic philosopher Willard van Orman Quine. As a student at the University of Oslo in the mid-fifties, Føllesdal wrote a master's thesis on Frege and the origins of the phenomenological movement, in which he hypothesized that Frege's 1894 review of Husserl's *Philosophy of Arithmetic* might have had a crucial impact on Husserl's thought at a critical point in his career. In addition, work that Føllesdal began publishing in the late sixties about similarities existing between Husserl's and Frege's theories of meaning attracted a great deal of interest. A 1983 bibliography prepared by Ethel Kersey inventoried no less than 104 English language items on the subject. Indeed, Føllesdal's work appeared at a time when philosophers were actively debating such theories of meaning and growing increasingly enthusiastic about Frege's contributions to their field. Føllesdal's hypotheses also surely played a significant role in stimulating research into the Austro-German origins of analytic philosophy.

In 1973, Guillermo E. Rosado Haddock, the co-author of the present work, argued in his "Edmund Husserls Philosophie der Logik und Mathematik im Lichte der gegenwärtigen Logik und Grundlagenforschung" that Frege had not had any major impact on Husserl's thought and that Føllesdal's theses were untenable. The following year, J.N. Mohanty published "Husserl and Frege: A New Look at their Relationship." In it, he argued at length that the basic change in Husserl's mode of thinking that could in itself have led to his explicit rejection of psychologism and to his advocacy of a theory of logic as a science of objective meanings had taken place before Frege published his review.

In my 1979 Sorbonne master's thesis on Husserl, Frege, and Jaakko Hintikka's Possible World Semantics, I advanced additional arguments to show that Frege had not influenced Husserl in the ways and to the extent that it was then fashionable to believe that he had. In spite of a wealth of objective evidence in favor of this thesis, much of it only available in German, analytic philosophers still found it congenial to believe the contrary. So such arguments still fell on deaf ears.

A new era in the study of Husserl's philosophy of logic and mathematics began opening up in 1982. In that year Mohanty published his book *Husserl and Frege*. In it he endeavored to show the extent to which phenomenology and analytic philosophy had grown out of common problems and concerns and often from a common set of distinctions and theses. He argued that despite the fact that there had been virtually no fruitful communication between Husserl's and Frege's followers, distinct conceptual routes linking phenomenology and analytic philosophy did indeed exist.

The same year, Professor Rosado Haddock's "Remarks on Sense and Reference in Frege and Husserl" (chapter 2 of this book) appeared in *Kant-Studien,* more than three years after it had been submitted. In it he concluded that the theory that Frege's influence on Husserl had both turned him away from psychologism and taught him to distinguish between sense and reference was completely unfounded.

Now analytic philosophers began to listen to arguments that Frege's influence on Husserl had been exaggerated. This might have put an end to any further inquiry. Some found it reassuring to believe that Frege had not influenced Husserl and considered that the matter should then be dropped. Indeed, if Husserl had not been influenced by Frege in any significant way, then why should his ideas be of interest to specialists in the field Frege had opened up? It was unfortunate that Husserl had not listened to Frege. He had had his chance and missed it. That line of reasoning was congenial in its way too, and many may have hoped that that was the proper attitude to adopt.

Those knowledgeable about Husserl's unique contribution, however, now saw the door finally opened to investigations into what Husserl's views truly were and how they truly connected up with mainstream philosophy of logic and mathematics. The field could now get off to a fair start.

The two authors of the present book fall into that last category of philosophers who saw the ground finally prepared for a fair assessment of Husserl's contribution to their field. The papers published here contain a very interesting progression and intertwining of ideas. They explore what for a long time has been terra incognita, territory that largely went unexplored, and even undetected, because of the misconceptions and obstacles (to which many more could be added) cited above.

However, the papers anthologized here are not merely expositions and assessments of neglected parts of Husserl's writings. A few of them are not primarily concerned with Husserl, but are actually the development of ideas barely sketched by him or not even present in his writings at all. But even those papers that may seem less dependent on Husserl's writings are "in the spirit" of the unknown Husserl that the rest of the papers are endeavoring to expose and critically assess. They serve to buttress both positions he took and the major arguments of this book.

1

Claire Ortiz Hill

HUSSERL AND FREGE ON SUBSTITUTIVITY

In the critical discussion of Gottlob Frege's logic in Edmund Husserl's *Philosophy of Arithmetic*,[1] Husserl outlines his objections to the use Frege makes of Leibniz's principle of the substitutivity of identicals in the *Foundations of Arithmetic*.[2] In the 1903 appendix to *Basic Laws II*,[3] Frege linked these same criticisms with Russell's paradox when, without mentioning Husserl's name, he traced the source of the paradox to points Husserl had made in the *Philosophy of Arithmetic*. For many philosophical, linguistic, and historical reasons[4] these two facts have gone virtually uncommented on. In the belief that Husserl's discussion of identity and substitutivity in Frege's theory of number may actually be able to shed light on some dark areas surrounding the significance of Russell's paradox for logic and epistemology I propose here to examine Husserl's criticisms and systematically tie his arguments in with observations made by Bertrand Russell and others who have studied Frege's work.

First, however, I must preface my discussion with a short historical digression aimed at showing how Husserl fit into Frege's intellectual world and his competency to deal with Frege's ideas. This is necessary because Husserl is not generally thought of as having been someone who could have understood Frege's work in 1891. Louis Couturat,[5] Alonzo Church,[6] and Dallas Willard[7] are among the very few people who seem to have noticed that Husserl wrote anything worthwhile or insightful at all about Frege's logic in the *Philosophy of Arithmetic*. Husserl is most often wrongly thought of as having been a kind of intellectual infant when he wrote it,[8] and for a long time it was thought that his intellectual awakening only began in 1894 with Frege's bitter review[9] of the book.[10]

The many people who still underrate Husserl's ability in 1891 to publish an insightful work concerning the philosophy of arithmetic are not, however, in possession of the facts, for during the years in which his philosophical ideas were developing, Husserl actually had the unusual privilege of directly participating in the early development of the very mathematical, logical, and philosophical ideas that would go on to determine the course of philosophical thought in English-speaking countries in the twentieth century. He was, in fact, directly and intimately involved in the earliest discussions of such pivotal issues in twentieth–century logic, mathematics, and analytic philosophy as number theory, the continuum problem, set theory, the axiomatic foundations of geometry, Russell's paradox, infinity, function theory, intentionality, intensionality, analyticity, identity, sense and reference, and completeness, all of which are philosophical issues which still, a hundred years later, present thorny problems for philosophers, filling the pages of the journals and books they read. Husserl's ideas now need to be knit back into the intellectual context that produced them.

1. Weierstrass, Brentano, Stumpf, and Cantor

It was Karl Weierstrass's courses on the theory of functions that, in the late 1870s and early 1880s, first awakened Edmund Husserl's interest in seeking radical foundations for mathematics. Husserl was impressed by his teacher's emphasis on clarity and logical stringency.[11] He was receptive to Weierstrass's efforts to further the work begun by Bernard Bolzano to instill rigor in mathematical analysis[12] and to transform the "mixture of reason and irrational instincts" it then was into a purely rational discipline. Weierstrass exercised a deep influence on Husserl and in 1883 Husserl became his assistant. It was from Weierstrass, Husserl once said, that he acquired the ethos of his scientific striving.[13]

After serving as Weierstrass's assistant for a year, Husserl traveled to Austria to study under Franz Brentano. Like Weierstrass, Brentano was working on Bolzano's ideas,[14] and under Brentano, Husserl studied Bolzano's writings and the *Paradoxes of the Infinite* in particular.[15] Brentano was then engaged in reforming logic[16] and was vigorously trying to revise old traditions, paying particular attention to matters of linguistic expression.[17] He was influenced by British empiricism[18] and Michael Dummett, for one, considers him to have been "roughly comparable to Russell and Moore" in England.[19] Russell himself actually explicitly acknowledged the kinship between his own ideas and those of Brentano, and there was enough superficial kinship between Russell's views on reference and those of Brentano's school for Russell to have at one point confused his ideas and theirs.[20]

Brentano sent Husserl to Halle to prepare his *Habilitationsschrift* on number theory under the direction of Carl Stumpf,[21] a man to whom Frege had appealed in 1882 for help in making his *Begriffsschrift* known. In his reply to Frege's request, Stumpf had mentioned how pleased he was that Frege was working on logical problems because it was an area where there was a great need for cooperation between mathematicians and philosophers. He agreed with Frege that arithmetical and algebraic judgments were analytic and expressed his own interest in working on that problem. He also suggested that Frege's ideas might be more favorably received if he first explained them in ordinary language. Frege appears to have taken Stumpf's advice by expressing his ideas in prose in the 1884 *Foundations of Arithmetic*.[22] Husserl began studying Frege's *Foundations* as Stumpf's colleague in the late 1880s, and he used it extensively as he worked on the *Philosophy of Arithmetic*.[23]

In Halle, Husserl befriended another man who, like him, had been profoundly influenced by Karl Weierstrass. This was the creator of the theory of sets, Georg Cantor.[24] Cantor, too, was carrying on the work Bolzano had begun,[25] and enough kinship is apparent between Husserl's and Cantor's work to have prompted scholars to speak of the influence Husserl may have had on Cantor's work,[26] and of Cantor's influence on Husserl's work.[27] Enough of a kinship exists between Frege's and Cantor's work to have prompted Michael Dummett to speak of Georg Cantor as "the mathematician whose pioneering work was closest to that of Frege . . ."[28] and as one "who ought, of all philosophers and mathematicians, to have been the most sympathetic" to Frege's work.[29]

Russell thought his own debt to Cantor was evident. In Russell's opinion Cantor had "conquered for the intellect a new and vast province which had been given over to Chaos and Night."[30] "In arithmetic and theory of series, our whole work is based on that of Georg Cantor," he wrote in the preface to *Principia Mathematica*.[31] And it was while studying Cantor's work that Russell found the paradoxes to which Frege's logic leads.[32] In *The Principles of Mathematics*, Russell pays homage to Weierstrass for the happy changes he, Dedekind, Cantor, and their followers had wrought in mathematics by adding "quite immeasurably to theoretical correctness" and thereby remedying "a diminution of logical precision and a loss in subtlety of distinction."[33] "No mathematical subject," he wrote there, "has made, in recent years, greater advances than the theory of Arithmetic. The movement in favour of correctness in deduction, inaugurated by Weierstrass, has been brilliantly continued by Dedekind, Cantor, Frege and Peano. . . ."[34] Through the labors of Weierstrass and Cantor, the fundamental problem of infinity and continuity had undergone a complete transformation, Russell considered.[35]

Husserl had actually rather fortuitously found himself in the right place at the right time, and by the time he published the *Philosophy of Arithmetic*, he had long been involved in philosophical investigations into the principles of mathematics and in the work to obtain greater precision in mathematics that ultimately made extensive formalization of mathematics possible and led to comprehensive formal systems like that of Russell's *Principia Mathematica*.[36]

In contrast, Frege remained aloof and apparently loath to undertake even the short train journey that would have taken him to Göttingen, Leipzig, or

Halle and the likes of Cantor, Zermelo, or Hilbert, or a bit farther to Berlin, Paris, Austria, or Cambridge where he could have met with Weierstrass, Brentano, Peano, or Russell.[37] And Frege actually devoted several sections of the second volume of the *Basic Laws of Arithmetic* to refuting Cantor's and Weierstrass's views.[38]

Russell didn't learn of Weierstrass's and Cantor's work until the mid-1890s and he first came into contact with Frege's work several years later than Husserl, Hilbert, Cantor, or Brentano's circle did. Most of these people already knew Frege's work in the 1880s. Russell was too young and too far-away actually to interact with the imposing figures whom Husserl regularly frequented over long periods.[39] So more than Frege, Russell, or Wittgenstein, Husserl was actually present and witnessed the very earliest stages of twentieth–century Anglo-American philosophy, and the *Philosophy of Arithmetic* was written under the influence of the same mathematicians and philosophers that ultimately played such a key role in determining the course of philosophy in English-speaking countries.

2. Husserl's Encounter with Frege's *Foundations of Arithmetic*

Husserl first obtained a copy of Frege's *Foundations of Arithmetic* in the late 1880s.[40] Though he did not use Frege's book at all in his *Habilitationsschrift* called "On the Concept of Number"[41] that he defended before Cantor and Stumpf at the University of Halle in 1887,[42] he thoroughly studied Frege's book in the *Philosophy of Arithmetic* published four years later. There he cites Frege more often than any other author. In a letter Frege himself once acknowledged Husserl's interest in his *Foundations*, noting that Husserl's study was perhaps the most thorough one that had been done up to that time. Husserl replied to Frege saying how much stimulation he had derived from Frege's work and acknowledged having "derived constant pleasure from the originality of mind, clarity . . . and honesty" of Frege's investigations which, he wrote, "nowhere stretch a point or hold back a doubt, to which all vagueness in thought and word is alien, and which everywhere try to penetrate to the ultimate foundations." While writing the *Philosophy of Arithmetic*, no other book, Husserl claimed, had provided him with nearly as much enjoyment as Frege's remarkable work had.[43]

Much of what Husserl had written about Frege's ideas in the *Philosophy of Arithmetic* was, though, critical and in the *Logical Investigations* Husserl would make a point of retracting certain of the objections he had voiced concerning Frege's views on analyticity and his opposition to psychologism there. A close look at Husserl's statement of retraction, however, shows that he only retracted three pages of his criticisms of Frege's logic (not eight as a typographical error in the English edition suggests), leaving most of his basic criticisms of Frege's logical project intact.[44] For instance, Husserl never retracted his statements that theories of number like Frege's are unjustified and scientifically useless, that all Frege's definitions become true and correct propositions when one substitutes extensions of concepts for the concepts, but that then they are absolutely self-evident and without value, and that the results of Frege's endeavors are such as to make one wonder how anyone could believe they were true other than temporarily.[45]

3. Substitutivity in Frege's *Foundations of Arithmetic*

"Now, it is actually the case that in universal substitutability all the laws of identity are contained," wrote Frege in §65 of *Foundations*. And in the brief summary of his views Frege offers in the last pages of that book, he repeats this conviction that: ". . . a certain condition has to be satisfied, namely that it must be possible in every judgement to substitute without loss of truth the right-hand side of our putative identity for the left side. Now at the outset, and until we bring in further definitions, we do not know of any other assertion concerning either side of such an identity except the one, that they are identical. We had only to show, therefore, that the substitution is possible in an identity" (§107). It is evident from this that substitutivity was destined to play a very central role in Frege's theories, so it is very important to examine the arguments of *Foundations* so as to understand exactly how substitutivity operated in Frege's philosophy of arithmetic.

Frege's *Foundations of Arithmetic* is divided into five parts, the first three of which are largely devoted to the refutation of views of number which Frege opposes. In part four he outlines his own theory and in part five he summarizes the results of his work. Frege begins outlining his own theories by affirming that numbers are independent objects (§55) which figure as such in identity statements like '1 + 1 = 2'. Though in everyday discourse numbers are often used as adjectives rather than as nouns, in arithmetic, he argues, their independent status is apparent at every turn and any appearance to the contrary "can always be got around," for example by rewriting the statement 'Jupiter has four moons' as 'the number of Jupiter's moons is four'. In the new version, Frege argues, the word 'is' is not the copula, but the 'is' of identity and means "is identical with" or "is the same as." "So," he concludes, ". . . what we have is an identity, stating that the expression 'the number of Jupiter's moons' signifies the same object as the word 'four'." Using the same reasoning he concludes that Columbus is identical with the discoverer of America for "it is the same man that we call Columbus and the discoverer of America" (§57). (Note that Frege here, as always, quite perspicuously distinguishes between words and objects. In his identity statements he is asserting the sameness of one object as given in two different ways by different linguistic expressions.)

Now that Frege is satisfied that he has established numbers as independent objects and acquired a set of meaningful statements in which a number is recognized as the same again, he turns to the question of establishing a criterion for deciding in all cases whether *b* is the same as *a*. For him this means defining the sense of the statement: 'the number which belongs to the concept F is the same as that which belongs to the concept G'. This, he believes, will provide a general criterion for the identity of numbers (§63).

Not wanting to introduce a special definition of identity for this, but wishing rather "to use the concept of identity, taken as already known, as a means for arriving at that which is to be regarded as being identical," Frege explicitly adopts Leibniz's principle that "things are the same as each other, of which one can be substituted for the other without loss of truth" (§65). However, even as he is writing Leibniz's formula right into the foundations of his logic, Frege modifies Leibniz's dictum in a way which, as I hope to show, has presented thorny problems for those who have tried to further Frege's insights and answer some of the really hard questions his logic raises. Although, as Husserl would point out in the *Philosophy of Arithmetic*, Leibniz's law defines identity, complete coincidence, Frege, here as elsewhere, [46] explicitly maintains that for him "whether we use 'the same' as Leibniz does, or 'equal' is not of any importance. 'The same' may indeed be thought to refer to complete agreement in all respects, 'equal' only to agreement in this respect or that" (§65).[47]

Frege believed that by rewriting the sentences of ordinary language, these differences between equality and identity could be made to vanish. So here he recommends rewriting the sentence 'the segments are equal in length' as 'the length of the segments are equal or the same' and 'the surfaces are identical in color' as 'the color of the surfaces is identical'. Since he believed all the laws of identity were contained in universal substitutivity, to justify his definition he believed he only needed "to show that it is possible, if line *a* is parallel to line *b*, to substitute 'the direction of line *b*' everywhere for 'the direction of line *a*'. This task is made simpler," he notes, "by the fact that we are being taken initially to know of nothing that can be asserted about the direction of a line except the one thing, that it coincides with the direction of some other line. We should thus have to show only that substitution was possible in an identity of this one type, or in judgement-contents containing such identities as constituent elements" (§65).

In these examples he has transformed statements about objects which are equal under a certain description into statements expressing complete identity. By erasing the difference between identity and equality he in fact is arguing that being the same in any one way is equivalent to being the same in all ways. While conceivably one could use this principle to stipulate substitution conditions for symbols, very few objects could satisfy its conditions and, outside of strictly mathematical contexts where differences between equality and identity seem not to apply in the same way as they do elsewhere, many of the inferences that could be made by appealing to such a principle would lead to evidently false and absurd conclusions.

Frege himself acknowledged that left unmodified this procedure was liable to produce nonsensical conclusions, or be sterile and unproductive. For him, the source of the nonsense lay in the fact that, as he himself points out, his definition provides no way of deciding whether, for example, England is or is not the same as the direction of the Earth's axis. Though he is certain that no one would be inclined to confuse England with the direction of the Earth's axis, this, he acknowledges, would not be owing to his definition which, he notes, "says nothing as to whether the proposition 'the direction of a is identical with q' should be affirmed or denied except for the one case where q is given in the form of 'the direction of b'" (§66).

As it stood, the definition was unproductive, according to him, because were we "to adopt this way out, we should have to be presupposing that an object can only be given in one single way. . . . All identities would then amount simply to this, that whatever is given to us in the same way is to be reckoned as the same. . . . We could not, in fact, draw from it any conclusion which was not the same as one of our premisses." Surely, he concludes, identities play such an important role in so many fields "because we are able to recognize something as the same again even although it is given in a different way" (§67; also §107).

Seeing that he could not by these methods alone "obtain any concept of direction with sharp limits to its application, nor therefore, for the same reasons, any satisfactory concept of Number either," Frege felt obliged to introduce extensions to guarantee that "if line a is parallel to line b, then the extension of the concept 'line parallel to line a' is identical with the extension of the concept 'line parallel to line b' and conversely, if the extensions of the two concepts just named are identical, then a is parallel to b" (§67; also §107).

While Frege maintained in *Foundations* that he attached "no decisive importance even to bringing in the extensions of concepts at all" (§107), by the time he wrote *Basic Laws* he felt obliged to accord them a fundamental role. There he would argue that the generality of an identity can always be transformed into an identity of courses-of-values and conversely, an identity of courses-of-values may always be transformed into the generality of an identity. By this he meant that if it is true that $(x)\ \phi(x) = \varphi(x)$, then those two functions have the same extension and that, vice versa, functions having the same extension are identical (*Basic Laws* §§9 and 21). "This possibility," he wrote then, "must be regarded as a law of logic, a law that is invariably employed, even if tacitly, whenever discourse is carried on about extension of concepts. The whole Leibniz-Boole calculus of logic rests upon it. One might perhaps regard this transformation as unimportant or even as dispensable. As against this, I recall the fact that in my *Grundlagen der Arithmetik* I defined a Number as the extension of a concept. . . ."[48]

4. Husserl's Criticisms

Husserl had the following remarks to make about Frege's theory of number as described above.[49] In his first objection to it, he appeals to common linguistic usage which distinguishes between the equality and the identity of two objects. Leibniz's definition, he points out, defines identity, not equality, so that as long as the least difference remains there will be propositions for which the elements in question will not be interchangeable salva veritate (p. 104). Here Husserl is appealing to the ordinary, nonmathematical use of the words 'equality' and 'identity'. For example, we commonly say that the United States of America was dedicated to the proposition that all men are created equal with respect to their legal rights, but I believe that no one has ever said, nor would be so foolish as to say, that all men are created identical. (It should be noted that Husserl's remarks never concern the identity or equality of signs, but only the equality or identity of objects and the properties that might be predicated of them.)

According to dictionaries, two things are identical when they are the same in every way. They are equal when they are the same under a specific description, as given in a particular way. The difference between equality and identity would then be the difference between sharing any given property or properties, or having all properties in common. Husserl's point is that if x were to have even one property that y does not have, then though they may be equal in one or in many respects, they are not identical and there will be statements in which substitution will fail, and so affect the truth value of statements made referring to them, or the outcome of one's inquiries regarding them.

In another argument, Husserl alludes to the problems that arise when one begins examining the grounds for determining the equality of two objects (pp. 108–9). One can declare two simple, unanalyzed objects equal without much further ado, he notes. But there is a certain ambiguity in ordinary language with regard to complex objects. If two objects are the same, then it follows that they must have all their properties in common. But the inverse does not seem to hold. Sometimes two objects have their properties in common and we still do not say that they are the same.

At first sight, Husserl's point may seem illogical for it seems that he is saying that x and y could be different without there being any discernible difference between them. Before condemning his analysis outright, however, it should be noted that, tangling with problems surrounding extensionality, identity, and classes, Bertrand Russell was moved to make the same observations. Writing in *Introduction to Mathematical Philosophy* on classes and problems connected with Leibniz's law of the identity of indiscernibles he argued that it was just "as it were, an accident, a fact about . . . this higgedly-piggedly job-lot of a world in which chance has imprisoned us" that no two particulars were precisely the same and he hypothesized that "there might quite well, as a matter of abstract logical possibility, be two things which had exactly the same predicates."[50] He also wrote in *Principia Mathematica* that: "It is plain that if x and y are identical, and φx is true, then φy is true . . . the statement must hold for any function. But we cannot say conversely: 'If, with all values of φ, φx implies φy, then x and y are identical'; . . . we cannot without the help of an axiom be sure that x and y are identical if they have the same predicates. Leibniz's 'identity of indiscernibles' supplied this axiom."[51]

We may in fact, Husserl continues his argument, declare the same objects to be equal in one case and different in another (pp. 108–9). He offered the following example to illustrate his point: two straight lines may in one case be said to be equal because they have the same length, but might otherwise be deemed unequal because to be equal two segments of a line must be parallel and have the same direction. Husserl tries to overcome the ambiguity involved by concluding that two objects are to be considered equal if they share the specific properties which constitute the main focus of interest of the investigation and these properties are the same.

An example will help illustrate Husserl's point. A few years ago a Jerusalem courtroom found a retired Ohio autoworker named John Demjanjuk guilty of being Ivan the Terrible, the murderer of hundreds of thousands of Jews. Throughout his fourteen-month trial, Demjanjuk had insisted that he was a victim of mistaken identity. For the Jerusalem courtroom that condemned him to death the only thing that mattered involved determining whether or not he was the same man who had operated gas chambers at Treblinka during World War II, any of the innumerably many other things that could be truthfully predicated of him were beside the point. Their reasoning was of the form: $F(x)$, and if $F(y)$ then $x \equiv y$. Killing hundreds of thousands of Jews was true of Ivan the Terrible and if the same were true of John Demjanjuk, then he would be Ivan the Terrible—and liable to hanging.

Numerous other examples can be found to illustrate the differences between equality and identity. For instance, is a person in an irreversible coma following an accident who is entirely dependent on machines to sustain her bodily functions identical to the person she was before the accident took place? Think of the innumerable things that could have been predicated of her before which are no longer true, and the truly macabre propositions that could result from substitution rules which do not take sufficient account of the difference between equality and identity. Certainly her family would never have entertained the thought of depriving her of the minimum means necessary to support her life before she was in the coma. And don't differences between equality and identity figure in many other dilemmas actually faced in medical practice today? Surely, some of the very real moral issues involved in abortion rights turn on whether a fetus is in all essential respects the same as the person that will develop from it if the pregnancy is not terminated. Such contexts make it hard to dismiss differences between equality and identity as being merely linguistic or psychological.

In another argument Husserl sides with those who hold that characteristics, properties, attributes, and concepts are not liable to the same identity conditions as objects are, so that talk of them cannot for him be reducible to talk of the objects they are about (pp. 134–35). He argues that Leibniz's definition turns the real state of affairs upside down (pp. 104–05). Assuming that, against all odds, one manages to find objects satisfying the conditions it sets down, then by what right can one replace one with the other in certain true propositions or all? The only precise answer, he replies, would be the identity or equality of the referents. However, here Husserl comes up against the same difficulty that Frege himself would encounter when confronted with Russell's paradox more than a decade later (see §5 of this text). Though characteristics which are the same form propositions which are the same, Husserl wrote, having equivalent propositions does not mean that the characteristics figuring in those propositions are the same. In other words, though two propositions may be formally equivalent by virtue of the fact that what is asserted of the reference in them is the same, from the fact that two propositions have the same reference it cannot be concluded that what has been asserted of their referent is the same. In *Introduction to Mathematical Philosophy*, Russell would give a reason for the problem Husserl perceived: "For many purposes, a class and a defining characteristic of it are practically interchangeable. The vital difference between the two consists in the fact that there is only one class having a given set of members, whereas there are always many different characteristics by which a given class may be defined."[52]

Husserl is making several interrelated, but different points which shed light on Frege's and Russell's difficulties. Quine's proposition that 'all bachelors are unmarried men'[53] provides a happy, relatively unproblematic example of equal properties coinciding in extension and so illustrates Husserl's point that if $F = G$, then if $(x) F(x)$ then $G(x)$. If being married to Jackie Kennedy is the same thing as being her husband, then anyone who is married to Jackie Kennedy is her husband. However, and this is Husserl's second point, were Jackie Kennedy's husband found out to be Marilyn Monroe's lover, it would not then follow that being Jackie Kennedy's husband and being Marilyn Monroe's lover are the same. The statement 'The husband of Jackie Kennedy was the lover of Marilyn Monroe' could be determined to be true if both the descriptions figuring in it were found to be true of the same man, John Kennedy. However, trivial, self-evident statements resulting from substitutions like 'John Kennedy was the husband of Jackie Kennedy' and 'John Kennedy is John Kennedy', though true, are not its equivalent. Likewise, if it could be determined that a certain retired Ohio autoworker and a certain sadistic concentration camp guard were the same person, this would not then mean that everything that is true of sadistic concentration camp guards is true of retired Ohio autoworkers. In the latter case it could be argued that the matter has been complicated by changes that have taken place over the course of time. That is not, however, the case in the example preceding it since a person can be married and have an extramarital affair at one and the same time.

Husserl's objection lies in the fact that predicates which are true of the same objects are not always themselves interchangeable salva veritate. Russell made the same point in 1918 to his audience at Gordon Square. The two propositional functions 'x is a man' and 'x is a featherless biped' are formally equivalent, he notes. When one is true, so is the other, and vice versa. However, he points out that there are a certain number of things that you can say about a propositional function which will not be true if you substitute another formally equivalent propositional function in its place. "For instance," he writes, "the propositional function 'x is a man' is one which has to do with the concept of humanity. That will not be true of 'x is a featherless biped'. Or if you say 'so-and-so asserts that such and such is a man' the propositional function 'x is a man' comes in there, but 'x is a featherless biped' does not."[54] In an age of organ transplants, the point made by appealing to the old example of the statements 'creature with a heart' and 'creature with kidneys' coinciding in extension, but not in intension is less abstract than it may have been earlier in the century. Although almost anyone would concede that everyone who has a beating heart has at least one kidney and vice versa, who would think of undergoing a heart or kidney transplant or operation with a surgeon who believed that having a heart and having a kidney were the same thing? We cannot conclude from $(x)\ F(x) = G\,(x)$ that $F \equiv G$.

Husserl further notes that arguing that two objects are equal because they are interchangeable obviously misses the point (p. 104). In a case like John Demjanjuk's, this requirement would in fact have quite disturbing consequences. For instance, appealing to it one might reason that if John Demjanjuk could take Ivan the Terrible's place at the gallows, then John Demjanjuk and Ivan the Terrible were the same man. Obviously if, as Demjanjuk claimed, he was not Ivan the Terrible, but rather a victim of mistaken identity, such a conclusion would constitute a very grave miscarriage of justice. It would also be quite unthinkable to write off the differences between being the retired Ohio autoworker and the sadistic concentration camp guard as being merely linguistic or psychological.

There is, in fact, no way of determining whether two things are interchangeable which does not presume knowledge of their identity, Husserl concluded (p. 105). If substitutivity is to serve as the criterion of identity, then establishing the identity of two things implies that one has already established their interchangeability, but this would imply undertaking innumerably many acts in which what is predicated of one object is established as being the same as what is predicated of the other object and establishing this would require once again establishing that the same things can be predicated of each one of these pairs and so on. Michael Dummett makes the same point when he writes that:

> the truth of an identity statement cannot be established by an appeal to Leibniz's law since, apart from the impossibility of running through the totality of first level concepts, there will often be no other way of ascertaining that a particular predicate which is true of the bearer of one name is also true of the bearer of the other except by establishing that the two names have the same bearer. There is, for example, no way of showing that the predicate 'is visible shortly before sunrise', which is plainly true of the Morning Star, is also true of the Evening Star, which does not depend upon showing that the Morning Star and the Evening Star are one and the same celestial body.[55]

Though Husserl provided the above criticisms of Frege's use of Leibniz's principle of substitutivity of identicals, it was not Husserl, but rather Frege himself who established the link between some of these very problems and the significance of Russell's paradox for the logical theories propounded in *Foundations* and *Basic Laws*. This is what I hope to show now.

5. Some of the Broader Philosophical Issues Involved

Knowing the central role Frege accorded to fixing the sense of an identity and of the link he made between substitutivity and identity in the *Foundations*,[56] and knowing that Frege ultimately concluded that Russell's paradox meant that there were irremediable flaws in the foundations proposed for arithmetic in *Foundations* and *Basic Laws*, it is important to take a close look at the connections between Frege's views on substitutivity and his reasons for despair regarding the tenability of his logical theories. The philosophical questions involving substitutivity and Russell's paradox are surely in this way tied into fundamental matters of vital concern to many philosophers in this century. In this section I want to look at what Frege himself thought was the fatal flaw in his reasoning and at how Russell would tackle the same problems. I examine these issues from another angle in chapter 5, a sister chapter to this one.

In writing the *Basic Laws of Arithmetic*, Frege set out to actually demonstrate the theory of number he had advanced in *Foundations*.[57] Once Russell informed him of the famous paradox, Frege immediately traced the source of the problem to Basic Law V (or Principle V) as promulgated in *Basic Laws*. [58]

Basic Law V was Frege's axiom of extensionality which codified, or rather mandated, the views regarding identity and substitutivity Frege believed his system required. Right from the beginning, the discovery of the paradox indicated to Frege that Basic Law V was false. Although transforming of the generality of an identity into an identity of ranges of values was allowable, he concluded, the converse is not always permissible.[59]

Frege initially thought a solution to the problems raised by Russell's paradox might be found,[60] and he finally proposed one which involved a modification of the problematic law.[61] By mid-1906, however, he had apparently decided that all efforts to repair his logical edifice were destined to failure.[62]

Frege was also very specific about precisely what it is about Basic Law V that leads to the paradoxes. In the several texts in which Frege pinpoints what he believed was the source of the difficulties, he consistently cites Basic Law V's transformation of concepts into objects for extensional treatment as being at fault.[63] In §§146–47 of the 1903 *Basic Laws II*, he is quite specific about the nature of the procedure he had come to advocate:

> If a (first-level) function (of one argument) and another function are such as always to have the same graph for the same argument, then we may say instead that the graph of the first is the same as that of the second. We are then recognizing something common to the two functions. . . . We must regard it as a fundamental law of logic that we are justified in thus recognizing something common, and that accordingly we may transform an equality holding generally into an equation (identity).[64]

An article on the logical paradoxes of set theory he was working on in 1906 gives further insight into his reasoning:

> Let the letters 'φ' and 'ψ' stand in for concept-words (nomina appellativa). Then we designate subordination in sentences of the form 'If something is a φ, then it is a ψ'. In sentences of the form 'If something is a φ, then it is a ψ and if something is a ψ then it is a φ' we designate mutual subordination, a second level relation which has strong affinities with the first level relation of equality (identity) . . . . And this compels us ineluctably to transform a sentence in which mutual subordination is asserted of concepts into a sentence expressing an equality. . . . Admittedly, to construe mutual subordination simply as equality is forbidden . . . .Only in the case of objects can there be a question of equality (identity). And so the said transformation can only occur by concepts being correlated with objects in such a way that concepts which are mutually subordinate are correlated with the same object.[65]

He gives as an example the sentence: 'Every square root of 1 is a binomial coefficient of the exponent -1 and every binomial coefficient of the exponent -1 is a square root of 1.' According to his theory this sentence is to be rewritten as 'The extension of the concept *square root of 1* is equal to (coincides with) the extension of the concept *binomial coefficient of the exponent − 1.*' The words 'the extension of the concept *square root of 1*' are now to be regarded as a proper name as, Frege claims, is indicated by the definite article. Such a transformation acknowledges that there is one and only one object designated by the proper name. Frege explains that:

> By permitting the transformation, you concede that such proper names have meanings. But by what right does such a transformation take place, in which concepts correspond to extensions of concepts, mutual subordination to equality? An actual proof can scarcely be furnished. We will have to assume an unprovable law here. Of course it isn't as self-evident as one would wish for a law of logic. And if it was possible for there to be doubts previously, these doubts have been reinforced by the shock the law has sustained from Russell's paradox.[66]

In 1912 he would write of how he had originally silenced his doubts and, in order to obtain objects out of concepts, had decided to admit the passage from concepts to their extensions, which until he died he maintained leads to Russell's paradox.[67] An article and a letter Frege wrote just prior to his death eighteen years later both clearly show that for the rest of his life he remained firm in his conviction that this flaw in his system was the precise cause of Russell's paradox and so undermined his whole life's work. He wrote:

> One feature of language that threatens to undermine the reliability of thinking is its tendency to form proper names to which no objects correspond. . . . A particularly noteworthy example of this is the formation of a proper name after the pattern of 'the extension of the concept *a*'. . . . Because of the definite article, this expression appears to designate an object; but there is no object for which this phrase could be a linguistically appropriate designation. From this has arisen the paradoxes of set theory which have dealt the deathblow to set theory itself.[68]

This procedure, Frege concluded, leads "into a thicket of contradictions. . . . Confusion is bound to arise if a concept word, as a result of its transformation into a proper name, comes to be in a place for which it is unsuited." People should be warned against changing a concept into an object.[69]

Bertrand Russell also eventually established a connection between certain puzzles involving descriptions containing the definite article and the contradictions, or paradoxes, connected with set theory.[70] So as he struggled to find solutions to puzzles and paradoxes connected with Frege's logic Russell was actually faced with resolving the problem Frege described above. Russell would even go so far as to write in his 1919 *Introduction to Mathematical Philosophy* that he considered analyses of the word 'the' to be of so very great importance to a correct understanding of descriptions and classes that he would give the doctrine of the word if he were dead from the waist down and not merely in prison (where he was at the time).[71] No account of Frege's problems with substitutivity is complete, then, without a look at Russell's attempts to solve them.

"The whole theory of definition, of identity, of classes, of symbolism is wrapped up in the theory of denoting," Russell declared in his 1903 *Principles of Mathematics*.[72] So, it is perhaps not surprising to find him recounting that during 1903 and 1904 his work was almost entirely devoted to denoting problems which he thought were probably relevant to his problems with the contradictions of set theory. His 1905 theory of denoting proved to him that they were and it represented his first major breakthrough in finding a solution to the paradoxes.[73]

One of the most significant problems associated with descriptions containing the definite article dealt with by Russell's new theory of denoting is directly tied in with Leibniz's principle of the substitutivity of identicals. If identity can only hold between x and y if they are different symbols for the same object it would not then seem to have much importance, Russell noted. However, he observed, identity statements containing descriptive phrases of the form 'the so-and-so' constitute an exception and lead to a puzzle he explained in these words:

> If a is identical with b, whatever is true of the one is true of the other, and either may be substituted for the other in any proposition without altering the truth or falsehood of that proposition. Now George IV wished to know whether Scott was the author of *Waverley*, and in fact Scott *was* the author of *Waverley*. Hence we may substitute *Scott* for *the author of 'Waverley'*, and thereby prove that George IV wished to know whether Scott was Scott. Yet an interest in the law of identity can hardly be attributed to the first gentleman of Europe.[74]

In Frege's logic descriptions like 'the discoverer of America' or 'the extension of the word "star"' were treated as objects. His Basic Law V would have made such descriptions subject to the same formal rules of identity as those governing objects, and so amenable to extensional treatment and substitution. Frege ultimately felt obliged to condemn this operation as illicit and judged his reliance upon it to be the fatal flaw in his logic and the source of the paradoxes associated with set theory.

Like Frege, Russell considered statements equating a term standing for an object with a description to be identities, and identities to be statements which equated objects. Russell had originally believed that all expressions denote directly, "that, if a word means something, there must be something that it means."[75] But his struggle to solve the paradoxes forced him to come

to terms with serious logical problems which seemed to him to be unavoidable when definite descriptions are regarded as standing for genuine constituents of the propositions in which they figure. For instance, if the expression 'the author of *Waverley*' really does denote some object *c*, Russell finally
reasoned, the proposition 'Scott is the author of *Waverley*' would be of the
form 'Scott is *c*'. But if *c* denoted any one other than Scott, this proposition
would be false; while if *c* denoted Scott, the resulting proposition would be
'Scott is Scott', which is self-evident and trivial and plainly different from
'Scott is the author of *Waverley*' which may be true or false. So, Russell concluded, *c* does not stand simply for Scott, nor for anything else[76] because

> No one outside a logic-book ever wishes to say that '*x* is *x*', and yet assertions of
> identity are often made in such forms as 'Scott was the author of *Waverley*'. . .
> The meaning of such propositions cannot be stated without the notion of iden-
> tity, although they are not simply statements that Scott is identical with another
> term, the author of *Waverley*. . . . The shortest statement of 'Scott is the author of
> *Waverley*' seems to be 'Scott wrote *Waverley*; and it is always true of *y* that if *y*
> wrote *Waverley*, *y* is identical with Scott'.[77]

Analyzed in this way, the description may be substituted for *y* in any
propositional function *fy* and a significant proposition will be the result.
These reflections on substitutivity and definite descriptions led Russell to the
conclusion that all phrases (other than propositions) containing the word
'the' (in the singular) are incomplete symbols which have a meaning in use
but when taken out of context do not actually denote anything at all.[78]

Russell also came to consider classes to be incomplete symbols in the
same sense descriptions are, and so this new way of analyzing away incomplete symbols represented for him a major breakthrough in resolving the
contradictions apt to result when descriptive phrases are wrongly assumed to
stand for an entity. In 1918 he told his listeners at Gordon Square:

> you find that all the formal properties that you desire of classes, all their formal
> uses in mathematics, can be obtained without supposing for a moment that there
> are such things as classes, without supposing, that is to say, that a proposition in
> which symbolically a class occurs, does in fact contain a constituent correspond-
> ing to that symbol, and when rightly analysed that symbol will disappear, in the
> same sort of way as descriptions disappear when the propositions are rightly
> analysed in which they occur.[79]

Russell considered this theory of definite descriptions to have been his
most valuable contribution to philosophy and frequently spoke in enthusiastic
terms of its role in resolving his logical problems, and the paradoxes in particular.[80] He once claimed that his success in his 1905 article "On Denoting"
was the source of all his subsequent progress. As a consequence of his new

theory of denoting, he said, he found at last that substitution would work, and all went swimmingly.[81] This new theory, he claimed, afforded a "clean shaven picture of reality" and "swept away a host of otherwise insoluble problems." It did not, however, sweep away all the problems and in spite of Russell's enthusiastic appraisals and the acclaim it has received, even Russell recognized that additional measures were needed[82] to guarantee that the unwanted whiskers wouldn't come back. So Russell marshalled the theory of types and the axiom of reducibility into his barbershop to try to finish the job.

However, even as he expounded the theory of types, Russell realized that it only solved some of "the paradoxes for the sake of which" it was "invented" and something more would be necessary to solve the other contradictions.[83] In 1917 he would go so far as to concede that ". . . the theory of types emphatically does not belong to the finished and certain part of our subject: much of the theory is still inchoate, confused and obscure."[84]

From the axiom of reducibility all the usual properties of identity and classes would follow. Two formally equivalent functions would determine the same class and, conversely, two functions which determine the same class would be formally equivalent. Without it or some equivalent axiom, "many of the proofs of *Principia* become fallacious," Russell believed, and "we should be compelled to regard identity as indefinable." [85]

Frege had introduced classes into his logical system to uphold the theory of substitutivity and identity his work on the foundations of arithmetic called for. Russell's paradox finally convinced him that this was an ill-fated move and he gave up. Russell devised a way of analyzing away classes and descriptions and proposed the very problematic axiom of reducibility to mandate the properties of identity, classes, and substitutivity the logic of *Principia* seemed to need. Neither he nor Frege ever felt he had ever solved the substitutivity problem.

6. Concluding Remarks

The chief objective of this paper has been to show Husserl's ability to evaluate Frege's logical work and pinpoint genuine problems in his reasoning. I have also tried to show that, independently of whether or not one is persuaded of the gravity of the problems with extensionality discussed here, Frege himself ultimately concluded that such problems were serious enough to sink his logic, and that in trying to free Frege's work from paradox, Russell felt obliged to come to terms with these same problems. A closer look at the Husserl–Frege relationship, in fact, reveals that the two men clashed swords on many more matters than have yet been discussed in the literature and that their ideas overlapped on many issues which still figure importantly in philosophical discussions today. Of course, the arguments of this paper but raise deeper questions regarding Russell's theory of definite descriptions and how Husserl's philosophy might deal with the questions raised. Naturally these are matters calling for in-depth study extending well beyond the limits of this paper and I have tried to begin to answer them elsewhere.[86] I would, however, like to make one more remark concerning them.

In his now classic article on Russell's mathematical logic[87] Kurt Gödel complained about *Principia Mathematica*'s lack of formal precision in the foundations and the fact that in it Russell omits certain syntactical considerations even in cases where they are necessary for the cogency of the proofs. In particular, he cites Russell's treatment of incomplete symbols, which he complains are introduced, not by explicit definitions, but by rules describing how to translate sentences containing them into sentences not containing them. The matter is especially doubtful, he notes, for the rule of substitution and it is chiefly this rule which would have to be proved.[88]

As an example of Russell's analysis of basic logical concepts, Gödel examines Russell's treatment of the definite article 'the' in connection with problems about what descriptive phrases signify (in deference to Frege, Gödel uses 'signify' and 'signification' in the place of Russell's 'denote' and 'denotation'). Gödel agrees with Russell that the apparently obvious answer that the description 'the author of *Waverley*' signifies Walter Scott leads to unexpected difficulties. For, Gödel reasons, if one admits an axiom of extensionality according to which the signification of a composite expression containing constituents which have themselves a signification depends only on the signification of these constituents, and not on the way in which this signification is expressed, then it follows that the statements 'Scott is the author of *Waverley*' and 'Scott is Scott' have the same signification. Of Russell's technique for solving this puzzle, Gödel writes that he could not help but feel that the problem raised by Frege's puzzling conclusion had only been evaded by Russell's theory of descriptions and that there was something behind it which is not yet completely understood.[89]

In a later paper Gödel wrote of the significance of Russell's paradox for set theory that "it might seem at first that the set-theoretical paradoxes would doom to failure such an undertaking, but closer examination shows that they cause no trouble at all. They are a very serious problem, not for mathematics, however, but rather for logic and epistemology."[90] Because of the connection both Frege and Russell saw between the paradoxes of set theory and the fact that descriptions are not as immediately translatable into the extensional language their theories required, I am inclined to think that Gödel's intriguing statements regarding Russell's theory of definite descriptions and epistemological and logical questions raised by Russell's paradox might be linked, and that with the theory of definite descriptions Russell artfully swept under the carpet a whole host of deep logical and epistemological problems the set-theoretical paradoxes raise for philosophy. I also believe that further work on Husserl's logic will show that he quite lucidly addressed these very issues. I myself pursue these questions further in my *Rethinking Identity and Metaphysics: On the Foundations of Analytic Philosophy*,[91] and in chapter 5 of the present book.

NOTES

1. E. Husserl, *Philosophie der Arithmetik* (Halle: Pfeffer, 1891).
2. G. Frege, *Foundations of Arithmetic* (Oxford: Blackwell, 1986 [1884]).
3. G. Frege, *The Basic Laws of Arithmetic* (Berkeley: University of California Press, 1964 [1893]), pp. 127–43.
4. I go into these in detail in my *Word and Object in Husserl, Frege and Russell* (Athens: Ohio University Press, 1991).
5. G. Frege, *Philosophical and Mathematical Correspondence* (Oxford: Blackwell, 1980), p. 7 (Couturat's July 1, 1899 letter to Frege).
6. A. Church, "Review of M. Farber, *The Foundations of Phenomenology*," *Journal of Symbolic Logic* 9 (1944): 63–65.
7. D. Willard, *Logic and the Objectivity of Knowledge, A Study in Husserl's Early Philosophy* (Athens: University of Ohio Press, 1984). Two comments on Husserl's psychological objections (as opposed to his logical objections) to Frege's views on identity appear in H. Ishiguro's *Leibniz's Philosophy of Logic and Language* (Cambridge: Cambridge University Press, 1990), p. 209, n. 43; and in D. Wiggins' *Sameness and Substance* (Oxford: Blackwell, 1980), pp. 20–21, n. 7.
8. Willard, pp. xii–xiv comments.
9. G. Frege, "Review of Dr. E. Husserl's *Philosophy of Arithmetic*," *Mind* 81, no. 323 (July 1972): 321–37. Translation of "Rezension von Dr. E. G. Husserl: *Philosophie der Arithmetik*," *Zeitschrift für Philosophie und philosophische Kritik* 103 (1894), 313–32.
10. M. Dummett, *Frege, Philosophy of Language*, 2nd ed. rev. (London: Duckworth, 1981), p. xlii; M. Dummett, *The Interpretation of Frege's Philosophy* (Cambridge, MA: Harvard University Press, 1981), p. 56; Frege, *Philosophical and Mathematical Correspondence*, pp. 60–61; Comments Willard, pp. xii–xiii, p. 118.
11. K. Schuhmann, *Husserl-Chronik* (The Hague: M. Nijhoff, 1977), pp. 6–11 on Weierstrass and Husserl.
12. M. Kline, *Mathematical Thought from Ancient to Modern Times* vol. 3 (New York: Oxford University Press, 1972), pp. 950–56, 960–66, 972 on Weierstrass and Bolzano.
13. Schuhmann, pp. 7, 11; Willard, pp. 3–4, 21–22; A. Osborn, *The Philosophy of E. Husserl in its Development to his First Conception of Phenomenology in the Logical Investigations* (New York: International Press, 1934), p. 12.
14. L. McAlister, *The Philosophy of Franz Brentano* (London: Duckworth, 1976), p. 49.
15. Osborn, p. 18.
16. McAlister, pp. 45, 53.
17. Osborn, p. 21.
18. Ibid., p. 17.

19. Dummett, *Interpretation of Frege's Philosophy*, pp. 72–73, 496–97; Dummett, *Frege, Philosophy of Language*, p. 683.
20. B. Russell, *My Philosophical Development* (London: Allen and Unwin, 1975 [1959]), p. 100. See my discussions of this in my *Word and Object in Husserl, Frege and Russell*, chapters 5 (§1) and 7.
21. Osborn, p. 29; Willard, pp. 32–34.
22. Frege, *Philosophical and Mathematical Correspondence*, pp. 171–72; Frege's letter to Marty, pp. 99–102, may have actually been addressed to Stumpf; see p. 99.
23. Schuhmann, p. 18; Frege, *Philosophical and Mathematical Correspondence*, p. 64.
24. A. Fraenkel, "Georg Cantor," *Jahresbericht der deutschen Mathematiker Vereinigung* 39 (1930), pp. 221, 253n., 257.
25. E. Husserl, *Introduction to the Logical Investigations. A Draft of a Preface to the Logical Investigations (1913)* (The Hague: M. Nijhoff, 1975), p. 37 and notes. Comparing the English edition with other editions one discovers the typographical error.
26. J. Cavaillès, *Philosophie mathématique* (Paris: Hermann, 1962), p. 182, in reference to the Fraenkel biography cited in note 24.
27. R. Schmit, *Husserls Philosophie der Mathematik* (Bonn: Bouvier, 1981, pp. 40–48; pp. 58–62; L. Eley, "Einleitung des Herausgebers," *Philosophie der Arithmetik* (The Hague: M. Nijhoff, 1970), pp. XXIII–XXV.
28. Dummett, *Frege, Philosophy of Language*, p. 630.
29. Dummett, *The Interpretation of Frege's Philosophy*, p. 21.
30. B. Russell, "The Study of Mathematics," *Mysticism and Logic* (London: Allen & Unwin, 1963 [1917]), p. 66; and B. Russell, *Principles of Mathematics* (New York: Norton, 1903), p. xviii.
31. B. Russell, *Principia Mathematica,* 2nd ed. rev. (Cambridge: Cambridge University Press, 1964 [1927]), p. viii.
32. Russell, *Principles of Mathematics*, §§100, 344, 500; Frege, *Philosophical and Mathematical Correspondence*, pp. 133, 147.
33. Russell, *Principles of Mathematics*, §149.
34. Ibid., §107.
35. Ibid., §249.
36. Schuhmann, p. 13.
37. Frege, *Philosophical and Mathematical Correspondence*, pp. 6–7, 52, 170; G. Frege, "Gottlob Frege: Briefe an Ludwig Wittgenstein," *Grazer philosophische Studien* 33/34 (1989) 11, 14.
38. G. Frege, *Grundgesetze der Arithmetik* (Hildesheim: Olms, 1966), §§68–85 of the second volume for Cantor and pp. 148–55 for Weierstrass. *Basic Laws II* is not available in English. §§56–67, 86–137, 139–44 are translated in G. Frege, *Translations from the Philosophical Writings* (Oxford: Blackwell, 1980), thus omitting Frege's criticisms of Cantor and Weierstrass.
39. B. Russell, "My Mental Development," *The Philosophy of Bertrand Russell*, Library of Living Philosophers, vol. 5, ed. P.A. Schilpp (Cambridge: Cambridge University Press, 1946), p. 11; C. Kilmister, *Russell* (Brighton: The Harvester Press, 1984), pp. 5–6, 54–66; I. Grattan-Guinness, *Dear Russell-Dear Jourdain* (London: Duckworth, 1977), pp. 132, 143–44.
40. Schuhmann, p. 18.

41. E. Husserl, *Über den Begriff der Zahl. Psychologische Analyse* (Halle: Heyne-mansche Buchdruckerei, 1887). In English in *Husserl: Shorter Works*, P. McCormick and F. Elliston, eds. (Notre Dame: Notre Dame University Press, 1981), pp. 92–119.

42. Schuhmann, p. 19.

43. Frege, *Philosophical and Mathematical Correspondence*, pp. 63–65.

44. E. Husserl, *The Logical Investigations* (London: Routledge and Kegan Paul, 1970), p. 179 n.

45. Husserl, *Philosophie der Arithmetik*, pp. 104–5, 134.

46. For example Frege, *Translations from the Philosophical Writings*, pp. 22–23, 56, 120–21, 141n., 146n., 159–61, 210; Frege, *Philosophical and Mathematical Correspondence*, p. 141; G. Frege, *Posthumous Writings* (Oxford: Blackwell, 1979), pp. 120–21, 182; Frege, "Review of Dr. E. Husserl's *Philosophy of Arithmetic*," pp. 327, 331.

47. I have had to alter Austin's translation of §65 to make it agree with other Frege texts and common usage according to which 'equal' means 'agreement in this respect or that' and 'identical' or 'the same' refers to complete agreement in all respects.

48. Frege, *Basic Laws of Arithmetic*, §9, see also p. 6 and Frege, *Translations from the Philosophical Writings*, pp. 159–61, 214; Frege, *Philosophical and Mathematical Correspondence*, pp. 140–41, 191.

49. I cite the page numbers to the original 1891 edition of Husserl's *Philosophie der Arithmetik* within the text.

50. B. Russell, *Introduction to Mathematical Philosophy* (London: Allen and Unwin, 1919), p. 192.

51. Russell, *Principia Mathematica*, p. 57.

52. Russell, *Introduction to Mathematical Philosophy*, p. 13.

53. W. Quine, "Two Dogmas of Empiricism," *From a Logical Point of View* (New York: Harper and Row, 1961 [1953]), pp. 27–32.

54. B. Russell, "The Philosophy of Logical Atomism," *Logic and Knowledge* (London: Allen and Unwin, 1956), pp. 265–66.

55. Dummett, *Frege, Philosophy of Language*, pp. 544–45.

56. Frege, *Foundations of Arithmetic*, p. x, §§62, 65, 107.

57. Frege, *Basic Laws of Arithmetic*, pp. 5, 7, 129, and §§38–47.

58. Ibid., p. 127; Frege, *Philosophical and Mathematical Correspondence*, pp. 130–32.

59. Frege, *Basic Laws of Arithmetic*, p. 132; Frege, *Philosophical and Mathematical Correspondence*, pp. 73, 132.

60. Ibid; ibid.

61. Frege, *Basic Laws of Arithmetic*, pp. 139–43.

62. Dummett, *The Interpretation of Frege's Philosophy*, pp. 21–27; Frege, *Posthumous Writings*, p. 176.

63. Frege, *Philosophical and Mathematical Correspondence*, pp. 54-56, 191; Frege, *Posthumous Writings*, pp. 181–82, 269–70.

64. Frege, *Translations from the Philosophical Writings*, pp. 159–60.

65. Frege, *Posthumous Writings*, pp. 181–82.

66. Ibid.

67. Frege, *Philosophical and Mathematical Correspondence*, p. 191.

68. Frege, *Posthumous Writings*, p. 269.

69. Frege, *Philosophical and Mathematical Correspondence*, p. 55.
70. Russell makes this connection in his *My Philosophical Development*, p. 60; "The Philosophy of Logical Atomism," *Logic and Knowledge*, pp. 262–66; "My Mental Development," pp. 13–14; and his *Essays in Analysis* (London: Allen & Unwin, 1973), p. 165.
71. Russell, *Introduction to Mathematical Philosophy*, p. 167.
72. Russell, *Principles of Mathematics*, p. 56.
73. Grattan-Guinness, pp. 79–80, 94; Russell, *My Philosophical Developr ent*, p. 60.
74. Russell, "On Denoting," *Logic and Knowledge*, pp. 47–48.
75. Russell, *My Philosophical Development*, pp. 62–63.
76. Russell, *Principia Mathematica*, p. 67; Russell, "The Philosophy of Logical Atomism," *Logic and Knowledge*, pp. 245–48.
77. Russell, "On Denoting," *Logic and Knowledge*, p. 55.
78. Russell, *Principia Mathematica*, pp. 67–68.
79. Russell, "The Philosophy of Logical Atomism," *Logic and Knowledge*, p. 266. Also Russell, *Principia Mathematica*, chapter III and the texts cited in note 70.
80. Kilmister, pp. 102, 108, 123, 138; Grattan-Guinness, pp. 70, 94 and the texts cited in note 70.
81. Grattan-Guinness, pp. 78–79.
82. Russell, *My Philosophical Development*, pp. 60–65.
83. Russell, "Mathematical Logic as Based on the Theory of Types," *Logic and Knowledge*, pp. 79–82.
84. Russell, *Introduction to Mathematical Philosophy*, p. 13.
85. Russell, *Principia Mathematica*, pp. 55–59, 75–77, 166–72.
86. My *Word and Object in Husserl, Frege and Russell*.
87. K. Gödel, "Russell's Mathematical Logic," *Collected Works*, vol. 2 (New York: Oxford University Press, 1990), pp. 119–41.
88. Ibid., p. 120.
89. Ibid., pp. 121–23.
90. Ibid., p. 258.
91. C. Hill, *Rethinking Identity and Metaphysics, On the Foundations of Analytic Philosophy* (New Haven: Yale University Press, 1997).

2

Guillermo E. Rosado Haddock

REMARKS ON SENSE AND REFERENCE IN FREGE AND HUSSERL

In this article I will deal with the theories of meaning of two of the most influential philosophers of the last hundred years, and ones whose seminal ideas on many philosophical issues will probably influence the generations to come. More specifically, I will be concerned with the theories on sense and reference of Gottlob Frege and Edmund Husserl.

1

My objectives in this article are threefold: (1) to give a relatively brief exposition of Frege's theories concerning the sense and reference of proper names, conceptual expressions (*Begriffswörter*), and assertive sentences; (2) to give a summary account of Husserl's corresponding theories and to compare them with Frege's, both in their contents and in their origins and possible influences; (3) to make some remarks on the sense and reference of assertive sentences. Related issues in both Husserl's and Frege's theories of meaning are excluded. Thus, for example, I will not deal with the fulfillment of meaningful expressions in Husserl, nor with the reference of sentences in contexts in which they do not have their usual reference in Frege (and Husserl).[1]

Before entering into the discussion properly, it is convenient to fix the terminology, especially since Husserl's and Frege's not only do not coincide, but can easily lead to confusion. In *Logische Untersuchungen* Husserl uses the words '*Sinn*' (sense) and '*Bedeutung*' (meaning) almost as synonyms,[2] and opposes them to '*Gegenstand*' (object) and to '*Gegenständlichkeit*' (objectuality). In what follows I will use the latter expression to name the more inclusive concept which includes not only objects but also states of affairs, whereas the term 'object' (*Gegenstand*) will be used to name the objectualities of names. Contrary to Husserl, Frege opposes in an unusual manner the terms

'*Sinn*' and '*Bedeutung*', which has led Frege's translators to render '*Bedeutung*' in Frege's sense as 'reference' or 'denotation'. Frege's *Bedeutung* comes near to *Gegenständlichkeit* in Husserl's terminology. Frege also uses the term '*Gegenstand*' to name the reference of names, although Husserl's and Frege's concepts of 'name' coincide only partially, since Frege—but not Husserl—considers sentences as names—even as proper names—but does not consider the so-called general (or common) names to be names.[3]

In this article I use the terms 'sense' (as a translation of '*Sinn*') and 'meaning' (as a translation of Husserl's '*Bedeutung*') almost as synonyms, and I oppose them to the terms 'reference' (as a translation of Frege's '*Bedeutung*'), 'denotation' and 'objectuality' (as a translation of '*Gegenständlichkeit*'), which I employ as almost identical expressions. I translate the term '*Gegenstand*' as 'object'.

2

Let us consider the following pairs of expressions: (1) 'the morning star' and 'the evening star', (2) 'the winner of Jena' and 'the loser of Waterloo', (3) 'Sir Walter Scott' and 'the author of Waverley', (4) '5 + 2' and '3 + 4', (5) 'Tarski's Ultrafilter Theorem' and 'Tarski's theorem on maximal dual ideals'. If we ask someone with the required knowledge about possible relations between the members of each pair, he will probably answer that in each pair both expressions refer to the same object. But this reference to the same object is established in a different way in each case. Not only are the expressions, or better, the proper names different in each case, but they refer also in a different way to the same object. (I understand, with Frege[4] and Husserl, the term 'proper name' to include not only proper names in the strict sense, like 'Napoleon' or 'Socrates', but also what Russell calls definite descriptions, or more generally, any word (minimal unity of meaning) or composite of words that can serve to determine univocally a singular object).

Consider now the following pairs of expressions: (1) '*Londres*' and 'London', (2) 'Napoleon' and 'Napoleon', (3) '7' and '7', (4) '*siete*' and 'seven', (5) 'France's capital' and '*die Hauptstadt Frankreichs*'. If we compare these examples with those of the first group, we find a striking difference. In the examples of the second group it is sufficient to know the language or languages to which the members of each pair belong to recognize immediately that they refer to the same object. (In the second example it is even unnecessary to know exactly whether 'Napoleon' is an expression of the English or of the German language.) On the contrary, in the examples of the first group we need some knowledge of astronomy, or history, or literature, or arithmetic, or logic, respectively, to recognize that in each pair of expressions both expressions have the same reference. Thus, equalities like '3 + 4 = 5 + 2' and '7 = 7' differ in their cognitive value, and so differ also sentences like 'The Evening Star is the Morning Star' and 'The Evening Star is the Evening Star'. (The notion of 'cognitive value' is decisive for Frege's rejection in "Über Sinn und Bedeutung"[5] of the interpretation of identity statements of the form '*a* = *b*' as expressing a relation between the references of '*a*' and '*b*'. According to him, on the basis of such an interpretation, if a statement of the form '*a* = *b*' is true, then there would not be any difference in cognitive value between the statements '*a* = *b*' and '*a* = *a*', since both statements would express a relation of an object with itself).[6]

We can say that in the examples of the second group both members of each pair refer in the same way to their denotation, and if there is still a noticeable difference—as in examples (1), (4), and (5)—it consists only in that each member of the pair is a perfect translation of the other into another language. On the contrary, in the examples of the first group both members of each pair refer in a different way to the object denoted. This particular way of referring to the objectuality, which is different both from the objectuality and from the expression in its physical appearance, and that is something objective, in opposition to the images that we use to associate with words, is what Frege (and Husserl also) calls the sense of an expression. This distinction is made by Frege for the first time in a lecture on January 9, 1891, which was published the same year with the title "Funktion und Begriff," and Frege formulates it in a more detailed way in his famous article "Über Sinn und Bedeutung" published the next year. In his letter to Husserl of May 24, 1891—to which I will refer in a more detailed manner later on—Frege presents a sort of sketch of his theory about sense and reference and compares it with Husserl's. It is in "Ausführungen über Sinn und Bedeutung," an article published posthumously,[7] but written between 1892 and 1895, that Frege discusses in a detailed manner his extension of the distinction to the so-called conceptual words (which I will discuss later on).

Thus, proper names have a reference, which is the object referred to by the proper name, and a sense, which is the way in which the proper name refers to this object. E.g., 'the winner of Jena' and 'the loser of Waterloo' are proper names which refer to Napoleon, but they refer to him in a different way. We can say that a proper name expresses a sense and through this sense refers to an object. (Of course, it is possible for a proper name, as for any expression, to have a sense but no denotation.)

The relation between the sense and the reference of expressions—whether they are proper names or not—is many-one, and the relation between the sign (i.e., the expression in its physical appearance) and the sense of expressions is also many-one. In other words, the sense univocally determines its reference, but the reference does not determine the sense univocally; and the sign determines the sense univocally, but the sense does not determine the sign univocally. Thus, the sense of the expression '5 + 2' determines univocally the number 7, but this does not determine univocally the sense of the expression '5 + 2'; whereas the sign 'London' determines univocally its sense, but this does not determine univocally the sign 'London', since we could have used the sign '*Londres*' to express the same sense.

Frege distinguishes the sense of an expression from the poetical elements which usually accompany it in colloquial language. These poetical elements of expressions, which I will call 'tone,' following Dummett's terminology,[8] do not add anything to the determination of the reference of an expression, and just serve to awaken in the listener (or reader) subjective images, which must be clearly distinguished both from the sense and from the reference of the expression concerned. As Dummett puts it,[9] the sense of an expression is the only part of its content that makes a (decisive) contribution to the determination of its reference, and of the reference of any sentence in which the expression occurs. According to Frege, the reference of a compound expression is determined by those of its constituents, and the sense of a compound expression is determined by those of its constituents.

But proper names (in the wider sense) are not the only constituents of a sentence. Frege distinguishes a second, radically different, constituent of sentences, which he calls 'conceptual words'. According to Frege we can divide essentially any (assertive) sentence into two components: a proper name and a (probably complex) conceptual word; and if there are two or more proper names in the sentence, there is more than one way in which we can divide the sentence into a proper name and a conceptual word. (Frege's motivation on this point—as on many others—obviously comes from logic). Before I consider the sense and the reference of sentences, I am going to consider Frege's conception of the sense and the reference of conceptual words. (On this point I do not follow Frege, since he first establishes the sense of sentences and then obtains from it the sense of its constituents.)

3

For Frege concepts are functions of one argument, whose value is a truth value. Let us consider briefly Frege's analysis of the concept of function. We will restrict ourselves to the analysis of functions of only one argument, as Frege himself does in "Funktion und Begriff"[10], but the generalization to functions of n arguments for an arbitrary positive integer n does not entail any further difficulties.

A sign for functions always contains at least one empty place for its argument, which is usually indicated by an 'x' or a 'y'. But this argument is not part of the function, and thus the letter 'x' (or 'y') is not part of the sign for functions. Hence, in the case of functions we can always speak of empty places, since what fills them does not belong properly to the function. If we consider, e.g., an expression for functions like '$2 \cdot x^3 + x$', where 'x' occupies the argument place, we can say that in the expression '$2 \cdot (\ )^3 + (\ )$' which we obtain from '$2 \cdot x^3 + x$' by deleting all occurrences of x and indicating all empty places with brackets, lies the 'essence' of the function.

The sign for functions is, thus, a sign that in itself does not denote any object, and requires an extension. Its reference is the function itself, which is called by Frege unsaturated (*ungesättigt*), since its sign requires an extension by a sign for its argument, and only when it has been so extended does it have a saturated reference. This reference of the function sign together with its argument sign is the value of the function for this argument, and since it is the reference of a saturated expression, Frege calls it an object. Hence, if in '$2 \cdot x^3 + x$' we substitute '4' for 'x' (and read '$2 \cdot x^3 + x$' as '$(2 \cdot x^3) + x$') we obtain the identity '$2 \cdot 4^3 + 4 = 132$'. In the particular case of a concept, the conceptual word, which is essentially unsaturated, refers, according to Frege, to a concept, which, in virtue of its predicative nature, is also unsaturated and essentially different from an object. If we fill the empty places of a conceptual word with an argument sign we obtain an expression with a saturated reference, i.e., with an object as reference. But in this particular case, this object, which is the value of the function which is a concept for a given argument, is a truth value: 'the true' if the argument falls under the concept, and 'the false' if the argument does not fall under the concept.

To fall under a concept is just to belong to the extension of the concept, i.e., to the set of objects to which that concept applies. It is somewhat surprising that for Frege the reference of a conceptual word is a concept and not the extension of a concept. It seems more astonishing if we remember that in "Ausführungen über Sinn und Bedeutung" Frege asserts that since an object which falls under a concept falls under all concepts of the same extension, we can substitute in a sentence a conceptual word for another conceptual word without changing the reference of the sentence, but changing its sense, whenever the corresponding conceptual words have the same extension of the concept. However, according to Fregean principles concerning the sense and reference of compound expressions, if the sense of a sentence, but not its reference, has been affected, there must have been a change in the sense of the conceptual words. But Frege does not clarify for us the sense of a conceptual word (or of a function sign). E.g., Frege should have shown that the sense of a conceptual word is different from both the conceptual word and the concept, and that the relation between conceptual words and their senses and the relation between these senses and their concepts are both many-one.

Furthermore, we can substitute in a sentence conceptual words which 'denote' different concepts with the same extension, e.g., 'is an even prime number' and 'is the successor of 1 in the sequence of natural numbers', without affecting the truth value of the sentence involved. Frege offers two reasons for his choice of the reference of conceptual words. (1) The extensions of concepts are objects and, thus, they cannot be the reference of unsaturated expressions. (2) If a concept did not apply to any object, i.e., if it had an empty extension, the conceptual word would have no reference.[11] Reason (1) is not completely clear, since we cannot conclude with complete certainty whether unsaturated expressions are unsaturated because they have an unsaturated reference, or whether the reference is unsaturated because it is the reference of an unsaturated expression. According to Frege's formulations in "Ausführungen über Sinn und Bedeutung"[12] the reference is unsaturated because its expression is unsaturated; but in "Funktion und Begriff"[13] Frege tells us that the expression for a function has to have an empty place, and then we can inquire about the grounds upon which this 'has to have' is based. On the other hand, (2) is based on Frege's requirement that in scientific usage all expressions (proper names, conceptual words, and sentences) must have a reference. An expression without any reference would deprive all compound expressions of which itself is a component part of having a reference. This requirement of Frege is reasonable, but it does not make a decisive point against the rival solution (i.e., that the reference of a conceptual word is the extension of the concept, whereas the concept is its sense). (This rival conception could stipulate that when a concept is absurd, i.e., there cannot be an object that falls under the concept, the reference of the conceptual word is the empty set, and when the concept is not absurd, but does not have a reference in the actual world, it could have a reference in other possible worlds, e.g., the conceptual word 'centaur'.) Since Frege's thesis has not attracted many followers, I prefer not to enter in a detailed discussion of (2).

Moreover, if the reference of a sentence is a truth value—as was Frege's opinion—then it is possible to make many more substitutions in sentences than the ones we have considered up to now. (Church's discussion of Frege's thesis in *Introduction to Mathematical Logic* illustrates this point very well.)

Frege makes an important point of the distinction between objects and concepts, telling us emphatically that neither they nor their respective linguistic expressions are interchangeable.[14] Thus, e.g., proper names cannot be used as predicates in sentences. Although this is an interesting problem (which could also be compared with some of Husserl's ideas), we cannot consider it here.

4

We have seen that when we extend a conceptual word with the help of an argument sign, we obtain an expression whose reference is a truth value: the true, if the object referred to by the argument sign falls under the concept, and the false, if this is not the case, i.e., if the object concerned does not fall under the concept. But according to Frege, the fundamental logical relation is that of an object falling under a concept. Assertive sentences (*Behauptungssätze*) have essentially two kinds of constituent parts: proper names and conceptual words. Proper names refer to objects and conceptual words refer to concepts. Assertive singular sentences express that the object concerned falls under the concept concerned. The sense of a sentence is a thought (*Gedanke*) (in the sense of the [theoretical content] of a judgment), and its reference is a truth value: the true or the false. Thus, all true sentences have the same reference (the true), and all false sentences have the same reference (the false). Both 'the true' and 'the false' are objects, since they are the references of expressions that do not need any further extension. Thus, Frege sometimes calls sentences 'names'.

According to Frege, a thought, all of whose component parts have a denotation, is either true or false, and it would be nonsense to speak about cases in which a thought is true and cases in which the same thought is false. In such situations we simply have different thoughts which are inadequately expressed by the same sentence. Such a sentence, which does not express a complete thought, is called by Frege an 'improper sentence'. Thus, only proper (or genuine) sentences express a thought unambiguously and have a truth value as reference. (An improper sentence, however, can be a component part of an aggregate of sentences [*Satzgefüge*], which expresses a complete thought.) Finally, it should be said that for Frege it is possible to grasp a thought without recognizing it as true (or as false).

Let us consider now Frege's arguments on behalf of his strange conception that the reference of a sentence is a truth value. I shall base my discussion not only on his famous article "Über Sinn und Bedeutung," but also on his little known sketch of a logic textbook "Einleitung in die Logik," written in 1906, but published only recently (in 1969).

In "Über Sinn und Bedeutung"[15] Frege argues in the following way: Let us suppose that a sentence has a reference. Let us substitute in this sentence one word with another having the same reference but a different sense. This cannot have any influence on the reference of the sentence. Under such transformations, however, the thought does not remain invariant. Thus since the thought cannot be the reference of the sentences, it will be its sense. We can exclude all considerations about the reference of expressions, and consider only their sense, i.e., the thought. Only our search for truth leads us from sense to reference. We have to look for a reference of a sentence if we are interested in the reference of its constituents. But this happens only when we ask about the truth value of the sentence. Thus, we are constrained to consider the truth value as the reference of the sentence.

Perhaps because he was not completely convinced of his argumentation, Frege adds that if he is correct, i.e., if the reference of a sentence is its truth value, then this has to remain invariant under substitutions in the sentence of expressions for expressions with different senses but the same reference. Moreover, he asks if there is anything else besides the truth value which belongs to every sentence in which we consider the reference of its components and which remains invariant under substitutions of expressions with the same reference, but with different senses. Husserl could answer Frege: yes, the state of affairs and the situation of affairs.

Frege's argumentation is not conclusive. He has shown only that, under the assumption—which we admit with Frege—that invariance under the kind of substitution mentioned immediately above should serve as an adequacy criterion for the reference of sentences, the thought cannot be the reference of a sentence.

In "Einleitung in die Logik"[16] the argumentation is also inconclusive. His argument runs as follows: Only when we are interested in the truth value of a sentence do we require of every proper name which occurs in the sentence that it possess a reference. Moreover, we know that for the sense of the sentence, i.e., for the thought, it is indifferent that all parts of the sentence have a reference or not. Thus, there must be something distinct from the thought and connected with the sentence, for which it is essential that all the components of the sentence have a reference. But the only thing for which it is essential that the components of the sentence have a reference is the truth value of the sentence. Thus, the reference of a sentence is its truth value.

In this case also Frege has not shown that the reference of the components of the sentence is essential only for the truth value of the sentence. The reference of the components of the sentence is essential both to the state of affairs and to the situation of affairs.

5

In his letter of May 24, 1891—the first of the Frege–Husserl correspondence—Frege thanks Husserl for sending him copies of *Philosophie der Arithmetik*, "Der Folgerungskalkül und die Inhaltslogik," and "Rezension von Schröder's Vorlesungen über die Algebra der Logik," all of them published the same year. Most of the letter consists of a commentary on Husserl's review of Schröder's book, and Frege tries to emphasize both the coincidences and the differences between his and Husserl's conceptions, with their respective evaluations of Schröder's work serving as a starting point. Frege comments that with respect to the relation between sense and reference they differ in their conceptions of the sense and the reference of conceptual words. He formulates his theory in the following way:

Sentence	Proper Name	Conceptual Word
↓	↓	↓
Sense of the Sentence (Thought)	Sense of the Proper Name	Sense of the Conceptual Word
↓	↓	↓
Reference of the Sentence (Truth Value)	Reference of the Proper Name (Object)	Reference of the Conceptual Word (Concept)	→ Object that falls under the Concept

Frege explains that for him, but not for Husserl, there is one more step between the conceptual word and the object than between the proper name and the object, and adds that in this way conceptual words have a reference even when the extension of the concept is empty. According to Frege, in the case of conceptual words, Husserl's conception is the following:

Conceptual Word

↓

Sense of the Conceptual Word
(Concept)

↓

Reference of the Conceptual Word
(Object that falls under the Concept)

In "Ausführungen über Sinn und Bedeutung" Frege comments on Husserl's review of Schröder's book and points out that Husserl accuses Schröder of not distinguishing between (1) the failure of a name to have a sense (or meaning), and (2) the failure of a name to have an object as reference. Moreover, in the letter mentioned above Frege says that by the time he published the *Grundlagen der Arithmetik* he had not made the distinction between sense and reference. Apparently the first work by Frege in which he explicitly establishes the distinction between sense and reference is "Funktion und Begriff," published in 1891 and of which Frege sent a copy with a dedication to Husserl more or less at the same time at which he sent him the letter, i.e., after having read and discussed in this letter Husserl's review of Schröder's book. It is just incredible that Føllesdal,[17] Solomon,[18] Angelelli,[19] and others could think that Husserl is indebted to Frege for this distinction and for his abandonment of psychologism. Remarkably, one of Angelelli's 'arguments' consists precisely in telling us that the copy referred to above of "Funktion und Begriff" was found in Husserl's private library after his death, and that it contained many remarks and comments by Husserl. Concerning his abandonment of his 'psychologistic position'—which as Frege remarked at the end of his review of *Philosophie der Arithmetik* was more a sort of conflicting mixture of a psychologist and an objectivist conception of logic—it is convenient to point out the following: (1) *Philosophie der Arithmetik,* although published in 1891, is an extension of Husserl's work of 1887 "Über den Begriff der Zahl," and contains mostly Husserl's conceptions before 1890. (2) *Logische Untersuchungen* took Husserl approximately ten years to think out and write, and that means from 1890 onwards.[20] (3) Husserl himself comments[21] in some manuscripts that his reading of Bolzano, Lotze, and Hume in the years 1890 and 1891 was the decisive factor for his antipsychologistic turn, and this he says before Frege's review. (4) There exist manuscripts by Husserl on the philosophy of mathematics written before 1894 (the year of Frege's review), which contain essentially the same conceptions on these matters that will be presented much later in the *Logische Untersuchungen* and in *Formale und transzendentale Logik.* (5) Moreover, any thorough examination of chapter 11 of the *Prolegomena* will reveal essential differences between Husserl's and Frege's conceptions of logic, mathematics, and the relationship between them.[22] (In many aspects Husserl's conception of mathematics is much nearer to Bourbaki's, but comes close also in some aspects to Hilbert's Program.)

Concerning the distinction between sense and reference, I should like to point out that Husserl established it—although in a somewhat unclear way—in 1890 in his article "Zur Logik der Zeichen," published in the Husserliana edition of *Philosophie der Arithmetik.*[23] In this article Husserl still admits Mill's conception for proper names in the strict sense (like 'Socrates' or 'Napoleon'), but establishes the distinction for descriptions ('indirect signs' in Husserl's terminology) and for 'general names'. Thus Husserl says:

> Bei indirekten Zeichen ist es notwendig zu trennen: dasjenige, was das Zeichen bedeutet und das, was es bezeichnet. Bei direkten Zeichen fällt beides zusammen. Die Bedeutung eines Eigennamens z.B. besteht darin, daß er eben diesen bestimmten Gegenstand benennt. Bei indirekten Zeichen hingegen bestehen Vermittlungen . . . und eben darum machen sie die Bedeutung aus.[24]

Husserl continues:

> Demgemäß besteht z.B. die Bedeutung des allgemeinen Namens darin, daß er irgendeinen Gegenstand aufgrund und vermittels gewisser begrifflicher Merkmale, die er besitzt, bezeichnet.[25]

Moreover, in the same article[26] Husserl defines two signs as (1) identical, if they designate the same object, or the same extension of objects, in the same way; and as (2) equivalent, if they designate the same object, or the same extension of objects, in a different way. These two definitions and the corresponding distinction between the two concepts defined are similar to Carnap's distinction between L-equivalent and equivalent designators.[27] (I must emphasize, however, that in the article referred to the distinction between sense and reference is not exempt from some confusion especially in some examples.)

Thus, we may conclude about Føllesdal's and others' statement that it was Frege's influence on Husserl that both turned Husserl away from psychologism and taught him to distinguish between sense and reference, that it is completely unfounded. Husserl did not generalize in *Ideen I* any theory of Frege on sense and reference[28] but in any case his own theory on sense and reference, and this generalization had been made before in the Fifth Investigation of the *Logische Untersuchungen*.

6

For Husserl, as for Frege, every expression not only expresses something, but also refers in this way to something; not only does it have a meaning (or sense), but it refers to an objectuality.[29] The expression refers to an objectuality through its meaning. It is precisely its meaning that solely constitutes the essence of an expression.[30] Since my interest here is limited to a comparison between Husserl's and Frege's theories of sense and reference, I am not going to consider the many other interesting distinctions which Husserl establishes, especially in chapter 1 of the First Investigation.

Thus, for Husserl also we have to distinguish between the sense (or meaning) and the reference of an expression. Two proper names can have a different sense (or meaning) but name the same object. The proper names 'the winner of Jena' and 'the loser of Waterloo' express different meanings, but name (in a different way) the same object. Hence, there is a perfect coincidence between Husserl and Frege concerning the sense and the reference of proper names.

Husserl would probably not accept as wide a use of his term 'universal name' as that given by Frege to his term 'conceptual word'. Moreover, since

Husserl's motivations are not exclusively logical, he would not admit Frege's classification of the components of singular assertive sentences in proper names and conceptual words, a classification primarily motivated by logical simplicity. Husserl's theory of a logical grammar pretends to serve as a foundation both for logic and for a sort of rational universal grammar, and cannot allow for a simplification like Frege's.[31] Since my interest is not primarily that of a historian, I can be content with saying that Husserl's concept of 'universal name' is in Husserl's theory what most nearly approximates Frege's concept of 'conceptual word'.

According to Husserl, the sense of a universal name is a concept, and its denotation is the extension of the concept, which Husserl calls a multiple extension, since it contains all objects that fall under the concept.[32] Thus, e.g., the universal name 'a horse' can refer (or better, be applied) in one case to Rocinante and in another case to Bucephalus, since both belong to the extension of the concept expressed by the universal name 'a horse'. Hence, universal names have a reference which changes from one concrete case to another in a fixed extension determined by a fixed concept. (Of course, the concept can change without any change in the extension.) But a universal name is not an equivocal name, since equivocity concerns fixed expressions with different meanings.

The sense of an assertive sentence (*Aussage*) is a proposition (*Satz*). Contrary to Frege, Husserl uses the word '*Satz*' to refer to the meaning of a sentence, i.e., as a synonym for '*Aussagebedeutung*', or to refer to the sentence together with its meaning. Sometimes Husserl even opposes the expressions '*Aussagesatz*' and '*Aussage*'.[33] In §11 of the Fourth Investigation Husserl even uses Frege's expression '*Gedanke*' as a synonym of '*Satz*' to refer to the sense of an assertive sentence.

Hence, an assertive sentence expresses a meaning (or sense), which is a proposition (or thought) and refers to an objectuality. This objectuality is, according to Husserl, not a truth value, but a state of affairs (*Sachverhalt*).[34] Although it is true that the predicates 'is true' and 'is false' apply to thoughts, this does not make them, according to Husserl, the reference of assertive sentences. Thus, the sentence 'The earth is round' refers to the state of affairs that the earth is round. Different sentences can refer to the same state of affairs by means of different propositions (thoughts). E.g., the sentences 'The morning star is a planet of our solar system' and 'The evening star is a planet of our solar system' refer to the same state of affairs by means of different thoughts. This and other similar examples convince us that substitution in assertive sentences of one expression by another with the same reference but different sense does not affect the state of affairs referred to by the sentence.

In the *Logische Untersuchungen* Husserl considers also the situation of affairs as the possible reference of an assertive sentence.[35] Although Husserl's discussion is not free from confusion, we use Husserl's example to make the distinction between state of affairs and situation of affairs clear. The sentences '*a* is larger than *b*' and '*b* is smaller than *a*' are assertive sentences that refer to different states of affairs but to the same underlying situation of affairs. Here also the corresponding propositions are different, but their difference is not reducible to a difference between component parts which have the same reference but a different sense. The following example is much clearer. Consider the expressions: (1) '5 + 2 < 6 + 3', (2) '4 + 3 < 6 + 3', and (3) '6 + 3 > 4 + 3'. Expression (2) differs from expression (1) since it expresses a different thought, it has a different sense, but differs from expression (3) not only because it expresses a different thought, but also because it refers to a different state of affairs. (The difference between (1) and (3) could be considered as a sort of composition of the differences between (1) and (2) and between (2) and (3).) However, all three expressions (1), (2), and (3) refer to the same situation of affairs. As in the case of the state of affairs, the situation of affairs is invariant under substitutions of expressions in an assertive sentence for other expressions having the same reference but a different sense.[36] Moreover, the following three relations are many-one: (a) the relation between propositions (i.e., thoughts) and states of affairs, (b) between states of affairs and situations of affairs, and (c) between situations of affairs and truth values.

In *Erfahrung und Urteil* Husserl tries to clarify the difference between states of affairs and situations of affairs. States of affairs are objects of the understanding, or better, categorial objects, which are constituted in judgments (as their reference), whereas situations of affairs are the prejudicative supports of states of affairs, a sort of 'reference base' for judgments. Husserl's scheme would be, thus, the following:

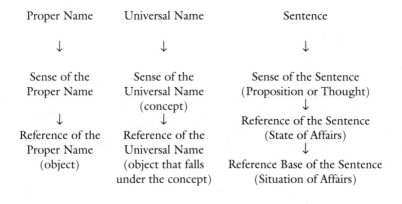

Proper Name	Universal Name	Sentence
↓	↓	↓
Sense of the Proper Name	Sense of the Universal Name (concept)	Sense of the Sentence (Proposition or Thought)
↓	↓	↓
		Reference of the Sentence (State of Affairs)
Reference of the Proper Name (object)	Reference of the Universal Name (object that falls under the concept)	↓
		Reference Base of the Sentence (Situation of Affairs)

↓

Truth Value

7

Although both the state of affairs and the situation of affairs remain invariant under substitutions that affect only the sense, but not the reference of partial expressions—contrary to Frege's belief that only the truth value would remain invariant—it is easy to imagine stronger sorts of transformations of sentences that affect the state of affairs and the situation of affairs without affecting the truth value of the sentence. This is precisely what Alonzo Church does in the introduction of his well-known *Introduction to Mathematical Logic*, although he erroneously thinks to be arguing on behalf of Frege's thesis that the reference of an assertive sentence is a truth value, and does not notice that he is assuming the validity of that same thesis.

In my opinion the preference for Frege's thesis or for any of the two Husserlian alternatives—or any other reasonable alternative—will probably be based on pragmatic considerations. I accept with Frege that invariancy under substitution of partial expressions with different senses but the same reference should be used as a criterion of adequacy in the sense that any proposed alternative which does not fulfill this requirement should be excluded. Furthermore, each of the three alternatives, i.e., invariance of truth value, invariance of situation of affairs, and invariance of state of affairs—and even the limit case of invariance of thought—determine partitions and corresponding equivalence classes in the set of all sentences of a language. Thus, under the partition determined by invariance of state of affairs, the sentences 'The morning star is a planet of our solar system' and 'The evening star is a planet of our solar system' belong to the same equivalence class, but they do not belong to the same equivalence class under the partition determined by invariance of thought. Moreover, the sentences 'Napoleon was defeated in the battle of Waterloo' and 'Paris is the capital of France' belong to the same equivalence class under the partition determined by invariance of truth value, but not under the partition determined by invariance of situations of affairs—and, hence, not under the partitions determined by invariance of state of affairs or invariance of thought. Finally, the sentences 'Sugar is soluble in water' and 'Solubility in water is a property of sugar' belong to the same equivalence class under the partition determined by invariance of situation of affairs, but not under the partition determined by invariance of state of affairs. (Moreover, all pairs of expressions of the form $(xRy, yR^{-1}x)$, where R^{-1} is the inverse relation of R, belong to the same equivalence class under the partition determined by invariance of situation of affairs, but not under the partition determined by invariance of state of affairs.)

In the case of each of the alternatives we can consider sentences as objects and the corresponding substitutions as transformations that preserve the reference of sentences and in each case we obtain (with appropriate definitions) a group of transformations. Thus, e.g., let us consider the transformations

$T_1, T_2, \ldots$ that preserve the state of affairs under variance of the sense of partial expressions. Let e be the transformation which applied to any sentence S transforms it in itself, i.e., $e(S) = S$. Let $T * T'$ be the transformation which consists of an application of T' to a sentence S and of T to the result of applying T' to S. Finally, let T^{-1} be the inverse of T. It is easily seen that the group axioms are satisfied. E.g., for any T, $T * e = T = e * T$ since e leaves any sentence unchanged, and, hence both $T * e$ and $e * T$ reduce to T. Associativity is trivial. It is easy to see that if $T(S) = S'$, where S' results from S by substituting an expression c' for an expression c in S with different sense but the same reference, then $T^{-1}(S') = S$, and this T^{-1} is trivially unique. On the other side, it is easily seen that commutativity is not satisfied. Thus the transformations form a nonabelian group. The other candidates for reference also give rise to corresponding nonabelian groups of transformations of sentences. Moreover, the group of transformations determined by invariance of state of affairs is isomorphic to a quotient subgroup of the group of transformations determined by invariance of situation of affairs, and both of them are isomorphic to quotient subgroups of the group determined by invariance of truth value. More generally, we can consider the map from any group of transformations determined by a 'stronger' invariance principle (e.g., invariance under state of affairs) to another group of transformations determined by a relatively 'weaker' invariance principle (e.g., invariance under situation of affairs) as a (partially) forgetful functor from a category of groups to a category of groups. An interesting problem consists in determining the structure of the set of all partitions determined by candidates for reference (which, thus, obey the criterion of adequacy). E.g., this set has a minimal element, namely, equivalence, i.e. sameness of truth value, and a trivial maximal element, namely, logical equivalence in Carnap's sense of invariance of proposition (or thought), whose corresponding group of transformations is the trivial group (e). (Here we assume an ideal language in which there are no mere differences of 'tone'. In the contrary case, there would be many Carnapian partitions, but only one of them would correspond to the trivial group.) Other members of the set are partitions determined by the two Husserlian candidates and partitions determined by truth value equivalences in appropriate classes of possible worlds.

Finally, I would like to make clear that most of the possible candidates (i.e., those that fulfill the criterion) can be excluded on pragmatic grounds. For instance, the 'maximal element' can be excluded on the grounds that it makes unnecessary the distinction between sense and reference, since there would be a trivial bijection between them. Candidates determined by truth value equivalence in an appropriate class of possible worlds have the handicap that preference for one of them would seem quite arbitrary. The Fregean alternative and the two Husserlian alternatives seem to be the more natural choices. Frege's candidate has the advantage of simplicity.

Actually, it is the simplest possible choice. But it has the disadvantage of not being very informative, since it obviates important semantic relations between sentences that are not reducible to sameness or difference of truth value (in the actual world). Both of Husserl's candidates are informative and nontrivial. Husserlian states of affairs have the advantage against situations of affairs that we usually learn to know these ones through them. On the other hand, situations of affairs are possibly more acceptable to philosophers not sympathetic to categorial objects. In any case, preference for one of the Husserlian candidates against the other is not easy to justify. Moreover, such an abstract metatheoretical study of semantic theories as has been sketched above could throw some light on the problem of selecting rival semantic theories. But before such a possibility can be taken seriously, it is necessary to determine a set of desirable properties that could serve, if not as a criterion of adequacy for semantic theories, at least as a guide for evaluating them. For instance, I conjecture that invariance of state of affairs is the strongest invariance principle that fulfills the Fregean criterion of adequacy, or equivalently, the group of transformations determined by invariance of state of affairs constitutes the smallest possible nontrivial group of transformations that obeys the criterion.

NOTES

1. See, e.g., E. Husserl, *Logische Untersuchungen*, 2nd ed. rev. (Halle: Niemeyer, 1913), Fourth Investigation, §11.
2. In E. Husserl, *Ideen zu einer reinen Phänomenologie und einer phänomenologischen Philosophie, Erstes Buch*, Husserliana, vol. III (The Hague: M. Nijhoff, 1950 [1913]). Husserl uses the term '*Sinn*' in a more general fashion to apply to all objectifying acts—approximately as a synonym of 'noematic nucleus'—whereas the word '*Bedeutung*' is used in the more restricted sense of the '*Sinn*' of meaning conferring acts.
3. See §§3 and 6 of the present chapter.
4. Frege uses the term 'proper name' also for sentences, but I will try to avoid this use, since it can lead to confusion.
5. G. Frege, "Über Sinn und Bedeutung," *Kleine Schriften*, ed. I. Angelelli (Darmstadt: Wissenschaftliche Buchgesellschaft, 1967; 2nd ed., Hildesheim: Olms, 1990), pp. 143–62.
6. Contrary to other commentators of Frege like Michael Dummett (see his *Frege, Philosophy of Language* [London: Duckworth, 1973]), I interpret Frege as considering that statements of the form '$a = b$' express a relation between senses, namely, the relation of having the same reference. More exactly, there is a (partial) function from the set of all senses to the set of all references, which establishes a partition in its domain, according to which all senses with the same reference belong to the same equivalence class. The relation of identity is then the congruence relation between two senses generated by the partition, and thus determined by the sameness of reference.

7. G. Frege, "Ausführungen über Sinn und Bedeutung," *Nachgelassene Schriften,* ed. H. Hermes et al. (Hamburg: Meiner, 1969; 2nd ed., 1983), pp. 128–36.
8. Dummett, *Frege, Philosophy of Language,* chapter 5.
9. Ibid., pp. 84–85.
10. G. Frege's, "Funktion und Begriff," *Kleine Schriften,* pp. 125–42.
11. See his letter to Husserl from May 24, 1891 in Frege's *Wissenschaftlicher Briefwechsel,* ed. G. Gabriel et al. (Hamburg: Meiner, 1976).
12. Frege, "Ausführungen über Sinn und Bedeutung," pp. 129–30.
13. Frege, "Funktion und Begriff," p. 128.
14. See, e.g., Frege's "Über Begriff und Gegenstand," *Kleine Schriften,* pp. 167–78.
15. Frege, "Über Sinn und Bedeutung," pp. 148–50.
16. Frege, "Einleitung in die Logik," *Nachgelassene Schriften,* pp. 210–11.
17. See D. Føllesdal, *Husserl und Frege* (Oslo: Ascheloug, 1958).
18. See R. Solomon, "Sense and Essence: Frege and Husserl," *International Philosophical Quarterly* 10 (1970): 387–401.
19. See Angelelli's remarks in Frege's *Kleine Schriften,* pp. 431–32.
20. See K. Schuhmann, *Husserl-Chronik* (The Hague: M. Nijhoff, 1977), p. 25.
21. Ibid., p. 26. See also Husserl's *Introduction to the Logical Investigations* (The Hague: M. Nijhoff, 1975), pp. 32–40.
22. This has been clearly shown in my dissertation "Edmund Husserls Philosophie der Logik und Mathematik im Lichte der gegenwärtigen Logik und Grundlagenforschung" (Ph.D. dissertation, Rheinische Friedrich-Wilhelms-Universität, Bonn, 1973). Concerning the relationship between Husserl and Frege see especially chapter 6. See also J. N. Mohanty's article "Husserl and Frege: A New Look at Their Relationship," *Research in Phenomenology* 4 (1975): 51–62 and Husserl's *Formale und transzendentale Logik,* Husserliana, vol. XVII (The Hague: M. Nijhoff, l974).
23. E. Husserl, "Zur Logik der Zeichen," *Philosophie der Arithmetic mit ergänzenden Texten,* Husserliana, vol. XII (The Hague: M. Nijhoff, 1970), pp. 340–73, translated in Husserl's *Early Writings in the Philosophy of Logic and Mathematics* (Dordrecht: Kluwer, 1994), pp. 20–51. The articles Husserl sent to Frege in 1891 are also anthologized here.
24. Ibid., p. 343.
25. Ibid. pp. 343–44.
26. Ibid., p. 344.
27. According to Carnap (see his *Meaning and Necessity* [Chicago: University of Chicago Press, 1947]) two designators are equivalent if and only if they have the same extension, and they are L-equivalent if and only if they have the same intension (p. 23). But for any expression, its ordinary (Fregean) reference is the same as its (Carnapian) extension (p. 125), whereas its ordinary (Fregean) sense is the same as its (Carnapian) intension (p. 126). Thus, two designators are equivalent if and only if they have the same (ordinary) reference, and two designators are L-equivalent (or identical) (p. 25) if and only if they have the same (ordinary) sense.
28. See D. Føllesdal, "An Introduction to Phenomenology for Analytic Philosophers," *Contemporary Philosophy in Scandinavia,* ed. R. Olson and A. Paul (Baltimore: The Johns Hopkins Press, 1972), pp. 417–29; and R. Solomon's "Husserl's Concept of the Noema" in *Husserl: Expositions and Appraisals,* ed. F. Elliston and P. McCormick (Notre Dame: Notre Dame University Press, 1977), pp. 168–81.

29. See, e.g., Husserl's *Logische Untersuchungen*, First Investigation, §12.
30. Ibid., §13.
31. See Husserl's *Logische Untersuchungen*, Fourth Investigation.
32. Husserl, *Logische Untersuchungen*, First Investigation §12.
33. See Husserl's *Logische Untersuchungen*, First Investigation §§11, 29, and Fourth Investigation §11.
34. Husserl, *Logische Untersuchungen*, Fourth Investigation §11.
35. Husserl, *Logische Untersuchungen*, First Investigation.
36. In both cases the invariance can be proved by induction on the complexity of sentences.

3

Guillermo E. Rosado Haddock

IDENTITY STATEMENTS IN THE SEMANTICS OF SENSE AND REFERENCE

By a semantic theory of sense and reference I understand an ordered quintuple $<E, S, R, <s_i>_{i\epsilon I}, <r_j>_{j\epsilon J}>$, where E is a set of syntactically well formed expressions of a given language—possibly divided in syntactic categories—S is a set of senses, R is a set of (possible) referents, the s_i are partial functions that assign to (at least some) members of E members of S, and the r_j are partial functions that assign to members of S members of R.[1]

Frege's semantic theory falls in a natural way under the definition just given. In Frege's semantics we have essentially two categories of expressions: the category of proper names and the category of functional expressions. The category of proper names can be further subdivided in two subcategories: the category of sentential proper names (or statements) and the category of nonsentential proper names (which includes both proper names in the strict sense and definite descriptions). The category of functional expressions, on the other hand, contains as a special case the conceptual (and relational) words. According to Frege, all (meaningful) expressions express a sense and through this sense refer to something. Proper names, e.g., have a sense and through it they refer to an object. In the particular case of sentential proper names, the sense expressed is a thought and the object referred to is a truth value, i.e., either the true or the false. Functional expressions also express a sense, but they refer to a function, not to an object. In particular, a conceptual word expresses a sense and through this it refers to a concept. This concept—that by being a function belongs to an ontological category different from that of objects—determines a (possibly empty) extension— that belongs to the ontological category of objects.[2]

1

Frege's semantics is not the only possible semantic theory of sense and reference,[3] and my discussion in this article is, thus, not restricted to it. I assume with Frege, however, that in identity statements the expressions at each side of the identity sign are proper names (in Frege's broad sense that includes both definite descriptions and strict proper names) and not conceptual words. Under this general assumption, I will try to show that in a semantic theory of sense and reference there is only one sound interpretation of identity statements. In my discussion we will benefit from some particularly interesting remarks of Frege in "Über Sinn und Bedeutung" on this issue, but I will not try to establish that my interpretation coincides with Frege's, since neither in the days when he wrote *Begriffsschrift* nor in his mature years did Frege have a clear notion of identity statements. I think, however, that my interpretation is the only one that does justice to those remarks of Frege in "Über Sinn und Bedeutung," and that it is, thus, implicit in that article.

2

Let us consider an identity statement of the form '$a = a$', e.g., 'Venus = Venus', and a true identity statement of the form '$a = b$', e.g., 'the morning star = the evening star' or 'Tully = Cicero'. Frege has clearly stated[4] that there is an important difference in cognitive value between statements of the form '$a = a$' and true statements of the form '$a = b$'. Sometimes complicated mathematical calculations or extensive empirical investigations are needed for establishing that a statement of the form '$a = b$' is true (or that it is false), whereas for knowing that a statement of the form '$a = a$' is true we do not need any such investigation. An interpretation of identity statements has to do justice to this difference in cognitive value between a statement of the form '$a = a$' and a true statement of the form '$a = b$', if it is to be considered an acceptable interpretation.

On the other hand, it seems that not all statements of the form '$a = b$' are empirical (or even synthetic), since, e.g., the statement 'the least even number = the only even prime number' seems to express an analytic truth. But it seems that some of the statements of the form '$a = b$' are empirical, e.g., 'the teacher of Alexander the Great = Aristotle' or 'the morning star = the evening star'. An acceptable interpretation of identity statements cannot, thus, exclude either the possibility that a true identity statement expresses an empirical truth or the possibility that it expresses a logical or mathematical necessity.

We arrive, thus, at the following two adequacy criteria for interpretations of identity statements.

Criterion I: An interpretation of identity statements must do justice to the difference in cognitive value between statements of the form '$a = a$' and true statements of the form '$a = b$'.

Criterion II: An interpretation of identity statements cannot imply either that all identity statements of the form '$a = b$' are empirical or that they are necessary (where—to fix concepts—we understand 'necessity' in Kripke's sense, i.e., a statement S is necessary if it is true in every possible world in which exist the objects referred to by denoting expressions in S).[5]

3

Now, how are we to interpret identity statements? First of all, we must distinguish two questions that are frequently taken as one and the same, namely: (1) What is expressed by an identity statement of the form '$a = b$'? and (2) Under what conditions is an identity statement of the form '$a = b$' true? The second question can be answered without difficulties by saying that in case that the expressions 'a' and 'b' have the same object as referent. This, however, does not answer the first question, which is the one under discussion when we try to interpret identity statements.

I shall consider the following six interpretations of identity statements: (I) They express a relation of identity between the two objects that are the referents of the expressions at each side of the identity sign. (II) They express the relation of identity with itself of an object that is the common reference of the expressions at each side of the identity sign. (III) They express the relation of identity between the expressions at each side of the identity sign. (IV) They express the congruence relation, determined by sameness of reference, between the expressions at each side of the identity sign. (V) They express the relation of identity between the senses of the expressions at each side of the identity sign. (VI) They express the congruence relation, determined by sameness of reference, between the senses of the expressions at each side of the identity sign.

In *Begriffsschrift* Frege seems to have fluctuated between interpretations (III) and (IV). When he introduces the identity sign, Frege states that identity statements express a relation between names or signs, which seems to indicate that he accepts interpretation (III).[6] However, when he tries to explain his conception in detail, Frege seems to favor interpretation (IV).[7] Caton's[8] and Schirn's[9] opinion that after all Frege's conception of identity statements in *Begriffsschrift* is not radically different from his conception in "Über Sinn und Bedeutung" seems to receive support therein. We will later see, however, that interpretation (IV) is essentially different from interpretation (VI), which, in my opinion, is the one implicit in "Über Sinn und Bedeutung." Dummett[10] and Khatchadourian[11] seem to interpret Frege's mature position in the sense of interpretation (II). Although—as remarked by Schirn[12]—some statements of Frege in *Grundgesetze der Arithmetik*[13] seem to give support to that interpretation, I think, however, with Schirn, that such an interpretation does not do justice to Frege's analysis of identity statements in "Über Sinn und Bedeutung." Some remarks of Frege in "Ausführungen über Sinn und Bedeutung"[14] as well as some remarks of Dummett[15] seem to favor interpretation (I). It seems, however, that in both cases such isolated remarks do not faithfully represent the opinions of their authors.

Let us consider now some difficulties that present themselves to interpretations (I) through (V).

4

First of all, if what is expressed by an identity statement were a relation of identity between expressions—i.e., if interpretation (III) were the case—then all statements of the form '$a = b$' would be false, since 'a' and 'b' are different expressions. Hence, e.g., the statements 'Londres = London', 'Hesperus = Phosphorus,' and 'the morning star = the evening star' would all be false. Only statements of identity of the form '$a = a$', like e.g., 'London = London,' could be true under such an interpretation (and even this only under the assumption that by an expression we mean an expression-type and not an expression-token).

On the other hand, if identity statements were to express a relation of identity between senses—as is stated by interpretation (V)—then all statements of the form '$a = b$', where 'a' and 'b' are expressions with different senses, would be false. Hence, e.g., the statements 'London = London' and 'Londres = London' would be true, but the statements 'Hesperus = Phosphorus' and 'the morning star = the evening star' would be false.

Let us consider now interpretation (I), according to which identity statements express a relation of identity between two objects that are the referents of the expressions at each side of the identity sign. But if they are two objects, then one cannot properly speak of identity. They could not have exactly the same properties since each one of them would have the property of not being the other, and this property could not be shared by this other, since an object cannot be different from itself. All that we could have under interpretation (I) would be a so-called relative identity, or, more precisely, an identity in a certain aspect. In such a case, however, one would have to establish in which aspect the objects under discussion are identical, and to establish this the corresponding identity statement is of no help, if we want to avoid an infinite regress. Moreover, it could happen that the object that is the referent of 'a' were identical in a certain respect to the object that is the referent of 'b', and that it were identical in a completely different respect to the object that is the referent of 'c'. In such a case, the statements '$a = b$' and '$c = a$' would both be true, but the statement '$c = b$' would possibly be false, since the referents of 'b' and 'c' could be non-identical in every aspect. Hence, under interpretation (I), the transitivity of identity would not be valid. Moreover, under such an interpretation, the identity sign would be equivocal. We would either (1) have to abandon the transitivity of identity, or (2) have as many relations of identity as there are aspects of reality in which two objects can be identical. Both (1) and (2) are inadmissible. Interpretation (I) is thus as unsound as interpretations (III) and (V).

According to interpretation (II), identity statements express the relation of identity of an object with itself. Under such an interpretation there would not be any difference in cognitive value between an identity statement of the form '$a = a$' and a true identity statement of the form '$a = b$'. More precisely, under this interpretation every true identity statement of the form '$a = b$' is necessary, since it is true in every possible world that every object is always identical with itself. Hence, under this interpretation not only 'London = London' and 'Londres = London' would be necessarily true, but also 'Hesperus = Phosphorus' and 'the morning star = the evening star'. Interpretation (II) violates both Criterion I and Criterion II. Moreover, one could ask the proponents of interpretation (II) if a false statement of the form '$a = b$' also expresses the relation of identity of an object with itself or something else. If it expresses the relation of identity of an object with itself, it could not be false, and, therefore, all statements of the form '$a = b$' would be necessarily true. If it expresses something else, then the sense of a statement of the form '$a = b$' would be a function of its truth value, but this is clearly nonsense. Interpretation (II) is, thus, unacceptable.

Finally, let us consider interpretation (IV). According to this interpretation, an identity statement expresses a congruence relation, determined by sameness of reference, between the expressions at each side of the identity sign. This interpretation clearly satisfies Criterion I, since it allows us to establish an essential difference between an identity statement of the form '$a = a$' and a true identity statement of the form '$a = b$'. As remarked by Schirn[16] and Kienzle,[17] however, the relation between an expression and its referent is arbitrary. Therefore, we could neither establish the truth nor the falsehood of a statement of the form '$a = b$' solely by the analysis of its component expressions. Hence, with the exception of linguistic conventions, all identity statements of the form '$a = b$' would be, as observed by Kienzle,[18] synthetic, and even a posteriori, under this interpretation. Thus, even statements like '$2^2 = 4$' or 'Londres = London' would be a posteriori under interpretation (IV). Such an interpretation clearly violates Criterion II.

5

On the other hand, the relation between a sense and its reference is not an arbitrary one. That the sense of, e.g., words such as 'the smallest even number', or 'the teacher of Alexander the Great', or 'the morning star' have the referent they have (i.e., the number 2, Aristotle, and Venus, respectively) is not the result of an arbitrary stipulation. It is an objective fact that two senses have the same referent, and, thus, that according to interpretation (VI) they belong to the same equivalence class of senses determined by sameness of reference. Moreover, interpretation (VI)—according to which identity statements express a congruence relation, determined by sameness of reference, between the senses of the expressions at each side of the identity sign—is the only one that: (I) satisfies Criteria I and II; (2) avoids the difficulties that present themselves to interpretations (I) through (V); and (3) does justice to Frege's remarks in "Über Sinn und Bedeutung" according to which not only the reference, but also the sense plays a decisive role in identity statements.

First of all, interpretation (VI) does justice to the difference in cognitive value that exists between identity statements of the form '$a = a$' and true identity statements of the form '$a = b$'. Whereas in an identity statement of the form '$a = a$', not only the reference but also the sense of the expressions at each side of the identity sign is the same, in a true identity statement of the form '$a = b$' the expressions 'a' and 'b' have the same reference, but usually have different senses. The cognitive value is in the two cases completely different. Whereas '$2 = 2$' and 'the teacher of Alexander the Great = the teacher of Alexander the Great' are true identity statements that do not add any new knowledge, the true identity statements 'the smallest even number = the only even prime number' and 'the teacher of Alexander the Great = the most famous disciple of Plato' do add new knowledge, and have, therefore, a greater cognitive value than the first two. According to interpretation (VI) both '$a = a$' and '$a = b$' express that the senses of the expressions at each side of the identity sign have the same reference, i.e., belong to the same equivalence class of senses, determined by sameness of reference. But whereas in the first case this is obvious, in the second case, it not only is not obvious, but could even not be the case, since '$a = b$' could be false. If, however, '$a = b$' is true, then we have learnt something that is not obvious, or, more exactly: that although the senses of 'a' and 'b' are different, they belong to the same equivalence class determined by sameness of reference, i.e., they have the same reference. Thus, interpretation (VI) satisfies Criterion I.

Moreover, in the first of the last two examples the surplus in cognitive value is obtained 'analytically', whereas in the second example the surplus in cognitive value is obtained by empirical means. Hence, one should clearly distinguish between the extensions of the concepts (and a fortiori between the concepts) 'statement with cognitive value' and 'synthetic statement' and, thus, between the extensions of the concepts 'statement with cognitive value' and 'empirical statement'. (More exactly, to have a cognitive value greater than zero is a mark (*Merkmal*) of the concept 'true synthetic statement' and a fortiori of the concept 'true empirical statement'.) To know the truth of the statement 'the smallest even number = the only even prime number' one needs some elementary knowledge of arithmetic, whereas to know the truth of the statement 'the teacher of Alexander the Great = the most famous disciple of Plato' one needs some knowledge of history of Ancient Greece. Moreover, the first statement is not only an a priori truth, but also a necessary one whereas the second statement is empirically true and contingent. Interpretation (VI) does not exclude any such possibility, and thus satisfies Criterion II.

It should by now be clear that interpretation (VI) not only does justice to the difference in cognitive value between an identity statement of the form '$a = a$' and a true identity statement of the form '$a = b$' (where the expressions 'a' and 'b' have different senses), and to the fact that some true identity statements are necessary and a priori, whereas others are contingent and a posteriori, but also avoids all the difficulties that present themselves to interpretations (I) through (V). Moreover, it is the only interpretation that does justice to each of the following theses more or less implicit in Frege's discussion of identity statements in "Über Sinn und Bedeutung": (1) An identity statement is true if and only if the expressions at each side of the identity sign have the same referent. (2) In a true identity statement the expressions at each side of the identity sign can either have the same sense or different senses. (3) True identity statements in which the expressions at each side of the identity sign have different senses possess a greater cognitive value than those in which the expressions at each side of the identity sign have the same sense. (4) Sameness of reference is not sufficient to do justice to the difference in cognitive value between identity statements of the form '$a = a$' and true identity statements of the form '$a = b$'. (5) That which is decisive for the difference in cognitive value existing between statements of the form '$a = a$' and true statements of the form '$a = b$' is the sense.[20]

6

Recently[21] Saul Kripke has offered an account of identity statements that has received wide acceptance in some circles of analytical philosophy. Although I have discussed this account somewhere else,[22] I would like to finish this article with some remarks about it.

Kripke divides the class of referring expressions into two disjoint subclasses, namely, the class of rigid designators and that of nonrigid ones. A designator is called rigid if it refers to the same object in every possible world, and is called nonrigid if that is not the case, i.e., if it refers to different objects in different possible worlds. According to Kripke, proper names in the strict sense (like 'Socrates' or 'Napoleon') are rigid designators— although not all rigid designators are strict proper names (e.g., 'the square root of 2' is a definite description and seems to be rigid). On the other hand, definite descriptions usually are nonrigid designators. Kripke follows Frege in attributing a sense to definite descriptions (like, e.g., 'the teacher of Alexander the Great' or 'the evening star') and in considering that through this sense they refer to an object in the actual world. He adds, however, that in other possible worlds definite descriptions could have a different referent from the one they have in the actual world. Hence, in another possible world the expression 'the teacher of Alexander the Great' could have had Socrates or Napoleon as referent. Strict proper names, however, do not express any sense, but only refer to some object, and this object, if it exists, is the referent of that strict proper name in every possible world. Therefore, Kripke concludes that a true identity statement of the form '$a = b$', where both 'a' and 'b' are strict proper names, is necessarily true, i.e., true in any possible world in which both 'a' and 'b' have a referent. On the other hand, if 'a' or 'b' or both are definite descriptions, the statement '$a = b$', if true, would probably be only contingently true. (A possible exception would be a statement like 'the smallest even number = the only prime even number'.)

Kripke acknowledges the possibility that a strict proper name may have had a different referent from the one it has in the actual world.[23] Thus, e.g., 'Hesperus' and 'Phosphorus' could have had different referents. But 'Hesperus = Phosphorus' is, after all, necessarily true. According to Kripke, this is based on the identity with itself in every possible world of the object Venus, which is the common referent of 'Hesperus' and 'Phosphorus' in the actual world. Khatchadourian,[24] who coincides with Kripke in believing in the rigidity of strict proper names, observes correctly that Kripke confuses the problem of the rigidity of designators with the completely different issue of the identity of an object with itself in every possible world. According to Khatchadourian, the sole (or at least principal) ground for the rigidity of strict proper names is that this is a condition for the intelligibility of any counterfactual statement in which they occur.

But this is not a distinct trait of strict proper names. In ordinary (or usual) discourse one assumes not only that strict proper names but also that definite descriptions always refer to the same object, and this assumption is a condition for the intelligibility of such a discourse. Both when someone says 'Tully could have been not a senator'[25] and when someone says 'The man who denounced Catiline could have been not a senator' usually the intelligibility of the statements presupposes that every strict proper name and every definite description that occurs in them refer to the object which is its referent in the actual world. Hence, in this respect strict proper names are not more rigid than definite descriptions. Certainly, however, there exist unusual situations in which someone intends to express by a statement like 'The person who denounced Catiline could have been not a senator' not that the man Cicero could have been not a senator, but that the person who denounced Catiline could have been a person different from the man Cicero and not a senator. E.g., it could have been a former collaborator of Catiline. In such a case the statement would express the possibility that the definite description has a different referent from the one it has in the actual world. But the statement 'Tully could have been not a senator' and even the statement 'Tully could have been not Cicero' admit similar interpretations. The first one could be expressing the possibility that the proper name 'Tully' could have a different referent from the one it has in the actual world, i.e., the man Cicero. The second statement could be expressing the possibility that the proper names 'Tully' and 'Cicero' have different referents, although in the actual world they have the same referent. There is no difference in rigidity between strict proper names and definite descriptions.

Moreover, Kripke's thesis that, e.g., 'Hesperus = Phosphorus' is necessarily true, whereas 'the morning star = the evening star' is only contingently true, is based not on any difference in rigidity between strict proper names and definite descriptions, but on the following assumptions of Kripke: (1) Strict proper names do not have any (Fregean) sense, whereas definite descriptions do have. (2) Identity statements between definite descriptions are interpreted according to interpretation (VI) above, i.e., as expressing a congruence relation, determined by sameness of reference, between the senses of the two definite descriptions. (3) Identity statements between strict proper names are interpreted according to interpretation (II), i.e., as expressing—if true—the identity of an object with itself. Now, if Kripke were to interpret identity statements between definite descriptions according to interpretation (II), then, as we have seen, all true identity statements between definite descriptions would be necessary, and there would be no modal difference between such identity statements and those between strict proper names. If, on the other hand, Kripke were to assume that strict proper names have a sense and were to interpret identity statements between strict proper names according to interpretation (VI), then there would be no modal difference either between such identity statements and identity statements between definite descriptions. Moreover, on the basis of (2) and (3), Kripke would have some difficulty in interpreting an identity statement between a strict proper name and a definite description. Such a statement would express a strange relation of identity between the sense of the definite description and the object referred to by the strict proper name. One would have either to consider all such logical monstrosities as semantical nonsense, or one would have to consider them all as false, since even in a case like

'Aristotle = the teacher of Alexander the Great'—where both expressions have the same referent—the sense of the definite description does not coincide with the referent of the strict proper name.

Finally, I would like to observe that since the relation between an expression and its referent is arbitrary, if we do not assume that strict proper names have a sense, then there is no way of showing that there exist rigid strict proper names. Moreover, I suspect that only a definite description that refers to a mathematical or other sort of nonreal (or ideal) entity could be a possible candidate for rigid designator. A discussion of this problem, however, would take us too far afield.

NOTES

1. A more exact definition could be given, but it is not necessary for our purposes.
2. For a detailed and critical exposition of Frege's semantics see either D. Shwayder's article "On the Determination of Reference by Sense," *Studies on Frege*, vol. 3, ed. M. Schirn (Stuttgart: Frommann-Holzboog, 1976), pp. 85–95 or my "Remarks on Sense and Reference in Frege and Husserl," chapter 2 of the present book. I have reviewed Schirn's *Studies on Frege* in *Diálogos* 38 (November 1981): 157–83.
3. See chapter 2 of the present book.
4. G. Frege, "Über Sinn und Bedeutung," *Kleine Schriften*, ed. I. Angelelli (Darmstadt: Wissenschaftliche Buchgesellschaft, 1967; 2nd ed., Hildesheim: Olms, 1990), pp. 143, 162.
5. See S. Kripke, "Naming and Necessity," *Semantics of Natural Language*, ed. D. Davidson and G. Harman (Dordrecht: Reidel, 1972), pp. 253–355 and pp. 763–69 and S. Kripke, "Identity and Necessity," *Identity and Individuation*, ed. M. K. Munitz (New York: New York University Press, 1971), pp. 135–64.
6. G. Frege, *Begriffsschrift* (Halle: Nebert, 1879; reprint, Hildesheim: Olms, 1964), p. 13.
7. Ibid., p. 14.
8. See C. Caton, "The Idea of Sameness Challenges Reflection," *Studies on Frege*, vol. 2, ed. M. Schirn (Stuttgart: Frommann-Holzboog, 1976), pp.167–80.
9. See M. Schirn, "Identität und Identitätsaussage bei Frege," *Studies on Frege*, vol. 2, ed. M. Schirn (Stuttgart: Frommann-Holzboog, 1976), pp. 181–82.
10. M. Dummett, *Frege, Philosophy of Language* (London: Duckworth, 1973).
11. See H. Khatchadourian, "Kripke and Frege on Identity Statements," vol. 2, *Studies on Frege*, ed. M. Schirn (Stuttgart: Frommann-Holzboog, 1976), pp. 269–98.
12. Schirn, p. 188.
13. G. Frege, *Grundgesetze der Arithmetik*, vol. 1 (Jena: Pohle, 1893; reprint, Hildesheim: Olms, 1966), p. 8.
14. G. Frege, "Ausführungen über Sinn und Bedeutung 1892–1895," *Nachgelassene Schriften*, ed. H. Hermes et al. (Hamburg: Meiner, 1969; 2nd ed., Hildesheim: Olms, 1990), pp. 130–31.

15. Dummett, p. 544.
16. Schirn, p. 184.
17. See B. Kienzle, "Notiz zu Freges Theorien der Identität," *Studies on Frege*, vol. 2, ed. M. Schirn (Stuttgart: Frommann-Holzboog, 1976), p. 218.
18. Ibid.
19. Frege, "Über Sinn und Bedeutung," pp. 143–44.
20. Ibid.
21. In Kripke's "Naming and Necessity" and "Identity and Necessity."
22. In my "Necessità a posteriori e contingenze a priori in Kripke: alcune note critiche," *Nominazione* 2 (June 1981): 205–17.
23. See Kripke, "Naming and Necessity," p. 270 and pp. 276–77.
24. Khatchadourian, p. 276.
25. Syntactically this and the following examples are instances of very bad English, but they express exactly what we mean.

4

Guillermo E. Rosado Haddock

ON FREGE'S TWO NOTIONS OF SENSE

1. Introduction

In my article "Remarks on Sense and Reference in Frege and Husserl"[1] (chapter 2 of the present book), I based my exposition of Frege's theory of sense in Frege's famous article "Über Sinn und Bedeutung"[2] and—since my interests were driven in another direction—tacitly assumed that Frege's notion of sense was unproblematic. Further acquaintance with Frege's philosophy has convinced me not only that Frege's notion of sense was far from being unproblematic, but that he really had two different notions of sense which he did not care to differentiate. One of these notions is the one presented in the papers cited above, as well as in, for example, Shwayder's paper "On the Determination of Reference by Sense." The second notion is scattered through different writings of Frege's mature period (after 1890), and can be traced back to his somewhat obscure notion of conceptual content.

2. Frege's Notion of Sense in "Über Sinn und Bedeutung"

According to Frege's discussion in "Über Sinn und Bedeutung," in each of the following pairs of expressions the members of the pair differ in sense but have the same reference: (i) {'7 + 1', '5 + 3'}; (ii) {'the morning star', 'the evening star'}; (iii) {'the author of *Meaning and Necessity*', 'the best known disciple of Frege'}; and (iv) {'the Ultrafilter Theorem', 'Tarski's theorem on maximal dual ideals'}. Since, according to the same source, the sense of a compound expression is completely determined by the senses of its component parts, in each of the following pairs of sentences the members of the pair differ in sense, i.e., they express different thoughts: (i) {'7 + 1 > 6', '5 + 3 > 6'}; (ii) {'The morning star is the morning star', 'The morning star is the evening star'}; (iii){'The author of *Meaning and Necessity* wrote *Der*

logische Aufbau der Welt', 'The best known disciple of Frege wrote *Der Logische Aufbau der Welt*'}; and (iv) 'The Ultrafilter Theorem depends on the Axiom of Choice', 'Tarski's theorem on maximal dual ideals depends on the Axiom of Choice'}. But since in each pair of sentences the references of the corresponding expressions are the same, and since the reference of a compound expression is completely determined by the references of its component expressions, in each of the above pairs of sentences the members of the pair have the same reference. According to Frege, the reference of any of those statements is a truth value, but this is inessential for our present purpose, since his principle of the determination of the reference of a compound expression both by its sense and by the references of its component expressions could still be defended—if at all—if we chose as the reference of statements one of the two Husserlian alternatives considered in our essay cited above and in sections 5 and 6 of the present essay.[4]

On the basis of this notion of sense, two different expressions would have the same sense, either if they belong to different languages and one is the exact translation of the other (e.g., 'France's capital' and 'die Hauptstadt Frankreichs' or 'l'auteur de *Meaning and Necessity*' and 'der Verfasser von *Meaning and Necessity*'), or if they belong to the same language but are perfect synonyms.

3. Frege's Notion of Conceptual Content

Although in *Begriffsschrift* the terms 'content', 'judgeable content', and 'conceptual content' seem to be important members of Frege's philosophical lexicon,[5] these terms were practically abandoned by Frege after 1890. Fregean scholars, however, have given sufficient attention neither to the concepts themselves nor to their disappearance from Frege's lexicon.

The term 'content' is the only one of the three that did not disappear completely from Frege's vocabulary after 1890. Curiously enough, it was used in a nontechnical and somewhat vague sense before 1890, but received a much more precise usage in some of his later writings. In "Der Gedanke" and elsewhere it appears as an essential constituent of statements which either coincides with the thought expressed or consists of the thought together with some coloring (*Färbung*) or illumination (*Beleuchtung*) of the thought.[6] In *Begriffsschrift* a judgeable content is characterized by contrasting it, on the one hand, with judgments and, on the other hand, with unjudgeable contents.[7] Whereas a judgment involves an assertion, a judgeable content has the linguistic form of a judgment, but does not involve any assertion. On the other hand, an unjudgeable content lacks the linguistic form of a judgment, and—according to Frege's discussion in *Begriffsschrift*—not only lacks an assertion, but cannot be asserted. (In *Grundgesetze der Arithmetik* Frege modifies this last point.[8]) Although, according to Frege, he later split the notion of judgeable content into thought and truth value, his notion of judgeable content in *Begriffsschrift* seems much closer to his notion of thought in "Über Sinn und Bedeutung" than to his notion of truth value.

The most enigmatic of the above cited notions—and the most important for our purposes—is the notion of conceptual content. In *Begriffsschrift* Frege characterizes this notion in the following confused way:[9]

(α) A sentence in the active mood and its corresponding sentence in the passive have the same conceptual content.

(β) Two sentences such that, if we maintain fixed the rest of the premisses, we can derive from them the same sentences, have the same conceptual content.

(γ) The conceptual notation does not need to differentiate between sentences having the same conceptual content. Only the conceptual content is relevant for the conceptual notation.

But (β) and (γ) are not easy to reconcile. (β) points to a sort of 'equiderivability' or 'equipollency'. Two sentences S and S′ have the same conceptual content if for any sentence S* and for any set of sentences Γ (of the same language as S and S′—e.g., the conceptual notation), $\Gamma \cup \{S\} \vdash S^*$ if and only if $\Gamma \cup \{S'\} \vdash S^*$. Hence, the sentences '$p \rightarrow q$', '$\neg p \vee q$' and '$\neg(p \wedge \neg q)$' of the language of propositional logic (with current notation) would have the same conceptual content—on the basis of (β). Moreover, in the language of the whole of mathematics based on Zermelo-Fraenkel set theory, the Axiom of Choice, Zermelo's Well-Ordering Principle, Zorn's Lemma, and many other seemingly very different mathematical statements would have the same conceptual content. However, according to (γ), the conceptual notation needs not differentiate between sentences having the same conceptual content. But any logician or mathematician would consider a language as completely inadequate for mathematics if it does not differentiate between the Axiom of Choice, Zermelo's Well-Ordering Principle, Zorn's Lemma, and the many other equivalent mathematical statements in set theory, logic, algebra, and general topology.

4. Frege's Second Notion of Sense

It is interesting to observe that in "Der Gedanke" Frege offers as examples of pairs of sentences expressing the same thought precisely a sentence in the active mood and its corresponding sentence in the passive, and also the similar case of a sentence containing the verb 'to give' and a sentence which is obtained from it by substituting the verb 'to receive' for the verb 'to give' and interchanging the nominative and the dative (together with the corresponding syntactic adjustments).[10] Thus, 'The Greeks defeated the Persians in the Battle of Platea' and 'The Persians were defeated by the Greeks in the Battle of Platea' would have the same sense—i.e., they would express the same thought—as would also the sentences 'John gave a present to Mary' and 'Mary received a present from John'. Analogously, on the basis of this characterization of sense, 9 > 3 and 3 < 9 would have the same sense—i.e., they would express the same thought. Moreover, according to Frege in "Der Gedanke," a language adequate for science does not need to differentiate between sentences—as those belonging to any of the pairs of sentences considered above—that have the same sense.[11] Hence, Frege's characterization of sense and of identity of sense in "Der Gedanke" is very similar to his characterization of conceptual content *Begriffsschrift*.

But Frege's notion of conceptual content was not free from confusion, since, as we observed above, (β) and (γ) are not easy to reconcile. Frege's conception of sense in "Der Gedanke" is also somewhat enigmatic in the same respect. In that paper, he says that facts are true thoughts.[12] But on the basis of "Über Sinn und Bedeutung," '5 + 2 < 9', '4 + 3 < 9', '8 − 1 < 9' and infinitely many other inequalities that correspond to the same fact, namely, that 7 < 9, express different thoughts. Since it would be groundless to identify a fact with some but not all its corresponding thoughts, the fact that 7 < 9 must be identified with all such corresponding inequalities.[13] But it is not clear at all why indefinitely many different true thoughts should be identical with the same fact. Moreover, that two thoughts are both identical to a given fact but not identical to each other would violate the transitivity of (absolute) identity.

Frege's identification of facts with true thoughts can be somewhat clarified if we take into account a letter of Frege to Husserl of 1906.[14] In this letter Frege takes the equipollency (or equiderivability) as a criterion for the identity of sense (i.e., of thought) of two sentences—hence, the similarity with the characterization of conceptual content is now complete. Thus, two sentences, S and S′ express the same thought if for any sentence S* and for any set of sentences Γ (of the same language as S and S′), $\Gamma \cup \{S\} \vdash S^*$ if and only if $\Gamma \cup \{S'\} \vdash S^*$. On the basis of this criterion for the identity of sense, Frege could say that S and S′ correspond to the same possible fact, and that, if true, the thought expressed by them would be identical to this fact. This is the (possible) rationale of Frege's identification of facts with true thoughts.

Thus, Frege's characterization of thought in "Der Gedanke" and in the letter to Husserl cited above coincides with his characterization of conceptual content in *Begriffsschrift* and is plagued by the same tensions discussed in section 3 above. But apart from such tensions, it should be clear by now that this notion of thought (and, more generally, of sense) is different from and even incompatible with Frege's notion of thought (correspondingly of sense) in "Über Sinn und Bedeutung." According to that paper, the equations '5 + 3 = 8' and '7 + 1 = 8' express different thoughts. But since they are equipollent, according to the letter to Husserl cited above, they express the same thought.

Moreover, on the basis of Frege's notion of sense in "Über Sinn und Bedeutung," it is not clear how it is that in each of the following pairs of sentences the sentences have the same sense, namely, in: {'The Greeks defeated the Persians in the Battle of Platea', 'The Persians were defeated by the Greeks in the Battle of Platea'}; {'John gave a present to Mary', 'Mary received a present from John'}; and {'9 > 3', '3 < 9'}. We will examine this last pair on the basis of Frege's notion of sense in "Über Sinn und Bedeutung," but the argument also applies to the other pairs.

If '9 > 3' and '3 < 9' express the same thought, then either (i) there is no difference in sense between '>' and '<', or (ii) the change in order of the arguments somehow 'compensates' for the difference in sense existing between '>' and '< '. But clearly '>' and '<' do not have the same sense. If they did have the same sense, then on the basis of "Über Sinn und Bedeutung," '9 > 3' and '9 < 3' would express the same thought. But these two inequalities do not even have the same truth value, and, therefore, cannot express the same thought. Hence, (i) is excluded. On the other hand, although not identical, the senses of ' >' and '<' are clearly related—as are the senses of 'to give' and 'to receive', and of 'to defeat' and 'to be defeated'. We could say that they are dual senses. Thus, for '9 > 3' and '3 < 9' to express the same thought, the change in order of the arguments '9' and '3' of '>' should compensate the difference in sense between '>' and '<'. But, on the basis of Frege's notion of sense in "Über Sinn und Bedeutung," there is no clue to an explanation of this compensating effect of the change in order of the arguments, and, therefore, no possible satisfactory explanation of the alleged identity of the senses of '9 > 3' and '3 < 9'—or of the identity of the senses of the members of any of the other two pairs of sentences considered above. On the basis of Frege's discussion in that article such an identity seems completely mysterious. A clear explanation can be immediately obtained on the basis of the notion of sense put forward in the letter to Husserl cited above. It is clear that in each of the three pairs of sentences mentioned above, the sentences can be considered equipollent. Thus, if equipollent sentences express the same thought, the sentences in each of those pairs will express the same thought, i.e., they will have the same sense.

We can conclude that Frege had two different notions of sense, namely: (i) the notion considered in "Über Sinn und Bedeutung" and presented in section 2 above; and (ii) a somewhat unclear notion that appears in "Der

Gedanke" and elsewhere, and which has its origin in Frege's old notion of conceptual content as characterized by (α) and (β) of section 3. On the other hand, Frege's remarks in "Der Gedanke" which are similar to (γ) of section 3 are difficult to reconcile with this last notion of sense—although not with that of "Über Sinn und Bedeutung"—and are most cogently rendered as a symptom of confusion in Frege that can also be traced back to his characterization of conceptual content in *Begriffsschrift*. In the next section we shall see that without Frege's second notion of sense the famous Principle V, or Basic Law V, of *Grundgesetze der Arithmetik* is completely unintelligible.[15]

5. On Interpreting Principle V

The famous Principle V of *Grundgesetze der Arithmetik*, namely,

$$\grave{\varepsilon}\,\Phi(\varepsilon)=\acute{\alpha}\,\Psi(\alpha)=\underset{\smile}{a}\!\!-\Phi(a)=\Psi(a)$$

expresses the identity of an equation between courses of values and the generalization for all arguments of the identity between the values of the corresponding functions. Thus, it is an identity statement between two statements. In my essay "Identity Statements in the Semantics of Sense and Reference" (chapter 2), I have shown that the only rendering of identity statements in such a semantics that avoids very serious difficulties is that which considers identity statements as expressing a congruence relation between the senses of the expressions at either side of the (principal) identity sign modulo sameness of reference.[16] A consequence of such a rendering— and of other renderings unacceptable on different grounds—is that for an identity statement to be true it is necessary that the expressions at either side of the (principal) identity sign have the same reference. Since Principle V is an identity statement between statements, and since for Frege the reference of a statement is a truth value, all that is needed for Principle V to be true is that the sentences at either side of the principal identity sign have the same truth value, i.e., that both have the True as reference or that both have the False as reference. But no matter if

$$\grave{\varepsilon}\,\Phi(\varepsilon)=\acute{\alpha}\Psi(\alpha)$$

has the True or the False as reference, there are denumerably—infinitely— many sentences (more exactly, thoughts or judgeable contents) expressible in conceptual notation that have as reference the same truth value as

$$\grave{\varepsilon}\,\Phi(\varepsilon)=\acute{\alpha}\Psi(\alpha).$$

Hence, on the basis of this rendering, Principle V rests completely undetermined and the (logically simple) notion of a course of values rests completely vague.

We could try to follow Sluga,[17] and consider such a logical principle as Principle V in such a way that the statements at either side of its principal identity sign not only have the same reference, but also the same sense, i.e., they express the same thought. But although Sluga's rendering seems to be based on remarks made by Frege himself,[18] it clearly conflicts with Frege's notion of sense in "Über Sinn und Bedeutung" and with the ensuing interpretation of identity statements mentioned above. On the basis of "Über Sinn und Bedeutung," identity statements in which the expressions at either side of the (principal) identity sign have the same sense are trivialities like 'The morning star is the morning star', or 'The morning star is *der Morgenstern*', or 'iv = 4'.[19] Moreover, as I have shown in chapter 2, if we interpret identity statements as Sluga wants us to interpret Principle V, '5 + 3 = 7 + 1' and 'The author of *Meaning and Necessity* is the best known disciple of Frege' would have the False as reference. But '5 + 3 = 7 + 1' not only has the True as reference, but was considered by Frege as an a priori truth, and even as an analytic one (i.e., derivable from logical axioms and definitions alone on the basis of logical rules of inference). Certainly Frege's notion of sense in "Über Sinn und Bedeutung" does not mix well with Sluga's rendering of Principle V.

But if we render Principle V in such a way that the statements at either side of the principal identity sign have not only the same truth value, but also the same conceptual content as characterized by (β), hence the same sense according to Frege's second notion of sense, Principle V becomes much more intelligible. Only a minor modification seems convenient, namely, that the notion of conceptual content should be rendered, not as a second notion of sense, but as an alternative notion of reference, so that it does not conflict with Frege's analysis of sense in "Über Sinn und Bedeutung" and the ensuing rendering of identity statements.

Under this new rendering of conceptual content, this notion comes close to Husserl's notion, not completely developed, of a situation of affairs (*Sachlage*). It is interesting to observe that, when contrasting situations of affairs with states of affairs (*Sachverhalt*) in *Erfahrung und Urteil*,[20] Husserl characterizes situations of affairs in a way that has some affinities with Frege's characterization (a) of conceptual content. Statements like 'The Greeks defeated the Persians in the Battle of Platea' and 'The Persians were defeated by the Greeks in the Battle of Platea' correspond to two different states of affairs but to the same situation of affairs. Similarly, '9 > 3' and '3 < 9' correspond to two different states of affairs but to the same situation of affairs. Both these notions were clearly differentiated by Husserl from the sense of statements, i.e. from the proposition or thought expressed by the statement.[21] That they are also different from the notion of truth value can be most easily seen by considering groups of transformations, as was sketched in my essay "Remarks on Sense and Reference in Frege and Husserl" (chapter 2).[22]

I am not interested in forcing further the analogy between Frege's notion of conceptual content and Husserl's notion of situation of affairs. Both these notions were not sufficiently developed by their originators, and—as we have seen—Frege's notion of conceptual content even received conflicting characterizations.

In the rest of this essay I argue on behalf of the introduction of the Husserlian distinctions in the semantic analysis of mathematics. In particular I will try to make precise a semantic notion which is a sort of 'explicans' of Frege's notion of conceptual content and of Husserl's notion of situation of affairs.

6. On the Semantics of Mathematics

I am going to use the expression 'abstract situation of affairs' for the notion that I am going to introduce. I prefer this more Husserlian terminology to Frege's conceptual content, since Frege's use of this expression involved some confusions discussed above that I want to avoid.[23] An alternative expression would be 'objective content'.

Following Frege in "Über Sinn und Bedeutung" and Husserl, let us say that the equations '$2^2 = 4$' and '$3 + 1 = 4$' express different thoughts, since '2^2' and '$3 + 1$' are expressions with the same reference, namely the number 4, but with different senses. Similarly, 'The teacher of Alexander the Great was born in Stagira' and 'The most famous disciple of Plato was born in Stagira' express different thoughts, since their respective constituent expressions 'the teacher of Alexander the Great' and 'the most famous disciple of Plato' have different senses although they have the same reference. Of course, the sentences in each of the two pairs have the same truth value, i.e., the True. But the statements 'The Ultrafilter Theorem was first proved by Tarski', 'Paris is the capital of France' and denumerably many other statements also have the truth value the True. However, the relation existing between '$2^2 = 4$' and '$3 + 1 = 4$' is much stronger than the relation existing between '$3 + 1 = 4$' and 'Paris is the capital of France'. Analogously, the relation between 'The teacher of Alexander the Great was born in Stagira' and 'The most famous disciple of Plato was born in Stagira' is much stronger than that existing between 'The teacher of Alexander the Great was born in Stagira' and 'The Ultrafilter Theorem was first proved by Tarski'. '$3 + 1 = 4$' is obtained from '$2^2 = 4$' by a very simple transformation, namely, the substitution of an expression for another expression having different sense—as conceived in

"Über Sinn und Bedeutung"—but the same reference, whereas we cannot obtain a sentence like 'Paris is the capital of France' from either equation by any such transformation, but only by a transformation that only preserves truth value. Similarly, 'The teacher of Alexander the Great was born in Stagira' is obtained from the sentence 'The most famous disciple of Plato was born in Stagira' by a transformation that consists of the substitution of an expression by another expression having different sense but the same reference, whereas 'Paris is the capital of France' and 'The Ultrafilter Theorem was first proved by Tarski' cannot be obtained from any of them or from each other by such a transformation. Following Husserl, I will say that '$2^2 = 4$' and '$3 + 1 = 4$', although having different senses, refer to the same state of affairs. Similarly, although 'The teacher of Alexander the Great was born in Stagira' and 'The most famous disciple of Plato was born in Stagira' have different senses, they refer to the same state of affairs. As I have argued in chapter 2, the invariance of state of affairs determines a group of transformations of sentences that is, on the one hand, much more inclusive than the one determined by the invariance of thought, and, on the other hand, much more restricted than the one determined by the mere invariance of truth value (considered by Church in *Introduction to Mathematical Logic*).[24] A semantic theory—like Frege's—that does not differentiate between these two sorts of invariance is completely inadequate both for natural language and for mathematics.[25]

But the above distinctions are not sufficient. Even in the case of natural languages, there are important invariance relations between sentences that are not adequately described either as relations of invariance of state of affairs, or as relations of invariance of truth value, nor as relations of invariance of thought. As Husserl has argued in *Erfahrung und Urteil*,[26] this is the case of the relation existing between a sentence in the active mood and its corresponding passive sentence. This is also the case of the relation existing between the arithmetical inequalities '$9 > 3$' and '$3 < 9$'. The states of affairs that correspond to '$9 > 3$' and '$3 < 9$' are different, but both inequalities correspond to the same situation of affairs.

In mathematics the need for a distinction between cases of invariance of state of affairs, like that prevailing between the equations '$3 + 1 = 4$' and '$2^2 = 4$', and cases in which the state of affairs changes but there still prevails a relation of invariance much stronger than the mere invariance of truth value is much more urgent than in natural languages. E.g., in many areas of mathematics there exist what we will call 'phenomena of duality' that cannot be adequately explained by a semantics based only on the notions of thought, state of affairs, and truth value.

As every mathematician knows, there is a sort of duality between the notions named in each of the following pairs: {ideal, filter}; {open set, closed set}; {lower bound, upper bound}; and {intersection, union}. Thus, a filter is sometimes called a 'dual ideal'. An ideal that is not properly contained in any other ideal is called a maximal ideal, and a filter that is not properly contained in any other filter is called an ultrafilter. Hence, an ultrafilter is a maximal dual ideal. Since an ultrafilter is not usually defined as a maximal dual ideal, but in a completely different way,[27] the expressions 'ultrafilter' and 'maximal dual ideal,' although naming the same mathematical entity, have different senses. Thus, the statements 'The Ultrafilter Theorem was first proved by Tarski' and 'The Maximal Dual Ideal Theorem was first proved by Tarski', although referring to the same state of affairs, express different thoughts.

A completely different relation is that existing between a theorem for filters and a corresponding dual theorem for ideals, e.g.: (i) 'Every filter can be extended to an ultrafilter' and (ii) 'Every ideal can be extended to a maximal ideal'. The maximal ideal theorem (ii) cannot be obtained from the ultrafilter theorem (i) by a simple substitution of an expression ε for an expression δ having the same reference but a different sense. Nevertheless, the relation between (i) and (ii) is much stronger than a mere coincidence of truth value. They are in some sense equivalent. Their correct semantical analysis requires some sort of notion that can give rise to an invariance relation intermediate between invariance of state of affairs and invariance of truth value.

Such an equivalence, however, is not a mathematical triviality. The notions of 'filter' and of 'ideal' are dual notions, but they do not seem to be equally fruitful in all areas of mathematics. Ideals are commonly used in algebra, whereas filters—although not a topological notion—have been very useful in general topology. 'Intersection' and 'union' of sets are in some sense dual notions and there are many results about unions that have their dual results about intersections. But these notions do not have the same behavior in any mathematical context. E.g., in the definition of a topological space based on open sets one considers only finite intersections but denumerably infinite unions. Hence, it is not completely trivial that in some contexts there exist corresponding dual results about these notions. Even in school arithmetic we have a similar situation with the operations of addition and subtraction (or multiplication and division), which are somehow dual operations with many corresponding 'dual' results, although if we restrict our considerations to the domain of natural numbers, subtraction (and also division) is not defined for all ordered pairs of numbers. (Hence, although for denumerably many ordered triples of natural numbers $<x, y, z>$, where $z = x + y$, there is a corresponding 'dual' ordered triple $<z, x, y>$, where $y = z - x$, there are also denumerably many such triples $<x, y, z>$ for which there is no corresponding 'dual' triple.) Therefore, situations of duality cannot be discharged as mathematical trivialities following immediately from the definitions of the concepts.

But not only situations of duality require a much finer analysis than can be given by means of the notions of thought, state of affairs, and truth value.

Particularly important is the case of seemingly unrelated results which have been shown to be mathematically equivalent. A very informative example is that of the Axiom of Choice and its equivalents. This axiom is equivalent (i.e., equipollent or interderivable) to many seemingly unrelated statements in different areas of mathematics, e.g., to Zermelo's Well-Ordering Principle, to Zorn's Lemma, and to the Principle of the Trichotomy of Cardinals in set theory, to the Upward and to the Downward Löwenheim-Skolem-Tarski theorems in logic, to Tychonoff's Compactness Theorem in general topology, to results about lattices and about vector spaces in algebra, and to many other mathematical results about the most diverse states of affairs.[28] It is clear that the relation between such statements is much stronger than mere coincidence of truth value, although not so strong as that existing between the statements 'The Ultrafilter Theorem was first proved by Tarski' and 'The Maximal Dual Ideal Theorem was first proved by Tarski'. The equivalence relation existing between, e.g., the Axiom of Choice and the Upward Löwenheim-Skolem-Tarski Theorem cannot be adequately described either by saying that they express different thoughts but refer to the same state of affairs (since they talk about seemingly unrelated things) or by saying that they have the same truth value (since '3 + 1 = 4' and 'Paris is the capital of France' have the same truth value, but are not equipollent).

Therefore, it seems unavoidable for an adequate semantic analysis of mathematics to distinguish the abstract situation of affairs of mathematical statements both from the state of affairs referred to by them and from their truth value. Thus, I will say that two mathematical statements have the same abstract situation of affairs if and only if they are (logico-mathematically) equipollent (or interderivable). (Hence, e.g., the Axiom of Choice, Zorn's Lemma, and the Upward Löwenheim-Skolem-Tarski Theorem have (or correspond to) the same abstract situation of affairs.) Invariance of abstract situation of affairs determines a group of transformations between sentences larger than the group of transformations determined by invariance of state of affairs—since transformations that preserve the state of affairs also preserve the abstract situation of affairs—and smaller than the group of transformations determined by invariance of truth value—since transformations that preserve the situation of affairs also preserve the truth value.[29]

Although the foregoing semantic distinctions between (i) thought, (ii) state of affairs, (iii) abstract situation of affairs, and (iv) truth value seem sufficient for the semantic treatment of natural languages, I am not asserting that they are sufficient for the semantic treatment of mathematics, but simply that they are necessary. In particular, we cannot decide here if all cases of mathematical equivalence can be adequately treated as cases of invariance of (a unique notion of) abstract situation of affairs. It could be the case that still finer distinctions have to be made, e.g., between different 'layers of abstraction' intermediate between states of affairs and truth values. But it could also be the case that any such new notion intermediate between those of state of affairs and abstract situation of affairs or between abstract situation of affairs and truth value, would collapse to one of them.

ACKNOWLEDGMENTS

Thanks are due to Professor Christian Thiel, Professor Michael Resnik, and an anonymous referee for some valuable comments on a first draft of this paper.

NOTES

1. G. E. Rosado Haddock, "Remarks on Sense and Reference in Frege and Husserl," *Kant-Studien* 73 (1982): 425–39, chapter 2 of the present book.
2. G. Frege, "Über Sinn und Bedeutung," *Kleine Schriften*, ed. I. Angelelli (Darmstadt: Wissenschaftliche Buchgesellschaft, 1967; 2nd ed., Hildesheim: Olms, 1990), pp. 143–62.
3. D. Shwayder, "On the Determination of Reference by Sense," *Studies on Frege*, vol. 3, ed. M. Schirn (Stuttgart: Frommann-Holzboog, 1976), pp. 85–95.
4. As illustrated by our examples, the coincidence of reference can sometimes be known a priori (e.g., in [i]), but others only a posteriori (e.g., in [ii] and [iii]).
5. See G. Frege, *Begriffsschrift* (Halle: Nebert, 1879; reprint, Hildesheim: Olms, 1964), chapter 1.
6. See G. Frege, "Der Gedanke," *Kleine Schriften*, pp. 342–62, especially p. 346. See also G. Frege, *Nachgelassene Schriften*, ed. H. Hermes et al. (Hamburg: Meiner, 1969; 2nd ed., 1983), pp. 213–14.
7. Frege, *Begriffsschrift*, p. 2.
8. See G. Frege, *Grundgesetze der Arithmetik*, vol. I, Pohle, Jena, 1893, reprint, Hildesheim: Olms, 1966, pp. 9–10.
9. Frege, *Begriffsschrift*, p. 3.
10. Frege, "Der Gedanke," p. 348.
11. Ibid., pp. 347–48: Frege does not say it explicitly, but it is clear from the context. See also his *Nachgelassene Schriften*, pp. 152–55 and *Wissenschaftlicher Briefwechsel*, ed. G. Gabriel et al. (Hamburg: Meiner, 1976), p. 102.
12. Frege, "Der Gedanke," p. 359.
13. The other possibility consists in identifying each of the above inequalities with a different fact, e.g., $5 + 2 < 9$ with the 'fact' that $5 + 2 < 9$. But this seems completely unjustified.
14. See Frege's *Wissenschaftlicher Briefwechsel*, pp. 102 and 105 *f.*, and also his *Nachgelassene Schriften*, pp. 213–14.
15. In his excellent classic on Frege's philosophy *Sinn und Bedeutung in der Logik Gottlob Freges* (Meisenheim am Glan: Anton Hain, 1965), Christian Thiel touched on the same difficulties with Frege's diverging utterances on

the notions of sense and of identity of sense (see pp. 134–45, especially pp. 139 and 143). This work has been translated into English: *Sense and Reference in Frege's Logic* (Dordrecht: Reidel, 1968).

16. In *Logique et Analyse* 25 (1982): 399–411, chapter 3 of the present book.

17. See H. Sluga, *Gottlob Frege* (London: Routledge and Kegan Paul, 1980), p. 156.

18. See G. Frege's 1891 "Funktion und Begriff," *Kleine Schriften*, pp. 125–42 (p. 130). It is not completely clear, however, which of his two notions of sense is Frege using in this context, although my ensuing discussion makes it more plausible that it was what I have called his second notion of sense (i.e., that derived from his notion of conceptual content).

19. Frege never considered his Principle V as evident as his other principles. See, e.g., his *Grundgesetze*, vol. I, p. VII and especially *Grundgesetze*, vol. 2 (Jena: Pohle, 1903; reprint, Hildesheim: Olms, 1966), p. 253.

20. See Husserl's *Erfahrung und Urteil*, 5th ed. (Hamburg: Meiner, 1976 [1939]), p. 285.

21. Husserl used the expression 'proposition' ('*Satz*') to refer to what Frege in "Über Sinn und Bedeutung" called 'thought' ('*Gedanke*'), but at least once in *Logische Untersuchungen* he used the expression 'thought' as a synonym for 'proposition' (see *Logische Untersuchungen*, 2nd ed. rev. [Halle: Niemeyer, 1913], Fourth Investigation § 11).

22. See also H. Weidemann's excellent article "Aussagesatz und Sachverhalt: ein Versuch zur Neubestimmung ihres Verhältnisses," *Grazer Philosophische Studien* 18 (1982): 75–99. Weidemann also argues against Frege that states of affairs remain invariant under transformations of sentences in which constituent expressions are substituted by expressions having different sense but the same reference, and concluded that states of affairs are a better candidate than truth values for the reference of sentences. However, Weidemann does not make the distinction between state of affairs and situation of affairs, and his notion of state of affairs seems (at least in some aspects) closer to my (Husserlian) notion of situation of affairs than to my (Husserlian) notion of state of affairs.

23. I hope that there does not arise any confusion with Barwise and Perry's recent usage of the expressions 'abstract situation' and 'situation' in semantics.

24. See A. Church, *Introduction to Mathematical Logic* (Princeton: Princeton University Press, 1956), pp. 24–25.

25. It would take me too far to explain the Husserlian notions of state of affairs and situation of affairs. Roughly, a state of affairs is a categorial objectuality that is the correlate or reference of one or more propositions. Thus, that $7 < 9$ is the state of affairs that is the reference, e.g., of '$5 + 2 < 9$' and of '$5 + 2 < 8 + 1$'. Situations of affairs are receptively apprehended objectualities that serve as basis of two or more states of affairs. For our purposes, however, it is sufficient to consider our 'explanations' of these notions as the equivalence classes to which belong a proposition S and any other proposition obtained from S by the two sorts of transformations under discussion. Thus, e.g., the state of affairs referred to by a statement S is the equivalence class of all statements that can be obtained from S by substituting one or more of its constituent expressions by expressions having different sense but the same reference. A more exact characterization can be given by an inductive definition.

26. Husserl, *Erfahrung und Urteil*, p. 285.

27. A proper filter **F** over a set **I** is defined as a family of subsets of **I**, i.e., **F** ⊆**P(I)**—the power set of **I**—such that (i) ∅∉**F**, (ii) if **M**∈**F** and **M***∈.**F**, then (**M**∩**M***)∈**F**, and (iii) if **M**∈**F** and **M**⊆**M***, then **M***∈**F**. An ultrafilter over **I** is then defined as a filter not properly contained in any other (proper) filter over **I**.

28. See, e.g., G. H. Moore's recent book *Zermelo's Axiom of Choice* (New York: Springer, 1982).

5

Claire Ortiz Hill

THE VARIED SORROWS OF LOGICAL ABSTRACTION

In 1894 Gottlob Frege published a scathing attack (Frege 1894) on the Cantorian theory of psychological abstraction Edmund Husserl espoused in the *Philosophy of Arithmetic* (Husserl 1891). Frege denounced that aspect of Husserl's attempt to provide sound foundations for arithmetic as "naive." Any view according to which a statement of number is not a statement about a concept or about the extension of a concept, Frege said, is naive for "when one first reflects on number, one is led by a certain necessity to such a conception" (p. 197). Had Husserl used the term 'extension of a concept' in the way, he, Frege, thought fit, he wrote, they "should hardly differ in opinion about the sense of a number statement" (pp. 201–2).

Yet despite that show of confidence, Frege's apparent self-assurance quickly evaporated less than ten years later upon learning of the contradiction of the class of all classes that are not members of themselves from Bertrand Russell. Frege immediately responded to Russell (Frege 1980a, 130–32) that it was the law about extensions that was at fault, and that its collapse seemed to undermine the foundations of arithmetic proposed in the *Basic Laws of Arithmetic*.

So it was a considerably less confident man who wrote in the appendix to the 1903 volume of *The Basic Laws of Arithmetic* that hardly anything more unfortunate could befall a scientific writer than to find the foundations of his work shaken after it was completed. He would have gladly dispensed with his law about extensions, Frege confessed, had he known of any substitute for it. At least, he consoled himself, he could draw comfort in the fact that misery loves company for he believed that everybody who had made use of extensions of concepts, classes, sets in proofs was in the same position he was (Frege 1980b, 214).

Frege never believed that his law about extensions recovered from the shock it had sustained from Russell's finding. And, as it happens, his review of Husserl's book actually contains some of the most forceful statements

Frege ever made in favor of them. For Frege actually expressed profound reservations about using extensions throughout his career, and many statements he made during his lifetime belie the aplomb with which he carried out his caustic attack upon Husserl's theories. Frege confessed several times that he had never actually become completely reconciled to introducing extensions into his reasoning. He alluded to having been "led by a certain necessity" (Frege 1894, 197) or to having been "constrained to overcome" his "resistance" to them (Frege 1980a, 191). By the end of his life he was warning that talk of extensions "easily . . . can get one into a morass" and that it leads "into a thicket of contradictions" (Frege 1979a, 55).

Here I propose to cast a cold eye on the logical moves which led Frege into that "morass" in an attempt to lay bare just what lie behind his conviction that everybody who made use of extensions of concepts, classes, sets in proofs was in the same position he was. I then study Russell's battle with the "thicket of contradictions." Finally, I call for lucidity regarding the differences between mere equality and full identity which are lost in the abstraction process. I must, of course, begin by defining logical abstraction.

1. What Logical Abstraction Is

Logical abstraction is a technique by which one singles out what is common among the members of a given set of objects. Through it a property is isolated and the particular equivalence obtaining between objects possessing that property comes to be regarded as an identity (on paper). A common predicate is interpreted as a common relation to a new term, the class of all those terms that are equal in terms of the property indicated by the predicate. The class of terms having that particular relation then replaces the common property inferred from the equivalence relation chosen, and all the other properties which might have normally served to distinguish those objects from each other, or from other objects equal to them in the same respect, are "abstracted" out of the picture. Once deleted on paper, the properties which originally might have marked any difference between the mere equality and the full identity of the objects are presumably expected to simplify matters by vanishing entirely from the reasoning.

Willard Van Orman Quine has provided the following description of the procedure:

> given a condition '---' upon x, we form the class $\hat{x}$--- whose members are just those objects x which satisfy the condition. The operator '$\hat{x}$' may be read 'the class of all objects x such that'. The class $\hat{x}$--- is definable, by description, as the class y to which any object x will belong if and only if --- (Quine 1961a, 87)

And Quine has provided this example of how that procedure might be applied in actual practice:

It may happen that a theory dealing with nothing but concrete individuals can conveniently be reconstrued as treating of universals, by the method of identifying indiscernibles. Thus consider a theory of bodies compared in point of length. The values of the bound variables are physical objects, and the only predicate is 'L', where 'Lxy' means 'x is longer than y'. Now where '~Lxy $\wedge$ ~Lyx', anything that can be truly said of x within this theory holds equally for y and vice versa. Hence it is convenient to treat '~Lxy $\wedge$ ~Lyx' as '$x = y$'. Such identification amounts to reconstruing the values of our variables as universals, namely lengths, instead of physical objects. (Quine 1961b, 117)

The talk of universals engaged in, Quine explains, may be regarded "merely as a manner of speaking—through the metaphorical use of the identity sign for what is really not identity but sameness of length. . . . In abstracting universals by identification of indiscernibles, we do no more than rephrase the same old system of particulars" (Quine 1961b, 118).

Many have found logical abstraction to be an attracive and convenient technique for translating various familiar expressions, and those having to do with numbers especially, into the notation of the extensional logic favored by so many logicians. Through logical abstraction one might introduce new objects into reasoning by translating many expressions into extensional language which would not on the surface seem to lend themselves to extensional treatment. Talk of properties is transformed into talk of classes as every monadic predicate comes to have a class as an extension—the class of all things of which the predicate is true. The notation Fy comes to mean y is a member of the class F, the class of all objects fulfilling a given condition—the class of those things that are equal in that particular respect (Quine 1961b, 120–21). Many things people want to accomplish can, in fact, be accomplished by leaving unwanted properties out of the picture (and, one might add, might not be accomplishable were they properly taken into account).

2. Frege Begins Abstracting Away Properties in *Foundations of Arithmetic*

Arithmetic was the point of departure for the ideas that led Frege to develop his logical theories. And the symbolic language he hoped might one day "become a useful tool for the philosopher" was modelled after the language of arithmetic (Frege 1879, Preface).

According to Frege's theory of arithmetic, as set out in the *Foundations of Arithmetic* (Frege 1884), numbers are always independent objects which as such are qualified to figure in identity statements, and any appearance to the contrary could "always be got around" because expressions which in everyday discourse do not seem to name independent objects can be rewritten so that they do (§§55–57).

Moreover, Frege held that with numbers it was "a matter of fixing the sense of an identity" (p. x, §§62, 106). And it became his aim "to construct the content of a judgement which can be taken as an identity such that each side of it is a number" (§63). So finding "a means for arriving at that which is to be regarded as being identical" (§63) became an integral part of his project to provide a deeper foundation for the theorems of arithmetic.

Not wanting to introduce a special definition of identity for this, but wishing rather "to use the concept of identity, taken as already known as a means for arriving at that which is to be regarded as being identical" (§63), Frege explicitly adopted Leibniz's principle that "things are the same as each other, of which one can be substituted for the other without loss of truth" (§65). Of his choice he wrote:

> This I propose to adopt as my own definition of equality. Whether we use 'the same', as Leibniz does, or 'equal', is not of any importance. 'The same', may indeed be thought to refer to complete agreement in all respects, 'equal', only to agreement in this respect or that; but we can adopt a form of expression such that this distinction vanishes. For example, instead of 'the segments are equal in length', we can say 'the length of the segments is equal', or 'the same', and instead of 'the surfaces are equal in color', 'the color of the surfaces is equal'. . . in universal substitutability all the laws of identity are contained (§65).

As Frege was writing Leibniz's formula right into the foundations of his logic, however, he modified Leibniz's dictum in an important way. Not one ever to adhere slavishly to usual linguistic practice, Frege advocated rephrasing statements in ways which eliminate distinctions obtaining in ordinary language. So, in the passage cited above he has recommended rephrasing the statement 'The segments are equal in length' as 'the length of the segments is equal or the same', and 'the surfaces are identical in color' as 'the color of the surfaces is identical'.

In so doing he adjusted Leibniz's principle to meet his own ends by deciding to translate sentences of natural languages into his symbolic language in a way he thought might do away with the differences between being identical (complete agreement in all respects) and equal (only agreement in this respect or that). For although Leibniz's law defines identity, complete coincidence, Frege, here as elsewhere, explicitly maintained that for him whether we use 'the same' as Leibniz does, or 'equal' is not of any importance. He would insist over and over again that for him there was no difference between equality and identity. For example in the preface to his 1893 *Basic Laws of Arithmetic* (p. ix), he explained that he had chosen to use the ordinary sign of equality in his symbolic language because he had convinced himself that it is used in arithmetic to mean the very thing that he wished to symbolize.

To achieve these goals Frege appealed to a process of logical abstraction (though he never used that term) by which statements in which certain objects are said to be equivalent in terms of a certain property predicated of them are transformed into statements affirming the identity-equality of abstract objects formed out of those properties. In the *Foundations of Arithmetic* (§65) he used the following examples to illustrate his move:

(5) 'The segments are equal in length'

and

(6) 'The surfaces are equal in color'

which he wished to see reformulated as

(7) 'The lengths of the segments are equal'

and

(8) 'The color of the surfaces is equal'.

Here, Frege has changed the statements of sameness of concrete properties predicated of concrete objects in (5) and (6) into statements (7) and (8) which affirm the equality-identity of abstract objects, in this case surfaces and lengths. He believes he has thus transformed statements about objects which are equal under a certain description into statements expressing a complete identity. By erasing the difference between identity and equality, he is in fact arguing that being the same in any one way is equivalent to being the same in all ways. However, he realized that many of the inferences that could be made by appealing to such a principle would lead to evidently false and absurd conclusions.

3. The Problems Frege Addressed from the Beginning

Frege himself acknowledged that left unmodified the procedure just described was liable to produce nonsensical conclusions, or be sterile and unproductive (Frege 1884, §66–67). For example, he realized that his definition of identity only afforded logicians a means of recognizing an object as the same again if determined in a different way, but it did not account for all the ways in which it could be determined.

To illustrate some nonsensical consequences of defining identity in this way, Frege carried the reasoning involved in his example of the identity of two lines one step further. However, the points he wished to make can be made more graphically by leaving the world of abstract geometrical figures for denizens of a more material one.

So, parroting Frege's reasoning in *Foundations of Arithmetic* §66, I propose to illustrate his point about possible nonsensical consequences of his definition by appealing to a more concrete case. Suppose it has finally been determined to be true that:

> the man who fired the shots from behind the grassy knoll is identical with the man who killed John Kennedy.

Frege's definition of identity would then afford us a means of identifying the man who fired the shots from behind the grassy knoll again in those cases in which he is referred to as the man who killed John Kennedy. But, as Frege recognized, this means does not provide for all the cases. For instance, it could not decide for us whether Lee Harvey Oswald was the man who fired the shots from behind the grassy knoll. While any informed person would consider it perfectly nonsensical to confuse Lee Harvey Oswald with the man who fired the shots from behind the grassy knoll, this would not, Frege would acknowledge, be owing to his definition. For it says nothing as to whether the statement:

> 'the man who fired the shots from behind the grassy knoll is identical with Lee Harvey Oswald'

should be affirmed or denied, except for the one case where Lee Harvey Oswald is given in the form of 'the man who killed John Kennedy'. Only if we could lay it down that if Lee Harvey Oswald was not the man who killed John Kennedy, (still following Frege's reasoning), could our statement be denied, while if he was that man, our original definition would decide whether it is to be affirmed or denied. But then we have obviously come around in a circle, Frege acknowledged. For in order to make use of this definition we should have to know already in every case whether the statement 'Lee Harvey Oswald is identical with the man who killed John Kennedy' was to be affirmed or denied.

Left as it was his definition was unproductive, Frege further judged, because in adopting this way out, we would be presupposing that an object can only be given in one single way. For otherwise, (still parroting Frege's reasoning, but using our more material example), it would not follow from the fact that Lee Harvey Oswald was not introduced by our definition that he could not have been by means of it. "All identities would then amount simply to this," Frege then wrote, "that whatever is given to us in the same way is to be reckoned as the same. This is, however, a principle so obvious and sterile as not to be worth stating. We could not, in fact, draw from it any conclusion which was not the same as one of our premisses." Surely though, he concluded, identities play such an important role in so many fields "because we are able to recognize something as the same again even although it is given in a different way" (Frege 1884, §67; §107).

4. Frege Espouses Extensions

So seeing that he could not by these methods alone obtain concepts with sharp limits to their application, nor therefore, for the same reasons, any satisfactory concept of number either, Frege was led to introduce extensions to guarantee that an identity holding between two concepts could be transformed into an identity of extensions, and conversely (Frege, 1884, §67; also §107). He hoped thereby to eliminate the undesirable consequences (§§66–67, 107) he saw accumulating around his theory of number and he devoted several sections of *Foundations of Arithmetic* to discussing the pros and cons of introducing them (§§68–73). However, in one of its concluding sections he wrote of extensions that: "This way of getting over the difficulty cannot be expected to meet universal approval, and many will prefer other methods of removing the doubt in question. I attach no decisive importance even to bringing in the extensions of concepts at all" (§107).

Although Frege ended the *Foundations of Arithmetic* claiming that he was not attaching any decisive importance to bringing in extensions, he did not propose any alternative and by the time it came to actually proving his theory of number in the *Basic Laws of Arithmetic*, he had managed to quiet the reservations he would later confess having had about introducing them. He explained in the preface to the book that extensions had taken on "great fundamental importance." Their introduction, he now maintained was "an important advance which makes for far greater flexibility" (pp. ix–x). "In fact," he wrote there, "I even define number itself as the extension of a concept, and extensions of concepts are, according to my definitions, graphs. So we just cannot do without graphs" (p. x). As far as he could see, he wrote, his basic law about extensions of concepts was the only place in which a dispute could arise. This, he believed, would be the place where the decision would have to be made (p. vii). Then, a year later he came out fighting for extensions in the 1894 review of Husserl.

In *Basic Laws*, Frege argued that the generality of an identity could always be transformed into an identity of courses of values and conversely, an identity of courses of values may always be transformed into the generality of an identity. By this he meant that if it is true that (x) F(x) = G(x), then those two functions have the same extension and that functions having the same extension are identical (Frege 1893, §§9 and 21). "This possibility," he wrote then, "must be regarded as a law of logic, a law that is invariably employed, even if tacitly, whenever discourse is carried on about extension of concepts. The whole Leibniz-Boole calculus of logic rests upon it. One might perhaps regard this transformation as unimportant or even as dispensable. As against this, I recall the fact that in my *Grundlagen der Arithmetik* I defined a Number as the extension of a concept . . ." (Frege 1893, §9).

In §§146–47 of the 1903 *Basic Laws II*, he characterized extensionality writing:

> If a (first-level) function (of one argument) and another function are such as always to have the same value for the same argument, then we may say instead that the graph of the first is the same as that of the second. We are then recognizing something common to the two functions . . . We must regard it as a fundamental law of logic that we are justified in thus recognizing something common to both, and that accordingly we may transform an equality holding generally into an equation (identity). (Frege 1980b, 159–60)

Frege never believed that any proof could be supplied that would sanction such a transformation. So he devised Basic Law V, or Principle V, to mandate the view of identity, equality, and substitutivity his system required. By transforming "a sentence in which mutual subordination is asserted of concepts into a sentence expressing an equality," Basic Law V would permit logicians to pass from a concept to its extension, a transformation which, Frege held, could "only occur by concepts being correlated with objects in such a way that concepts which are mutually subordinate are correlated with the same object" (Frege 1979, 182).

5. Strong Extensionality

Now that we have looked at Frege's characterization of extensionality, it is very important to look at how extensions could help Frege solve the particular problems with identity and substitutivity in connection with which they were introduced in the *Foundations of Arithmetic*.

First of all, Frege was conscious of the fact that the logician's job was initially made simpler because she initially only had to recognize the object as given in the particular way stipulated by her identity statement. But Frege realized that his definition afforded no means of recognizing that object as the same again when given in a different way. So returning to our Kennedy example, suppose that the wife of a man who fired at Kennedy from behind the grassy knoll has come forward with incontrovertible evidence that her husband, one Mr. Knoll, fired shots at Kennedy and killed him. She was actually with her husband behind the grassy knoll that day, but was too blinded by love for him to tell the police what she knew. She also wanted to protect her family. But Mr. Knoll has since died and her conscience is tormenting her. In addition to her eyewitness account she has produced authentic diaries in which her husband gave details of his plans. So, (changing the descriptions to expressions that name directly so as to avoid any problems deriving solely from the fact that descriptions appear in the putative identity statement):

'Mr. Knoll is identical with Kennedy's assassin'

is a true statement. Initially, it looks as if the case is closed. We can now substitute Mr. Knoll's name every time we find a reference to Kennedy's assassin. However, our identity statement is completely blind as concerns other alternatives. And once we give the matter further thought we find that Kennedy's assassin has so often been identified with Lee Harvey Oswald that substitution in most contexts would yield nonsense. For example, most of the Warren Report would become complete nonsense.

Now as concerns Frege's second point regarding the sterility of the procedure. It is certain that the single fact given by our identity is a highly informative statement. But just because our *x* is both F and G, this does not mean that if something is F it is G. We have only learned that Mr. Knoll killed Kennedy. But only on paper has Basic Law V ruled out the possibility that Lee Harvey Oswald, or someone else, might also have fired at Kennedy, and so might also have been Kennedy's assassin. Kennedy did not die instantly, and numerous factors may have finally conspired to bring about his death.

However, with Frege's principle of extensionality, we put on logical blinders. It mandates that since Mr. Knoll is identical to Kennedy's assassin, then anyone who was Kennedy's assassin was Mr. Knoll. Our concepts have acquired the sharp limits our quest for knowledge and substitution requires (Frege 1884, §67; also §107). Which is all fine on paper, but extensionality will not of itself keep Lee Harvey Oswald from slipping back into the picture as, for example, someone who might also have been Kennedy's assassin, but was definitely not Mr. Knoll. It may seem at first attractively simple to obliterate distinctions between identity and equality, but the differences between *x* and *y* when they are joined together by the equals sign to make an informative statement do not just go away because we have a rule stipulating that equality is to function as identity. For informative identity statements are the breeding ground of contradictions in extensional systems. The seed of contradictions derivable in extensional systems lies buried in them—something, of course, wholly unacceptable to Frege, a man who developed a symbolic language whose stated first purpose was to to provide "the most reliable test of validity for a chain of inferences and to point out every presupposition that tries to sneak in unnoticed" (Frege 1879, 6).

In his most confident moments, however, Frege believed that Basic Law V could bring logicians out of what he called the "queer twilight" of identity in which he saw mathematicians performing their logical conjuring tricks (Frege 1980b, 151–52). But the procedure he had hoped would bring identity out of the penumbra into the clear light of day had actually authorized various illicit logical moves which yielded contradictory results by letting logicians put their symbols to wrong uses and allowing type ambiguities to creep into reasoning unnoticed, with the result that, for example, a class might seem to be a member of itself.

6. The Shaking of Frege's Foundations for Arithmetic

When informed of Russell's paradox of the class of all classes which are not members of themselves in 1902 (Frege 1980a, 130–31), Frege immediately designated the logical transformations legitimized by Basic Law V as being the source of the problem. Basic Law V was false, he recounted in the appendix to *Basic Laws of Arithmetic II* (Frege 1980b, 214–24). The way he had introduced extensions was not legitimate (p. 219), and the interpretation he had so far put on the words 'extension of a concept' needed to be corrected.

In the several texts in which he pinpoints what he believed was the source of the difficulties, he consistently cites Basic Law V's transformation of concepts into objects for extensional treatment as being at fault (Frege1980a, 54–56, 191; 1979, 181–82, 269–70). There was nothing, he decided, to stop him from transforming an equality holding between two concepts into an equality of extensions in conformity with the first part of his law, but from the fact that concepts are equal in extension we cannot infer that whatever falls under one falls under the other. The extension may fall under only one of the two concepts whose extension it is. This can in no way be avoided and so the second part of his law fails (Frege 1980b, 214n. f, 218–23). "If in general, for any first-level concept, we may speak of its extension, then the case arises of concepts having the same extension, although not all objects that fall under one fall under the other as well. This, however, really abolishes the extension of the concept" he concluded (Frege 1980b, 221).

An example will make Frege's problem clearer. To illustrate the point he is making we might turn to the modern analogue of his 'the number of Jupiter's moons is 4', i.e., the putative identity statement Quine made famous:

(1) '9 is the number of the planets'.

According to Basic Law V, supposing this to be a true identity, the number which belongs to the concept 9 is the same as that which belongs to the concept the number of the planets. But the converse does not necessarily hold. For though 9 may be the number of the planets, we cannot infer that whatever falls under one falls under the other. There is nothing other than Basic Law V to guarantee that all objects that fall under '9' fall under 'the number of the planets' as well. For instance, according to materialistic astronomy:

(2) The ninth planet, Pluto, was discovered in 1930.

(3) The planet Pluto may be an anomaly and not a planet at all, but a giant asteroid flung into its present position when it had a close gravitational encounter with one of the outer planets.

(4) A Planet X may exist far beyond Pluto, which would explain apparent irregularities in Neptune's orbit.

(5) According to astronomers the dark matter accounting for more than 90% of the total mass of the universe could be made of giant planets.

So, while Frege's Basic Law V would secure for us that all objects that fall under '9' fall under 'the number of the planets' as well, according to materialistic astronomy, by substitution into (1) 9 may equal 8, or 10, or an as yet undetermined number of planets. And this indeterminacy causes Frege's system to go haywire.

Frege finally decided that all efforts to repair his logical edifice were destined to failure. By 1912 he had laid down his extensions and conceded defeat. As he wrote for an article by Philip Jourdain:

> And now we know that when classes are introduced, a difficulty, (Russell's contradiction) arises. . . . Only with difficulty did I resolve to introduce classes (or extents of concepts) because the matter did not appear to me to be quite secure —and rightly so as it turned out. The laws of numbers are to be developed in a purely logical manner. But numbers are objects. . . . Our first aim was to obtain objects out of concepts, namely extents of concepts or classes. By this I was constrained to overcome my resistance and to admit the passage from concepts to their extents . . . I confess . . . I fell into the error of letting go too easily my initial doubts (Frege 1980a, 191).

When specifically asked about the causes of the paradoxes of set theory, Frege explained that the "essence of the procedure which leads us into a thicket of contradictions" consisted in regarding the objects falling under F as a whole, as an object designated by the name 'set of Fs', 'extension of 'F', or 'class of Fs' etc. (Frege 1980a, 55). He wrote that the paradoxes of set theory

> arise because a concept e.g. fixed star, is connected with something that is called the set of fixed stars, which appears to be determined by the concept—and determined as an object. I thus think of the objects falling under the concept fixed star combined into a whole, which I construe as an object and designate by a proper name, 'the set of fixed stars'. This transformation of a concept into an object is inadmissible, for the set of fixed stars only seems to be an object, in truth there is no such object at all. (Frege 1980a, 54; 55)

"The definite article," he explained, "creates the impression that this phrase is meant to designate an object, or, what amounts to the same thing, that 'the concept star' is a proper name, whereas 'concept star' is surely a designation of a concept and thus could not be more different from a proper name. The difficulties which this idiosyncrasy of language entangles us in are incalculable" (Frege 1979, 270).

"From this," Frege wrote, "has arisen the paradoxes of set theory which have dealt the death blow to set theory itself" (Frege 1979, 269).

7. Intellectual Sorrow Descends upon Bertrand Russell

Intellectual sorrow, Bertrand Russell has said, descended upon him in full measure when the contradiction about classes which are not members of themselves put an end to the logical honeymoon he was having when he began writing the *Principles of Mathematics* (Russell 1959, 56). One of the things he found once he set out to find out exactly how and why Frege's theories could have given rise to the contradiction was that one could not generally suppose that objects which all have a certain property form a class which is in some sense a new entity distinct from the objects making it up (Russell 1973, 171).

Unbridled use of the principle of logical abstraction was producing fake objects, which in turn were causing the worrisome contradictions.

In *Mathematical Logic*, Quine showed how easily class abstraction leads to Russell's contradiction:

> The usual way of specifying a class is by citing a necessary and sufficient condition for membership in it. Such is the method when one speaks of "the class of all entities x such that . . .," appending one or another matrix. The class of all entities x such that x writes poems, e.g., is the class of poets . . . Despite its sanction from the side of usage and common sense, however, this method of specifying classes leads to trouble. Applied to certain matrices, the prefix 'the class of all entities

x such that' produces expressions which cannot consistently be regarded as designating any class whatever. One matrix of this kind, discovered by Russell, is '$\sim(x \in x)$'; there is no such thing as the class of all entities *x* such that $\sim(x \in x)$. For suppose *w* were such a class. For every entity *x*, then, $x \in w = \sim(x \in x)$. Taking *x* in particular as *w* itself, we are led to the contradiction: $(w \in w) = \sim(w \in w)$ (Quine 1940, §24).

9. The Distinction of Logical Types

Early in his search for a solution to the problem of the paradoxes, Russell believed that "the key to the whole mystery" would be found in the distinguishing of logical types (Russell 1903, §104). So he established a hierarchy of classes according to which the first type of classes would be composed of classes made up entirely of particulars, the second type composed of classes whose members are classes of the first type, the third type composed of classes whose members are classes of the second type, and so on. The types obtained would be mutually exclusive, making the notion of a class being a member of itself meaningless. No totality of any kind could be a member of itself. "In this way," Russell believed, "we obtain a series of types, such that, in all cases where formerly a paradox might have emerged, we now have a difference of type rendering the paradoxical statement meaningless" (Russell 1973, 201).

Russell believed that the theory of types he developed led "both to the avoidance of contradictions, and to the detection of the precise fallacy which has given rise to them" (Russell 1927, 1). And he believed that no solution to the contradictions was technically possible without it. However, he ultimately became aware that it was not "the key to the whole mystery" (see Russell 1919, 135; 1956, 333; Quine, 1961a, 91–92).

For one thing, he realized that the theory only solved some of the paradoxes for the sake of which he had invented it. Deeper problems caused the old contradiction to break out afresh and he realized that "further subtleties" would be needed to solve them. For, though Russell's theory of types restores some of the logical structure Frege had eclipsed when talk of classes and their extensions opened the door to the illegitimate inferences in the first place, the theory just treats the symptoms of the malady. It does not come to terms with the problem as to how and why the logical types had become confused to begin with—as to what had made Frege try to obtain objects out of concepts in the first place. It does not adequately address the deeper logical problems concerning identity and equality which had tempted Frege to introduce classes and a law saying that a class could be predicated of its own extension. The quest to uproot the paradoxes would require further investigations into what classes were.

10. Whither Classes?

Having evaded some of the contradictions by distinguishing between various types of objects, and having proposed a hierarchy of types, Russell was obliged to come to some conclusions regarding the ontological status of classes. He decided that classes could not be independent entities. Regarding them as such leads inescapably to the contradiction about the class of classes which are not members of themselves. He had originally believed that:

> When we say that a number of objects all have a certain property, we naturally suppose that the property is a definite object, which can be considered apart from any or all of the objects, which have, or may be supposed to have, the property in question. We also naturally suppose that the objects which have the property form a *class*, and that the class is in some sense a new single entity, distinct, in general, from each member of the class. (Russell 1973, 163–64)

According to Frege's theory, Russell explained,

> Whatever a class may be, it seems obvious that any propositional function ϕx determines a class, namely the class of objects satisfying ϕx. Thus 'x is human' defines the class of human beings, 'x is an even prime' defines the class whose only member is 2, and so on. We can then (so it would seem) define what we mean by 'x is a member of the class u', or 'x is a u' as we may say more shortly. This will mean: 'There is some function ϕ which defines the class u and is satisfied by x'. We then need an assumption to the effect that two functions define the same class when they are equivalent, i.e. such that for any value of x both are true or both false. Thus 'x is human' and 'x is featherless and two-legged' will define the same class. From this basis the whole theory of classes can be developed. (Russell 1973, 171)

Any such object that might be proposed, he believed, presupposed the notion of class, i.e., an object uniquely determined by a propositional function, and equally determined by any equivalent propositional function (Russell, 1903, §489). However, he had become convinced that this was precisely the kind of reasoning that had gotten himself, Frege, and others involved in the contradictions. For we cannot, then, escape the contradiction, Russell explained:

> For it is essential to an entity that it is a possible determination of x in any propositional function ϕx; that is, if ϕx is any propositional function, and a any entity, ϕa must be a significant expression. Now if a class is an entity, 'x is a u' will be a propositional function of u; hence, 'x is an x' must be significant. But if 'x is an x' is significant, the best hope of avoiding the contradiction is extinguished. (Russell 1973, 171)

The idea that classes were not entities shed some light on the ontological nature of classes by saying what they were not, but Russell had to do more than that. For the contradictions were, however, a distressing after-effect of introducing the extensions Frege had reluctantly appealed to because he saw that without them the theory of abstraction he was prescribing could produce nonsensical conclusions, or be sterile and unproductive (Frege 1884, §§66–67). Frege had resorted to extensions out of necessity and they served a definite purpose in his system.

And Russell was well aware of the need for classes. "The reason," he wrote in the closing pages of the *Principles of Mathematics*,

> which led me, against my inclination to adopt an extensional view of classes, was the necessity of discovering some entity determinate for a given propositional function, and the same for any equivalent propositional function. Thus "*x* is a man" is equivalent (we will suppose) to "*x* is a featherless biped", and we wish to discover some one entity which is determined in the same way by both these propositional functions. The only single entity I have been able to discover is the class as one. . . . (Russell 1903, §486)

And he believed that

> without a single object to represent an extension, Mathematics crumbles. Two propositional functions which are equivalent for all values of the variable may not be identical, but it is necessary there should be some object determined by both. Any object that may be proposed, however, presupposes the notion of *class*... an object uniquely determined by a propositional function, and determined equally by any equivalent propositional function. (Russell 1903, §489)

So Russell could not very well just demolish classes. The trick was to find a way of making them disappear from the reasoning in which they were present without really completely letting go of them. If, he reasoned, "we can find any way of dealing with them as symbolic fictions, we increase the logical security of our position, since we avoid the need of assuming that there are classes without being compelled to make the opposite assumption that there are no classes" (Russell 1919, 184).

11. Having Your Classes and Deleting Them Too

To "lay hold upon the extension of a concept," Frege had proposed transforming "a sentence in which mutual subordination is asserted of concepts into a sentence expressing an identity." Since only objects could figure in identity statements, he realized he would have to find a way of correlating objects and concepts in a way which correlated mutually subordinate concepts with the same object. He suggested this might be achieved by translating language asserting mutual subordination into statements of the form 'the extension of the concept X is the same as the extension of the concept Υ' in which the descriptions would then be regarded as proper names as indicated by the presence of the definite article. But by permitting such a transformation, Frege realized, one is conceding that such proper names have meanings (Frege 1979, 181–82).

But it was precisely that sort of recipe for making what was incomplete behave as if it were complete in identity statements that Russell's paradox had cast doubt upon, and the problems that procedure caused were precisely the ones Russell hoped to circumvent through a new theory of classes based on his theory of definite descriptions by which he hoped to realize Frege's goal of correlating classes with extensions in such a way that concepts which are mutually subordinate would be correlated with the same objects (re. Frege 1980b, 214).

While struggling to get to the bottom of the problem of the fake objects created by logical abstraction, Russell had discovered some parallels existing between the problems which arise when classes are treated as objects and problems which come up when descriptions are treated as names. These analogies plus the success he had with his 1905 theory of definite descriptions gave him an idea as to how classes might be analyzed away much as descriptions had been, and thus gave him a concrete idea as to how he might sweep his problems away.

Statements containing descriptions had proven amenable to further analysis. So once Russell was satisfied that classes and descriptions both fell into the same logical category of nonentities represented by incomplete symbols, he decided to extend his ideas about analyzing away descriptions to include class symbols. He reasoned that since:

> we cannot accept "class" as a primitive idea. We must seek a definition on the same lines as the definition of descriptions, i.e. a definition which will assign a meaning to propositions in whose verbal or symbolic expression words or symbols apparently representing classes occur, but which will assign a meaning that altogether eliminates all mention of classes from a right analysis of such propositions. We shall then be able to say that the symbols for classes are mere conveniences, not representing objects called "classes," and that classes are in fact, like descriptions, logical fictions, or (as we say) "incomplete symbols." (Russell 1919, 181–82)

According to Russell's theory of definite descriptions: "There is a term c such that $\phi\, x$ is always equivalent to 'x is c'" (Russell 1919, 178). That being so the putative identity statement 'Scott is the author of *Waverley*' might be rewritten 'Scott wrote *Waverley*, and it is always true of c that if c wrote

Waverley, *c* is identical with Scott' (Russell 1956, 55). In this construal of the sentence, what was not identical has been made to be equivalent. A symbol deemed equatable with 'Scott' has gone proxy for the description and this symbol will generally be "obedient to the same formal rules of identity as symbols which directly represent objects" (Russell 1927, 83).

This means of drawing objects out of descriptions provided Russell with a practical model of how to make nonentities function as entities without incurring contradictory results. Unadulterated class abstraction was generating contradictions, but the theory of definite descriptions was a different way of making an object fit to go proxy for what was said about it. By adapting the theory to class symbols, he thought he might acquire a "method of obtaining an extensional function from any given function of a function," which was precisely what the theory of classes needed (Russell 1927, 187).

12. Problems with Equivalent Propositional Functions

Russell, however, also realized that resolving the paradoxes would mean coming to terms with problems with classes caused by the fact that it was "quite self-evident that equivalent propositional functions are often not identical" (Russell 1903, §500). And these problems display certain formal similarities with the puzzles about descriptions which had impressed upon Russell the need to remove descriptions from statements in which they were present in the first place.

One of the reasons Russell had had to find a way to get rid of descriptions was that, while an informative statement of the form 'the author of *Waverley* is the author of *Marmion*' could be true and seem to be a genuine identity statement in virtue of the fact that both descriptions were true of the same individual, the two descriptions were obviously not themselves identical, nor were they always intersubstitutable salva veritate. If the only thing that mattered were that both descriptions are true of the same person, Russell observed, then any other phrase true of Scott would yield the same statement. Then 'Scott is the author of *Marmion*' would be the same as 'Scott is the author of *Waverley*', which is obviously not so, since from the one we learn that Scott wrote *Marmion* and from the other that he wrote *Waverley*, but the former statement tells us nothing about *Waverley* and the latter nothing about *Marmion*. Properly reconstrued, the descriptions would disappear and the statement transformed into: 'Someone wrote *Waverley* and no one else did, and that someone else wrote *Marmion* and no one else did' (Russell 1917, 217–18).

This way of correlating mutually subordinate descriptions with the same object could help solve a parallel problem in the paradox plagued theory of classes, e.g., the problem of two propositional functions having the same graph without everything that is true of one being true of the other. For just as descriptions are tied to a particular characterization of an object, so classes are formed by specifying the definite property giving the class. Moreover, just as two different descriptions might be true of the very same object, so a single class of objects might be defined in different ways, each one corresponding to a different sense of the class name. Yet as Russell observed "if *a* and *b* be different class-concepts such that $x \in a$ and $x \in b$ are equivalent for all values of *x*, the class-concept under which *a* falls and nothing else will not be identical with that under which falls *b* and nothing else" (Russell 1903, §488).

Russell was pleased to point out, however, that the theory of classes inspired by his theory of descriptions would leave intact all the fundamental properties desired of classes, the principal one of these being that "two classes are identical when, and only when, their defining functions are formally equivalent" (Russell 1927, 189). "The incomplete symbols which take the place of classes serve the purpose of technically providing something identical in the case of two functions having the same extension," he believed (Russell 1927, 187).

Presto chango! By an act of logical prestidigitation, the theory of classes has been rendered "symbolically satisfactory." We can have our classes and delete them too. Statements verbally concerned with classes have been reduced to statements that are concerned with propositions and propositional functions. Logicians can effectively pass from a class to its extension on paper without incurring contradictory results. We have avoided contradictions arising from supposing that classes are entities and acquired a technique for laying hold of the extension of a class. In addition, functions having the same extension would be identical. And all our efforts have brought us ever nearer to the deep reasons why Frege's introducing extensions into his logic caused the paradoxes in the first place.

13. The Ultimate Source of the Contradictions

Russell considered the technique he devised for making incomplete symbols obey the same formal rules of identity as symbols which directly represent objects to be a breakthrough in solving the paradoxes and a host of other problems (Russell 1959, 49, 60; Grattan-Guinness 1972, 106–7; 1977, 70, 79–80, 94 and note; 1975, 475–88; Kilmister 1984, 102, 108, 123, 138; Hill 1997). However, he realized that incomplete symbols could only obey the same formal rules of identity as symbols referring to objects in so far as "we only consider the *equivalence* of the resulting variable (or constant) values of propositional functions and not their identity" (Russell 1927, 83), an observation which brings us right back to the problems concerning logical abstraction, identity and substitutivity which first obliged Frege to introduce extensions.

So to assess the ultimate effectiveness of the various techniques Russell invented for evading the paradoxes, it is important to take a close look at what Russell once called the "ultimate source" of the contradictions. In a statement which brings us all the way back to Frege's original idea that "the existence of different names for the same content is the very heart of the matter if each is associated with a different way of determining the content" (Frege 1879, §8) and to his original reasons for introducing extensions, Russell affirmed that the cause of his and Burali-Forti's contradiction was to be found in that:

> if *x* and *y* are identical, φ*x* implies φ*y*. This holds in each particular case, but we cannot say it holds *always*, because the various particular cases have not enough in common. This distinction is difficult and subtle . . . the neglect of it is the ultimate source of all the contradictions which have hitherto beset the theory of the transfinite. (Russell 1973, 188)

These contradictions most intimately concerned with identity and the formal equivalence of functions proved especially hard to stamp out. For them Russell had to develop a new tactic more directly aimed at the theory of identity which had brought about the confusion of types and the reification of incomplete symbols in the first place.

In particular, Russell still faced the following predicament: The contradictions had taught him that there was a hierarchy of logical types. If contradiction producing vicious circle fallacies were to be avoided, functions would have to be divided into types, and all talk of functions would then necessarily be limited to some one type, which would effectively make statements about all functions true with a given argument, or all properties of a some given object, meaningless.

However, Russell was perfectly aware that "it is not difficult to show that the various functions which can take a given object *a* as argument are not all of one type" (Russell 1919, 189), and, even "that the functions which can take a given argument are of an infinite series of types" (Russell 1919, 190). By various technical devices we could, he once reminded readers, "construct a variable which would run through the first *n* of these types, where *n* is finite, but we cannot construct a variable which will run through them all, and, if we could, that mere fact would at once generate a new type of function with the same arguments, and would set the whole process going again" (Russell 1919, 190). So whatever selection of functions one makes there will always be other functions which will not be included in the selection.

Russell nonetheless believed that "it must be possible to make propositions about *all* the classes that are composed of individuals, or about *all* the classes that are composed of objects of any one logical 'type.' If this were not the case, many uses of classes would go astray. . . ." (Russell 1919, 185). "If mathematics is to be possible," he believed, "it is absolutely necessary . . . that we should have some statements which will usually be equivalent to what we have in mind when we (inaccurately) speak of 'all properties of *x*'. . . . Hence we must find, if possible, some method of reducing the order of a propositional function without affecting the truth or falsehood of its values" (Russell 1927, 166; 1956, 80).

14. The Axiom of Reducibility

So to cope with contradictions arising from necessary talk of 'all properties,' or 'all functions,' Russell introduced the axiom of reducibility. This specially designed axiom would be "equivalent to the assumption that 'any combination or disjunction of predicates is equivalent to a single predicate'" (Russell 1973, 250; 1927, 58–59), and would provide a way of dealing with any function of a particular argument by means of some formally equivalent function of a particular type. It would thus yield most of the results which would otherwise require recourse to the problematical notions of all functions or all properties, and so legitimize a great mass of reasoning apparently dependent on such notions (Russell 1927, 56).

For Russell, the axiom embodied all that was really essential in his theory of classes (Russell 1919, 191; 1927, 58). "By the help of the axiom of reducibility," Russell affirmed, "we find that the usual properties of classes result. For example, two formally equivalent functions determine the same class, and conversely, two functions which determine the same class are formally equivalent" (Russell 1973, 248–49). He came to believe classes themselves to be mainly useful as a technical means of achieving what the axiom of reducibility would effect (Russell 1919, 191). It seemed to him "that the sole purpose which classes serve, and one main reason which makes them linguistically convenient, is that they provide a method of reducing the order of a propositional function" (Russell 1927, 166). Classes were producing contradictions. They should be expunged and replaced with this axiom which seemed to him "to be the essence of the usual assumption of classes" and to retain "as much of classes as we have any use for, and little enough to avoid the contradictions which a less grudging admission of classes is apt to entail" (Russell 1927, 166–67; 1956, 82; 1919, 191). Russell leaned on the axiom of reducibility at every crucial point in his definition of classes in *Principia Mathematica* (pp. 75–81).

Russell considered that many of the proofs of *Principia* "become fallacious when the axiom of reducibility is not assumed, and in some cases new proofs can only be obtained with considerable labour" (Russell 1927, xliii).

He also believed that without the axiom, or its equivalent, one would be compelled to regard identity as indefinable and to admit that two objects might agree in all their predicates without being identical (Russell, 1927, 58). In particular, by resorting to the axiom of reducibility one might avoid a difficulty with the definition of identity which Russell explained as follows:

> We might attempt to define "x is identical with y" as meaning "whatever is true of x is true of y," i.e., ϕx always implies ϕy." But here, since we are concerned to assert all values of "ϕx implies ϕy" regarded as a function of ϕ, we shall be compelled to impose upon ϕ some limitation which will prevent us from including among values of ϕ values in which "all possible values of ϕ" are referred to. Thus for example "x" is identical with "a" is a function of x; hence, if it is a legitimate value of ϕ in "ϕx always implies ϕy," we shall be able to infer, by means of the above definition, that if x is identical with a, and x is identical with y, then y is identical with a. Although the conclusion is sound, the reasoning embodies a vicious-circle fallacy, since we have taken "$(\phi)(\phi x$ implies $\phi a)$" as a possible value of ϕx, which it cannot be. If, however, we impose any limitation upon ϕ it may happen, so far as appears at present, that with other values of ϕ we might have ϕx true and ϕy false, so that our proposed definition of identity would plainly be wrong. (Russell 1927, 49)

"But in virtue of the axiom of reducibility," Russell writes in *Principia Mathematica* *13, "it follows that, if $x = y$ and x satisfies ψx, where ψ is any function . . . then y also satisfies ψy." And this effectively made his definition of identity as powerful as if he had been able to appeal to all functions of x (Russell 1927, 168). For if one assumes the axiom of reducibility, then

> every property belongs to the same collection of objects as is defined by some predicate. Hence there is some predicate common and peculiar to the objects which are identical with x. This predicate belongs to x, since x is identical with itself; hence it belongs to y, since y has all the predicates of x; hence y is identical with x. It follows that we may *define* x and y as identical when all the predicates of x belong to y . . . (Russell 1927, 57; 1973, 243)

So by virtue of the axiom of reducibility, one might have the properties of identity and equality upon which the logic of *Principia Mathematica* is built. Russell thought of it as a generalized form of Leibniz's principle of the identity of indiscernibles (Russell 1919, 192; 1927, 57; 1973, 242). So he was finally back to square one, i.e., to the reasons why Frege's theory of identity had made him appeal to extensions in the first place.

15. Why All the Logical Acrobatics?

What had made such a radical measure necessary?

The axiom of reducibility was another attempt to rub out the differences between equality and full identity. In his search for a criterion for deciding whether in all cases *x* is the same as *y*, Frege had turned to Leibniz's principle of substitutivity of identicals (Frege 1884, §65). Then he adopted a form of expression by which being the same in one way would be the same as being the same in all ways, making the differences between equality and identity seem to go away. After that he reluctantly introduced extensions as a way of artificially rectifying the problems with substitutivity which that attempt to equate equality and identity had caused (Frege 1884, §§66–67). Basic Law V would guarantee that an identity holding between two concepts could be transformed into an identity of extensions and conversely, that functions having the same extensions were identical—and it leads to Russell's paradox.

Russell struggled long and hard with the problems buried in Frege's theory of identity, and he devised some sharp logical instruments to erase the differences between equality and identity which had not vanished from reasoning as obligingly as hoped. Wielding his theory of types and his technique for analyzing away classes, Russell began sweeping away the tangled web Frege began to weave when he adopted the inference wrecking practice of equating identity with lesser forms of equivalence. By adroitly wiping out a wealth of intensions and casting them into logical oblivion, the axiom of reducibility would sweep *Principia Mathematica* nearly clean of intensions and so win new territory for extensional ontology.

Yet in spite of all that might be achieved by means of the axiom, Russell expressed reservations about it reminiscent of those Frege had expressed regarding Basic Law V. Russell deemed the axiom "only convenient, not necessary" (Russell 1919, 192), and even called it "a dubious assumption" and a "defect" (p. 193). "This axiom," he admitted, "has a purely pragmatic justification: it leads to the desired results, and to no others. But clearly it is not the sort of axiom with which we can rest content" (Russell 1927, xiv).

Russell, who had once written that "Mathematics, rightly viewed, possesses not only truth, but supreme beauty," concluded that the "solution of the contradictions . . . seemed to be only possible by adopting theories which might be true but were not beautiful" and that the "splendid certainty" he had "always hope to find in mathematics had become lost in a bewildering maze" (Russell 1959, 155–57). What had appeared so convenient, simple, and austerely beautiful had spawned error, contradiction, ugliness, and messiness.

16. Frege's and Russell's Problems Live On

But Russell's struggle with the paradoxes was not the end of the story. No matter how convenient and attractive abstraction may seem to be as a technique for translating expressions into the popular notation of extensional logic, the properties marking the difference between equality and identity do not docilely submit to logical measures designed to wipe them out. They easily slip back into reasoning unawares to sow surprise, antinomy, nonsense, confusion, and contradiction, frustrating the aims of the brave new logic which it was hoped might wipe them out with the stroke of a pen. So they have survived the campaign to extirpate them which has finally served to demonstrate the reality and ineradicability of what it was trying to remove from logical reasoning.

In particular, evidence of the logical violence done began to surface again when philosophers working on modal and intensional logics dared to try to bring light to some dark areas of logic where philosophers had been warned not to tread. Taking a bolder attitude toward limning the true and ultimate structure of reality than many of their contemporaries, philosophers like Jaakko Hintikka and Ruth Barcan Marcus ventured beyond the narrow confines strong extensional calculi impose on philosophical reasoning and began working to increase the depth and utility of the standard languages and to develop intensional languages capable of investigating epistemic and deontic contexts and of analyzing the many nonextensional statements which figure significantly in the empirical sciences, law, medicine, ethics, engineering, politics, and ordinary philosophy but which have been deemed unfit for study because they complicate matters by not conforming to the rigid standards for admission into the stark, sterile logical world Quine and so many others have found so beautiful. And their work has been instrumental in pulling the deep issues underlying the puzzles, contradictions, and paradoxes haunting logical arguments out of the shadowy netherworld to which they were being consigned and into the clear light of day. So even as opponents of intensional and modal logics battled to contain logical reasoning within the narrow confines of strong extensional calculi, more and more reasons for not shoving reasoning into an extensional mold began gathering right in the "beautiful" world they were so intent upon preserving.

17. Marcus on Distinguishing between Identity and Lesser Forms of Equivalence

Founder of quantified modal logic, Ruth Barcan Marcus has been one of the staunchest and most eloquent advocates of lucidity regarding the differences between identity and weaker forms of equivalence which explicit or implicit extensionalizing principles would extinguish. As part of her ongoing campaign to expand classical logic to deal with larger areas of discourse, she has drawn attention to ambiguities regarding equality and identity that have slipped into logical reasoning and are present there now. In particular, she has drawn attention to the extent to which extensional logical systems are dependent on (1) directly, or indirectly imposing restrictions prohibiting some intensional functions, and (2) equating identity with a weaker form of equivalence (Marcus 1960).

"The usual reason given for reducing identity to equality," she has written, "is that it provides a simpler base for mathematics, mathematics being concerned with aggregates discussed in truth functional contexts, not with predicates in intensional contexts. Under such restrictive conditions, the substitution theorem can generally be proved for equal (formally equivalent) classes, with the result that equality functions as identity" (Marcus, 1960, 58).

Extensionality, Marcus explains, has acquired the undeserved reputation of being a clear, unambiguous concept, and as such well-suited to the needs of mathematics and the empirical sciences where, it is claimed, there is no need to traffic in fuzzy, troublesome nonextensional notions. However, strongly extensional functional calculi, she notes, are "inadequate for the dissection of most ordinary types of empirical statement" (Marcus 1993, 5). "Establishing the foundations of mathematics," she points out, "is not the only purpose of logic, particularly if the assumptions deemed convenient for mathematics do violence to both ordinary and philosophical usage" (Marcus 1960, 58). No single well-defined theory of extensionality exists, she argues, but only stronger and weaker principles of extensionality. She has urged "that the distinctions between stronger and weaker equivalences be made explicit before, for one avowed reason or another, they are obliterated" (Marcus 1960, 55), a request which would seem to be perfectly in keeping with the requirements of a logic which prides itself on its clarity, and was devised to keep ambiguity from creeping unawares into reasoning.

18. Abiding Differences

It is easy to find puzzles, contradictions, paradoxes, failures of basic logical principles, etc. which illustrate the need for lucidity regarding the differences between identity and weaker forms of equivalence. Now I want to draw attention to some undesirable consequences of failing to distinguish between identity and lesser forms of equivalence which I believe demonstrate the real need for consciously trafficking in intensional notions in order to control confusion and to draw the fine distinctions which are both germane and indispensable to many undertakings, scientific and otherwise.

Significant differences between equality and identity come to the fore in a much more tangible way and the issues at stake become clearer once we leave the realm of x's and y's, and a's and b's and turn to the reference of the signs. Since the logical ideas we are examining are supposed to be a useful tool for scientists and philosophers and since identity statements are only to be true or false on the basis of what they stand for, this is surely what we are supposed to do. There are, after all, few cases in which the difference between being called 'x' or being called 'y' really matters.

While strong extensional principles may prove appropriate in certain contexts, e.g., in criminal investigations or judicial proceedings where a person's guilt or innocence may be the sole determinant factor, relying upon them could have disastrous consequences in other contexts. For instance, in medical research and practice, extensionality could unnecessarily complicate situations and generate confusion, and even make the difference between life and death, sickness and health.

Consider this example. Doctors at Toronto's Hospital for Sick Children have discovered that the immune system of certain diabetics identifies a protein present on the surface of their insulin-producing cells as being the same as a protein present in cow's milk with which it is in many respects almost identical. Unable to distinguish between the two proteins, the immune system stimulates the body to attack and destroy its own insulin-producing cells in the pancreas causing juvenile onset diabetes which may lead to blindness, kidney failure, and heart disease.

Marcus's point about identity and equivalence is most apropos here for the immune systems in question are identifying two proteins as being the same on the basis of compelling similarities; they are not picking out essential differences between the milk protein and the protein on the insulin-producing cells. In a case like this equating identity with lesser forms of equivalence may be having disastrous consequences for diabetics.

Intensional factors marking the differences between identity and lesser forms of equivalence also come into play in organ transplant operations where what is sought are organs which are alike in certain respects, but surely not identically the same as the diseased or defective organs they are meant to replace. A strong enough equivalence may obtain between different organs of the same type so that the host organism will not reject the transplanted organ as foreign because it is sufficiently like the organ it supplants. It is precisely intensional considerations that mark the difference between unwanted strict identity and the sought after equivalence powerful enough to keep the substitutivity of the organs from breaking down. To say of organ x that it has all the properties in common with organ y essential for a successful transplant is to make a weaker claim than to claim they are identical. The situation with bone-marrow transplants appears to be more complex than for organs because six different genetic identity markers are involved in matching tissue types.

And don't differences between equality and identity figure in many other dilemmas actually faced in medical practice today? Surely, many of the really challenging moral issues involved in abortion, euthanasia, medically assisted suicide, etc. turn on whether the person to be killed is in all essential respects the same as that person once was, or in the case of abortion the same as the person who will be if the pregnancy is not terminated. For instance, are patients in apparently irreversible comas identical to the people they were before they were in that unfortunate situation? Think of the innumerable things that could been have been predicated of them before which are no longer true, and the truly macabre propositions that could result from sub-stitution rules which do not take sufficient account of the difference between equality and identity. Surely, such considerations are involved drawing the line between a mercy killing and a premeditated murder.

It is plain to see that in the above cases doctors, judges, medical researchers, and concerned individuals could not even seriously consider resorting to any conceptual tool which was so blunt, crude, and blind as one which systematically disregards the differences between identity and lesser forms of equivalence. However inconvenient they may be on paper, and however uncongenial they may be to some philosophers, intensions can be the decisive factor in real life situations where failure to resort to them is sure to increase perplexity, engender confusion, and complicate matters instead of bringing clarity and precision. Anyone insisting upon deintensionalizing the situations just cited would surely inspire contempt in the scientific commu-nity which fortunately would not rely on so crude a conceptual instrument in situations which cry out for the conscious intensional adjustment and pro-gressive refinement of a conceptual tool which less finely tuned could have disastrous consequences. The blithe use of extensional notions could, in the case of the organ transplant operation, be dangerous enough to cause a patient to die of post-operative complications due to simplistic ideas about extensionality and scientific thinking.

I would now like to close with a warning Edmund Husserl made in *For-mal and Transcendental Logic* in 1929, almost forty years after having con-demned Frege's theory of equality and identity in *Philosophy of Arithmetic*. Mathematicians, Husserl observed, are not in the least interested in the dif-ferent ways objects may be given. For them objects are the same which have been correlated together in some self-evident manner. However, he con-tended, logicians who do not bewail the lack of clarity involved in this, or who claim that the differences do not matter are not philosophers since here it is a matter of insights into the fundamental nature of formal logic, and without a clear grasp of the fundamental nature of formal logic, one is obvi-ously cut off from the great questions that must be asked about logic and its role in philosophy (Husserl 1929, 147–48).

REFERENCES

Dummett, M. 1991. *Frege, Philosophy of Mathematics*. Cambridge, MA: Harvard University Press.

Frege, G. 1879 (1967). "*Begriffsschrift*, a formula language, modeled upon that of arithmetic for pure thought." *From Frege to Gödel*. Ed. J. van Heijenoort. Cam-bridge, MA: Harvard University Press. 5–82.

———.1884 (1986). *Foundations of Arithmetic*. Oxford: Blackwell.

———.1893 (1964). *Basic Laws of Arithmetic*. Berkeley: University of California Press.

————.1894 (1984). "Review of E.G. Husserl's Philosophy of Arithmetic." *Collected Papers on Mathematics, Logic and Philosophy.* Oxford: Blackwell. 195–209.

————.1979. *Posthumous Writings.* Oxford: Blackwell.

————.1980a. *Philosophical and Mathematical Correspondence.* Oxford: Blackwell.

————.1980b. *Translations from the Philosophical Writings.* Ed. P. Geach and M. Black. Oxford: Blackwell.

Garciadiego, A. 1992. *Bertrand Russell and the Origins of the Set-theoretic 'Paradoxes.'* Basel: Birkhäuser.

Grattan-Guinness, I. 1972. "Bertrand Russell on His Paradox and the Multiplicative Axiom:An Unpublished Letter to Philip Jourdain." *Journal of Philosophical Logic* 1:103–10.

————. 1975. "Preliminary Notes on the Historical Significance of Quantification and the Axiom of Choice in Mathematical Analysis." *Historia Mathematica* 2: 475–88.

————. 1977. *Dear Russell–Dear Jourdain, A Commentary on Russell's Logic Based on his Correspondence with Philip Jourdain.* London: Duckworth.

Hill, C. O. 1991. *Word and Object in Husserl, Frege and Russell, the Roots of Twentieth Century Philosophy.* Athens, OH: Ohio University Press.

————. 1997. *Rethinking Identity and Metaphysics.* New Haven: Yale University Press.

Hintikka, J. 1962. *Knowledge and Belief.* Ithaca, NY: Cornell University Press.

————. 1969. *Models for Modalities.* Dordrecht: Reidel.

————. 1975. *The Intentions of Intentionality and Other New Models for Modalities.* Dordrecht: Reidel.

Hintikka, J., and D. Davidson, eds. 1969. *Words and Objections, Essays on the Work of W.V. Quine.* Boston: Reidel.

Husserl, E. 1891 (1970). *Philosophie der Arithmetik, mit ergänzenden Texten (1890–1901).* Husserliana, vol. XII. Dordrecht: Kluwer.

————. 1929 (1978). *Formal and Transcendental Logic.* The Hague: M. Nijhoff.

Kaplan, D., and R. Montague. 1960. "A Paradox Regained." *Notre Dame Journal of Formal Logic* 1, 3 (July): 79–90.

Kilmister, C.W. 1984. *Russell.* London: The Harvester Press.

Marcus, R. 1960. "Extensionality." *Mind* 69: 55–62.

————. 1993. *Modalities.* New York: Oxford University Press.

Montague, R. 1963. "Syntactical Treatments of Modality, with Corollaries on Reflexion Principles and Finite Axiomatizability." Proceedings of a Colloquium on Modal and Many-Valued Logics, Helsinki, 23–26 August, 1962, *Acta Philosophica Fennica* XVI: 153–67.

Quine, W. 1940. *Mathematical Logic.* New York: Norton.

————. 1947. "The Problem of Interpreting Modal Logic." *Journal of Symbolic Logic* 12, no. 2 (June): 43–48.

————. 1952 (1962). *Methods of Logic.* London: Routledge and Kegan Paul.

————. 1960. *Word and Object.* Cambridge, MA: M.I.T. Press.

————. 1961a (1953). "New Foundations For Mathematical Logic." *From a Logical Point of View.* New York: Harper and Row, 80–101.

————. 1961b (1953). "Logic and the Reification of Universals." *From a Logical Point of View.* New York: Harper and Row, 102–29.

————. 1969. *Set Theory and Its Logic.* Cambridge, MA: Harvard University Press.

————. 1976. *Ways of Paradox.* Cambridge, MA: Harvard University Press.

————. 1994. "Promoting Extensionality." *Synthese* 98: 143–51.

Rodriguez-Consuerga, F. 1991. *The Mathematical Philosophy of Bertrand Russell: Origins and Development.* Basel: Birkhäuser.

Russell, B. 1903. *Principles of Mathematics.* London: Norton.
———. 1917 (1986). *Mysticism and Logic.* London: Allen & Unwin.
———. 1919. *Introduction to Mathematical Philosophy.* London: Allen & Unwin.
———. 1927 (1964). *Principia Mathematica to *56,* 2nd ed. Cambridge UK: Cambridge University Press.
———. 1944. "My Mental Development." *The Philosophy of Bertrand Russell.* Ed. P. Schilpp. Library of Living Philosophers, vol. 5. Evanston, IL: Northwestern University Press, 3–20.
———. 1956. *Logic and Knowledge.* London: Allen & Unwin.
———. 1959 (1975). *My Philosophical Development.* London: Allen & Unwin.
———. 1973. *Essays on Analysis.* Ed. D. Lackey. London: Allen & Unwin.

6

Claire Ortiz Hill

FREGE'S ATTACK ON HUSSERL AND CANTOR

One hundred years ago Gottlob Frege published a damaging, abusive review (Frege 1894)[1] of Edmund Husserl's *Philosophy of Arithmetic* (Husserl 1891).[2] Although rather a lot has now been written about Frege's review and the role it might have played in the development of Husserl's thought,[3] much still remains to be rectified regarding Frege's assessment of the book and the credence his review has been accorded. Philosophers have generally been all too willing to trust Frege's judgment, and so all too ready to dismiss Husserl's book as irredeemably muddled.

Up until now it has been easy to underrate the *Philosophy of Arithmetic* because, although philosophers have long been familiar with the tenor of Frege's review, Husserl's book is only now beginning to receive the attention it deserves,[4] and numerous misconceptions still abound concerning it. Those who have examined the issues closely, however, have often risen to its defense.[5] Some have drawn attention to the fact that Frege attacked views Husserl had not espoused.[6]

Moreover, Frege himself actually seemed to indicate something of the kind when he wrote at the end of his review that he had taken it to be his task to exhibit the damage caused by the irruption of psychology into logic "in the proper light," but conceded that the errors he believed he "was compelled to point out are to be laid to the charge, not so much of the author, as of a widespread philosophical disease" (Frege 1894, 209).

Frege's statement raises an interesting question as to the extent to which he might actually have been condemning someone else's abuses. Here I will argue that a close look at his criticisms of Husserl's views on abstraction, extensionality, and psychologism yields important clues as to the answer to that question. I will endeavor to show the extent to which Frege used his review of the *Philosophy of Arithmetic* as a forum for attacking Georg Cantor's theory of number. By so doing I hope to help put Frege's objections "in the proper light," and so undo some of the damage done to Husserl's book.

1. Abstraction

In his review, Frege repeatedly criticized Husserl's use of a theory of number abstraction in the *Philosophy of Arithmetic*. It is worthwhile to examine Frege's objections closely because Husserl's use of a theory of abstraction cannot just be dismissed off hand. Husserl was in quite respectable company there. Richard Dedekind, Georg Cantor, Giuseppe Peano, and Bertrand Russell all had theories of abstraction. In fact, Frege himself is often said to have espoused a theory of abstraction.[7] So it is very important to understand what exactly Frege and Husserl each meant when they used the word.

Husserl clearly describes the theory of abstraction he had come to advocate in chapter 4 of the *Philosophy of Arithmetic* where he defines number abstraction as a procedure by which, while actually engaged in the counting process, the counter "abstracts from" the particular properties of the individual members of the multiplicity, or set of items being counted, the particular way in which they are given, and any relations obtaining among them, only regarding the set as being composed of distinct featureless items to be counted (pp. 85–88). Anticipating the objection that in so doing the items themselves and any relations obtaining between them would naturally disappear (p. 84), Husserl adds that abstracting from the particular properties of the items to be counted merely means not directing one's attention toward them while actually counting. "That absolutely does not have the effect of making these contents and the relations obtaining among them disappear" (p. 85). He ends his chapter by condemning Aristotle's, John Locke's, and J. S. Mill's theories of abstraction (pp. 91–92).

This being Husserl's view, Frege was perfectly correct to write that Husserl would have one "abstract from the peculiar constitution of the individual contents that make up the multiplicity and retain each one only in so far as it is a something or a one." Frege was also correct in writing that Husserl's "process of abstracting the number goes hand in hand with a process of emptying all content" (Frege 1894, 196).

And by so characterizing Husserl's theory of abstraction, Frege rather neatly put Husserl into the category of thinkers to which he belonged when writing the *Philosophy of Arithmetic*, i.e., into the school of Karl Weierstrass and Georg Cantor, two men whose ideas Frege also vehemently opposed.[8] Lecture notes which have survived from Husserl's student years indicate that Husserl had been pointed in this direction by his teacher Weierstrass[9] who exercised a profound influence on him.[10]

Georg Cantor, also a Weierstrass student, was Husserl's close friend, colleague, and mentor at the University of Halle from 1886 to 1900.[11] And Husserl's description of the abstraction process in the *Philosophy of Arithmetic* is actually quite similar to descriptions of the same process Cantor made during those years. A look at Husserl's personal copies of Cantor's

Contributions to the Theory of the Transfinite (1886–1890) in fact shows that Husserl marked and underlined precisely those passages (and almost exclusively those passages) in which Cantor defined the abstraction process. For instance, Husserl marked and underlined the passage in which Cantor wrote:

> By the **power** or **cardinal number** of a set M . . . I understand the general concept or species concept (universal) which one obtains when one abstracts from the set both the nature of its elements as well as all relationships which the elements have either between themselves or to other things, in particular to the order that may prevail between the elements, and only reflects on that which is common with all sets which are **equivalent** with M.[12]

But, unwilling just to condemn Husserl for holding views he really did hold in the *Philosophy of Arithmetic*, Frege quite unfairly went on to charge in his review that Husserl's procedure would "cleanse things of their peculiarities . . . in the wash-tub of the mind" where things "assume a quite peculiar pliancy. . . ." There, Frege charges, "we can easily change objects by directing our attention towards them or away from them. . . . We attend less to a property, and it disappears. By thus making one characteristic mark after another disappear, we obtain more and more abstract concepts" (Frege 1894, 197). Then Frege went on to provide this caricature of the procedure Husserl had advocated:

> Suppose, e.g., that there are a black and white cat sitting side by side before us. We do not attend to their colour, and they become colourless—but they still sit side by side. We do not attend to their posture, and they cease to sit . . . but each of them is still in its place. We no longer attend to the place and they cease to occupy one—but they continue presumably to be separate. We have thus perhaps obtained from each of them a general concept of cat. By continued application of this procedure, each object is transformed into a more and more bloodless phantom. (Frege 1894, 197–98)

In *Foundations of Arithmetic* (§34), Frege had explicitly stated that in the abstraction process the things themselves do not lose any of their characteristics. One might disregard the properties which distinguish a white cat and a black cat, but the white cat would still remain white, and the black one black whether or not one thought about their colors, or made any inference from their difference in this regard. The cats would not become colorless. They would remain precisely as before.

However, ten years later, he apparently could no longer imagine that abstraction did not actually change objects because in his review of the *Philosophy of Arithmetic* he maintained that it must surely be assumed that the process of abstraction effects some change in the objects and that they become different from the original objects which are either transformed or actually created by the abstraction process (Frege 1894, 204).

Nonetheless, in spite of the force of the charges he directed against Husserl in his review, Frege was honest enough to admit that Husserl himself did not hold that the mind creates new objects or changes old ones, and actually acknowledged that Husserl "disputes this in the most vehement terms (p. 139)" (Frege 1894, 205).

When, however, Frege goes on to charge that Husserl had taken "the road of magic rather than of science" (Frege 1894, 205), we have a good initial clue as to whom else Frege wishes to attack, for he had accused Cantor of the very same thing in reviewing the *Contributions to the Theory of the Transfinite*. There Frege complained about Cantor's use of the verb 'abstract' which Frege branded "a psychological expression and, as such, to be avoided in mathematics" (Frege 1984, 180; 181).

A posthumously published draft of this review (Frege 1979, 68–71) is more revealing and considerably less tame than the one actually published. There Frege likens mathematicians who like Cantor talk of abstraction to "negroes from the heart of Africa." For these mathematicians, Frege contends, words like "abstraction" are supposed to have "the kind of magical effects" that enable them to abstract from any properties of things which bother them (p. 69). In the spirit of the cat example in his review of Husserl, Frege complains that mathematicians like Cantor find a whole host of things in mice which are unworthy to form a part of the number. "Nothing simpler," Frege writes:

> one abstracts from the whole lot. Indeed when you get down to it everything in the mice is out of place: the beadiness of their eyes no less than the length of their tails and the sharpness of their teeth. So one abstracts from the nature of the mice . . . one abstracts presumably from all their properties, even from those in virtue of which we call them animals, three-dimensional beings. . . . (p. 70)

In this review, Frege even alludes to someone who he says he suspects is one of Cantor's pupils and who when asked what general concept he arrives at when given a pencil exerts himself "to the utmost in abstracting from the nature of the pencil and the order in which its elements are given," to answer 'the cardinal number one' (p. 71).

2. Misery Loves Company

After summarizing the fundamental ideas of the *Philosophy of Arithmetic*, Frege characterizes Husserl's endeavor as "an attempt to justify a naïve conception of number." Frege calls naïve "any view on which a statement of number is not a statement about a concept or about the extension of a concept." "For," he contends, "when one first reflects on number, one is led by a certain necessity to such a conception" (Frege 1894, 197). "If the author had used the word 'extension of a concept' in the same sense as I," Frege declares, "we should hardly differ in opinion about the sense of a number statement" (Frege 1894, 201–2).

In the *Philosophy of Arithmetic*, however, Frege complained, multiplicities, sets, are more indeterminate and more general than numbers. Husserl would first analyze the concept of multiplicity and use it in determining definite numbers and the generic concept of number presupposing them. He would go from the general to the particular and back again (Frege 1894, 195). Husserl had even argued that the relationship between the number and the generic concept of what is numbered was in certain respects the opposite of what Frege supposed (Husserl 1891, 186).

Frege's criticism can only be fairly assessed in connection with his own struggles with extensionality, because on several occasions he confessed that he had never become completely reconciled to using extensions, and he expressed profound reservations about them throughout his career. As it happens the 1894 review of Husserl's book actually contains some of the most forceful statements Frege ever made in their favor.

Frege first appealed to extensions in his 1884 *Foundations of Arithmetic* (§68) in an attempt to eliminate certain undesirable consequences (§§66–67, 107) of the theory of number he advocated there. And he devoted several sections of that work to discussing the pros and cons of introducing them (§§68–73). In one of its concluding sections he wrote of extensions that: "This way of getting over the difficulty cannot be expected to meet with universal approval, and many will prefer other methods of removing the doubt in question. I attach no decisive importance even to bringing in the extensions of concepts at all (§107).

Although Frege had ended the *Foundations of Arithmetic* claiming that he did not attach any decisive importance to bringing in extensions, he did not propose any alternative. And when Georg Cantor wrote in an 1885 review of the book that it was unfortunate that Frege had taken extensions of concepts as the foundation of the number concept,[13] Frege immediately rose to defend himself against Cantor's charge that it was "a reversal of the proper order when one undertakes to base the latter concepts on the concept of the extension of a concept" because the extension of a concept was generally something quantitatively completely undetermined, and that for such quantitative determination, the concept of number would have to have been given from somewhere else (Frege 1984, 122). So the Frege and Husserl exchange on the relationship between numbers and extensions of concepts described in the first two paragraphs of this section was actually an echo of the earlier exchange Frege and Cantor had had on the same subject.

In the *Philosophy of Arithmetic*, Husserl surely further incurred Frege's wrath by writing that all Frege's definitions become true and correct propositions when extensions of concepts are substituted for concepts, but then they are absolutely self-evident and without value. The results of Frege's endeavors, Husserl wrote, were such as to make one wonder how anyone could believe them to be true other than temporarily (Husserl 1891,134). Husserl in fact advanced several specific arguments against Frege's theory of identity, substitutivity, and extensionality, and even pointed to the problem regarding equivalent propositional functions which years later Frege would blame for the collapse of his logical edifice.[15]

Frege explained in the preface to the 1893 *Basic Laws of Arithmetic*[16] that he was trying to prove the theory of number he had expressed in *Foundations of Arithmetic*. The extensions he had introduced, he now explained to his readers had made for "far greater flexibility" and "also have a great fundamental importance" (p. ix). "In fact," he writes, "I even define number itself as the extension of a concept. . . ." (p. x). As far as he could see, he wrote, his basic law about extensions (Basic Law V or Principle V) was the only place in which a dispute could arise. This would be the place where the decision would have to be made (p. vii).

Then, a year later Frege came out fighting for extensions in his review of Husserl. In seeming reply to Cantor's charge regarding the indeterminacy of extensions, Frege defiantly wrote in his review: "A concept under which only one object falls has a determinate extension, as does a concept under which no object falls, or a concept under which infinitely many objects fall" (Frege 1894, 202).

Nonetheless, in spite of the sureness regarding extensions Frege expressed in that review, upon receiving the news of the paradox of the class of all classes that are not members of themselves in a letter from Russell in 1902, Frege immediately wrote back that it was the basic law about extensions that was at fault, and that its collapse seemed to undermine the foundations of arithmetic he had proposed for arithmetic.[17] "I have never disguised from myself," Frege wrote of his law in his 1903 appendix to *Basic Laws of Arithmetic II*,

> its lack of the self-evidence that belongs to the other axioms. . . . I should have gladly dispensed with this foundation if I had known of any substitute for it. And even now I do not see how arithmetic can be scientifically established . . . unless we are permitted . . . to pass from a concept to its extension . . . *Solatium miseris socios habuisse malorum.*. . . Everybody who in his proofs has made use of extensions of concepts, classes, sets, is in the same position as I.[18]

So, he concluded there, the way he had introduced extensions was not legitimate (p. 216), and the interpretation he had so far put on the words 'extension of a concept' needed to be corrected. Then he set out to track down the origin of the contradiction (p. 217). There was nothing, he decided, to stop him from transforming an equality holding between two concepts into an equality of extensions in conformity with the first part of his law, but from the fact that concepts are equal in extension we cannot infer that whatever falls under one falls under the other. The extension may fall under only one of the two concepts whose extension it is. This can in no way be avoided and so the second part of the law fails. This, however, really abolishes the extension of the concept, he concluded (p. 214 note f; pp. 218–23).

Frege never believed that his law about extensions recovered from the shock it had sustained from Russell's paradox (Frege 1979, 182), and came to rue having used the expression 'extension of a concept' which he finally concluded "leads us into a thicket of contradictions" (Frege 1980, 55).[19] By criticizing Frege's use of extensions, Husserl and Cantor had obviously struck a sensitive chord and so provoked an angry reaction by Frege.

3. Confusing Logic, Psychologism, and Empiricism

In his destructive review of *Philosophy of Arithmetic*, Frege further charged that Husserl's theory of number was conceived in the sin of confusing logic and psychologism. Frege's principal charges turn on Husserl's use of the word '*Vorstellung*', or 'presentation',[20] which Frege had decided to use only to designate the subjective, psychological phenomena (Frege 1884, §27 n.; Frege 1979, 72-76) which in "On Sense and Reference" he explicitly identified with the traditional empirical sources of knowledge, i.e., sense impressions, direct experiences, internal images arising from memories of sense impressions, and any object insofar as it was sensibly perceptible or spatial.[21] For Frege psychologism and empiricism were closely linked. Arithmetic, however, was a branch of logic for which no ground of proof need be drawn from experience or from intuition (ex. Frege 1984, 180; Frege 1884, v, §§8, 14).

As far as I know no one has ever used the word '*Vorstellung*' in Frege's special way, and he was most certainly correct to say that Husserl did not. In the *Philosophy of Arithmetic*, Husserl incurred Frege's wrath by using the word to designate both what Frege had decided was subjective and what Frege thought of as objective. In so doing, Frege complained, Husserl had transposed everything into the subjective mode (Frege 1894, 198). He had turned concepts, objects, and meanings of words (Frege 1894, 197) and numbers into "*Vorstellungen*, the results of mental processes or activities. . . ." (Frege 1894, 207). And, Frege further charges that since in Husserl's book "all things are *Vorstellungen* we can easily change objects by directing our attention towards them or away from them" (Frege 1894, 197). We can make properties disappear one after another, and actually essentially change objects (Frege 1894, 197–98).

Husserl's use of the word *'Vorstellung'* in the *Philosophy of Arithmetic* is ambiguous. A close look at the context, however, usually makes the intended reference clear, and several passages from the *Philosophy of Arithmetic* clearly indicate that Husserl was not guilty of confusing logic and psychology in the sense of obliterating the differences between mental processes and objects.[22] For example, the following passages from the book clearly show that Husserl did not change meanings into mental entities:

> Suppose we have a *Vorstellung* of a set of objects A, B, C, D. According to the order of succession in which the whole is formed finally, only D is given as a sensory *Vorstellung*. The remaining contents are, however, merely given as imaginary *Vorstellungen* altered in time and content. If we go in the opposite direction, from D to A, the phenomenon is different. The logical meaning nullifies all these differences. . . . While we make the *Vorstellung* of the set, we pay no attention to the fact that, in the grouping process, changes have occurred in the content. It is our intention to hold them together and unify them and so the logical content of the *Vorstellung* is not, for example, some D, then C just before, B even earlier, until we get to the most radically changed, A, but rather: (A, B, C, D). (Husserl 1891, 28–29)

This passage shows that Husserl did not reduce everything to subjectivity. He considered the theory that mental acts can engender relations to be untenable. In *Philosophy of Arithmetic* he wrote:

> Our mental activity does not make relations; they are simply there, and when interest is directed toward them they are noticed just like any other content. Genuinely creative acts that would produce any new content . . . are absurd from the psychological point of view. . . the act can in no way generate its content. . . . (Husserl 1891, 42)

Husserl even plainly wrote: "Isn't it obvious that a 'number' and the 'presentation of a number' are not the same thing?" (Husserl 1891, 30).

Once again, Cantor's presence is felt when Frege writes that according to Husserl "numbers are supposed to be *Vorstellungen*, the results of mental processes or activities," and charges Husserl with actually removing elements. Frege had also charged Cantor with "creating" numbers (Frege 1884, §96 note), and with engaging in the "psychological and hence empirical" activity of psychically removing elements.[23] Moreover, when criticizing Husserl for writing that when the number of items to be counted is beyond our capacity for presentation we are then to "idealize" our capacity for presentation (Frege 1894, 207, re. Husserl 1891, 251), Frege was surely also thinking of Cantor's statement in the *Mannigfaltigkeitslehre* that whenever "one comes to no greater number, one imagines a new one."[24]

The conflict between what is mental and the ideal realm of pure logic is plainly manifest and unresolved in Cantor's writings[25] where subjective and objective are mixed together in a puzzling way that cries out for clarification. In the same passage of the *Mannigfaltigkeitslehre* in which Cantor explicitly rejects the belief that "the source of knowledge and certainty is located in the senses or in the so-called form of pure intuition of the world of *Vorstellungen*," he goes to write that "certain knowledge . . . can only be obtained through concepts and ideas (*Ideen*), which are at best only stimulated by outer experience, but which are principally formed through inner induction, like something which, so to speak, already lay within us and is only awakened and brought to consciousness."[26] In *Contributions to the Theory of the Transfinite* he would write that "the act of abstraction with respect to nature and order . . . effects or rather awakens in my intellect the concept 'five',[27] and that the cardinal number belonging to a set was "an abstract image in our intellect."[28] Frege was perfectly justified in qualifying such appeals to "inner intuition" as "rather mysterious" (Frege 1884, §86). Cantor's philosophy was mystical to say the least.[29]

Though Husserl often used similar language, a close examination of the context shows that his statements are almost always qualified and explicated,[30] which makes them philosophically more sophisticated than Cantor's words are.[31] In particular, if in his famous review Frege seems to be attacking a more extreme form of psychologism than Husserl ever espoused, this is because the psychologism Husserl actually tried out in the *Philosophy of Arithmetic* was a variation on a theme by the realist and empiricist Franz Brentano.[32] By obliterating the differences between Husserl's empiricistic use of the word '*Vorstellung*' and Cantor's more Platonic ideas, Frege rather unfairly glossed over the differences between Husserl's and Cantor's ideas in this regard, making Husserl's ideas appear less reasoned than they actually were. In many respects the *Philosophy of Arithmetic* can be read as a first attempt on Husserl's part to, as he termed it, "banish the metaphysical fog and mysticism from mathematical investigations like Cantor's,"[33] and to redress some of the very weaknesses in accounts of the relationship between the subjectivity of knowing and the objectivity of mathematics[34] which Frege found so irritating.

Husserl was finally "deeply dissatisfied" with the analyses of the *Philosophy of Arithmetic*.[35] He later wrote of being "tormented" by "the incredibly strange worlds . . . of pure logic and the world of act-consciousness" while struggling to outline the logic of mathematical thought and calculation.[36] The idea of set, he recalled having reasoned, was supposed to arise out of the collective combination, and since everything which could be grasped intuitively was either physical or psychical, the collective had to be psychical. So it had to arise through psychological reflection in Brentano's sense, through reflection upon the act of collecting. But Husserl was finally compelled to conclude that

Brentano's theories of presentation "could not help,"[37] and that there was "an essential, quite unbridgeable difference between the sciences of the ideal and the sciences of the real," the correct assessment of which presupposed "the complete abandonment of the empiricistic theory of abstraction, whose present dominance renders all logical matters unintelligible."[38]

4. Conclusion

In the 1893 *Basic Laws of Arithmetic*, Frege accorded extensions "great fundamental importance" (p. x). A year later, he published an angry attack on the ideas of two Halle mathematicians who had criticized his use of extensions. His famous review of Husserl was partly a veiled assault on Cantor's ideas as reflected in the first book of a younger critic. Frege actually directly incorporated into his review of Husserl several specific criticisms he had already made of Cantor's work. The review was also partly a reflection of certain personal psychological problems Frege had regarding extensions and of his inability to accept criticism gracefully.

By drawing attention to these facts and to the relationship between Cantor's and Husserl's ideas, I have tried to contribute to putting Frege's attack on Husserl "in the proper light" by providing some insight into some of the issues underlying criticisms which Frege himself suggested were not purely aimed at Husserl's book. I have tried to undermine the popular idea that Frege's review of the *Philosophy of Arithmetic* is a straightforward, objective assessment of Husserl's book, and to give some specific reasons for thinking that the uncritical reading of Frege's review has unfairly distorted philosophers' perception of a work they do not know very well.

To put Frege's objections "in the proper light," the *Philosophy of Arithmetic* also needs to be evaluated in connection with kindred attempts "to provide a more detailed analysis of the concepts of arithmetic and a deeper foundation for its theorems."[39] For instance, Husserl was not the only mathematician to try to marry Brentano's ideas on presentation and Cantor's theory of arithmetic. The logical empiricist Bertrand Russell did likewise.[40] "In Arithmetic . . . our whole work is based on that of Georg Cantor" Russell wrote in the preface to *Principia Mathematica*.[41] Russell even for a time confused his own ideas on *Vorstellungen* with Frege's ideas, and those of Brentano's student Alexius Meinong.[42]

NOTES

1. G. Frege, "Rezension von E. Husserl: Philosophie der Arithmetik," *Zeitschrift für Philosophie und philosophische Kritik* 103 (1894): 313–32. In the text I cite as (Frege 1894) the translation in Frege's *Collected Papers on*

Mathematics, Logic and Philosophy (Oxford: Blackwell, 1984), pp. 195–209. Other articles in the *Collected Papers* are cited in the text as (Frege 1984).

2. E. Husserl, *Philosophie der Arithmetik* (Halle: Pfeffer, 1891), cited in the text as (Husserl 1891).

3. For example: D. Føllesdal's, "Husserl and Frege: A Contribution to Elucidating the Origins of Phenomenological Philosophy," *Mind, Meaning and Mathematics*, ed. L. Haaparanta (Dordrecht: Kluwer, 1994), a translation of his 1958 Norwegian master's thesis; J.N. Mohanty, *Husserl and Frege* (Bloomington, IN: Indiana University Press, 1982); D. W. Smith and R. McIntyre, *Husserl on Intentionality* (Dordrecht: Reidel, 1982); C. O. Hill, *Word and Object in Husserl, Frege, and Russell* (Athens, OH: Ohio University Press, 1991); M. Dummett, *Frege, Philosophy of Mathematics* (Cambridge, MA: Harvard University Press, 1991), chapter 12.

4. For example in J. P. Miller, *Numbers in Presence and Absence* (The Hague: M. Nijhoff, 1982); D. Willard, *Logic and the Objectivity of Knowledge* (Athens, OH: Ohio University Press, 1984); D. Bell, *Husserl* (London: Routledge, 1990).

5. M. Farber, *The Foundations of Phenomenology* (Cambridge, MA: Harvard University Press, 1943), pp. 16–17, 25–60; J. N. Findlay, "Translator's Introduction" to Husserl's *Logical Investigations* (London: Routledge and Kegan Paul, 1970), p. 1; A. Church, "Review of M. Farber's *The Foundations of Phenomenology*," *Journal of Symbolic Logic* 9 (1944): 64; G. E. Rosado Haddock, "Remarks on Sense and Reference in Frege and Husserl," chapter 2 of the present book; Bell pp. 67, 69, 77.

6. Hill, *Word and Object in Husserl, Frege and Russell*, chapters 2 and 5; Bell, pp. 79–81; Willard, pp. 43, 62–63, 69–70, 118–21; Mohanty, p. 26.

7. R. Dedekind, "The Nature and Meaning of Numbers" §VI, *Essays on the Theory of Numbers* (New York: Dover, 1963 [1887]); G. Cantor, *Gesammelte Abhandlungen*, ed. E. Zermelo (New York: Springer, 1932), pp. 379, 387, 411–12; G. Frege, *Foundations of Arithmetic* (Oxford: Blackwell, 1986 [1884]), §§29–44, cited in the text as (Frege 1884); B. Russell, *Principles of Mathematics* (London: Norton, 1903), pp. 115, 166, 219, 285, 305, 314–15, 497, 519. See also J. Dauben, *Georg Cantor, His Mathematics and Philosophy of the Infinite* (Princeton, NJ: Princeton University Press, 1979), pp. 151, 170–71, 176–77, 220–28; M. Hallett, *Cantorian Set Theory and Limitation of Size* (Oxford: Clarendon, 1984), pp. 54–85, 119–64; I. Grattan-Guinness, "Georg Cantor's Influence on Bertrand Russell," *History and Philosophy of Logic* 1, 1980, 68–71; C. Thiel, "Gottlob Frege: Die Abstraktion," *Studies on Frege*, vol. l, ed. M. Schirn (Stuttgart: Frommann-Holzboog, 1976), pp. 243–64; Dummett, pp. 50–52, 82–85, 145, 167–68.

8. Frege criticizes Cantor and Weierstrass in *Grundgesetze der Arithmetik*, vol. 2 (Hildesheim: Olms, 1966), §§68–86 and 143–55; also see Frege's "Draft Towards a Review of Cantor's *Gesammelte Abhandlungen zur Lehre vom Transfiniten*," in Frege's *Posthumous Writings* (Oxford: Blackwell, 1980), pp. 68–71, cited in the text as (Frege 1980); "Review of Georg Cantor, *Zur Lehre vom Transfiniten: Gesammelte Abhandlungen*," in Frege's *Collected Papers on Mathematics, Logic and Philosophy*, pp. 178–81; Dummett, chapter 21 and p. 243 where Dummett notes that in criticizing Weierstrass, Frege descended "rapidly into the grossest abuse"; Dauben, pp. 220–28.

9. L. Eley, "Einleitung des Herausgebers," *Philosophie der Arithmetik mit ergänzenden Texten*, Husserliana, vol. XII (The Hague: M. Nijhoff, 1970); Miller, pp. 1–6, 19.

10. K. Schuhmann, *Husserl-Chronik* (The Hague: M. Nijhoff, 1977), pp. 6–9; Miller, pp. 2–10; M. Kusch, *Language as Calculus vs. Language as Universal Medium* (Dordrecht: Kluwer, 1989), pp. 14–15.
11. Schuhmann, pp. 19, 22; A. Fraenkel, "Georg Cantor," *Jahresbericht der deutschen Mathematiker Vereinigung* 39, pp. 221, 253 note, 257 (abridged in Cantor's *Gesammelte Abhandlungen*); E. Husserl, *Introduction to the Logical Investigations, A Draft of a Preface to the Logical Investigations 1913* (The Hague: M. Nijhoff, 1975), p. 37 and notes; Eley, pp. XXIII–XXV; R. Schmit, *Husserls Philosophie der Mathematik* (Bonn: Bouvier, 1981), pp. 44, 58; M. Husserl, "Skizze eines Lebensbildes von E. Husserl," *Husserl Studies* 5 (1988): 114; W. Illemann, *Husserls vorphänomenologische Philosophie* (Leipzig: Hirzel, 1932), p. 50; I must thank Ivor Grattan-Guinness for bringing to my attention the letters published in *Georg Cantor Briefe*, ed. H. Meschkowski and W. Nilson (New York: Springer, 1991), pp. 321, 373–74, 379–80, 423–24, and W. Purkert and H. Ilgauds, *Georg Cantor 1845–1918* (Basel: Birkhäuser, 1991), pp. 206-7.
12. Cantor, p. 387; Dauben, p. 221; Grattan-Guinness, pp. 68–69. I personally examined Husserl's copies of Cantor's works at the Husserl Archives in Leuven in June 1993.
13. Cantor, "Rezension von Freges *Grundlagen*," *Gesammelte Abhandlungen*, pp. 440–41.
14. G. Frege, "Erwiderung auf Cantors Rezension der *Grundlagen der Arithmetik*," *Deutsche Literaturzeitung* 6 Nr. 28 (1885), Sp. 1030; Frege, *Collected Papers on Mathematics, Logic and Philosophy*, p. 122.
15. I discuss Husserl's arguments at length in *Word and Object in Husserl, Frege and Russell*, chapter 4, and "Husserl and Frege on Substitutivity," chapter 1 of the present book.
16. G. Frege, *Basic Laws of Arithmetic* (Berkeley, CA: University of California Press, 1964 [1893]), pp. vii–x.
17. G. Frege, *Philosophical and Mathematical Correspondence* (Oxford: Blackwell, 1979), pp. 131–32, cited in the text as (Frege 1979).
18. G. Frege, *Translations from the Philosophical Writings* 3rd ed. (Oxford: Blackwell, 1980), p. 214.
19. I document this more fully in chapter 6 of *Word and Object in Husserl, Frege and Russell* and in "Frege's Letters," *From Dedekind to Gödel, Essays on the Development of the Foundations of Mathematics*, ed. J. Hintikka (Dordrecht: Kluwer, 1995), pp. 97–118. Dummett discusses problems with extensionality at length in *Frege, Philosophy of Mathematics* cited above.
20. Following Russell's translation in "On the Nature of Acquaintance," *Logic and Knowledge* (London: Allen and Unwin, 1956), pp. 169–70. Frege's translators have often misleadingly opted for "idea."
21. Frege, *Translations from the Philosophical Writings*, pp. 59–64. I discuss this at length in *Word and Object in Husserl, Frege and Russell*, chapters 5 and 7.
22. See Miller, pp. 7–8, 21–23.
23. Frege, *Collected Papers on Mathematics, Logic and Philosophy*, pp. 180–81; Dauben, pp. 220–28, 239.
24. Cantor, p. 195; Dauben, p. 206.
25. Hallett, pp. 16–18, 34–35, 121, 128–33, 146–58; Dauben, p. 132.
26. Cantor, p. 207 note 6; Hallett, p. 15.
27. Cantor, p. 418 note 1; Hallett p. 128.
28. Cantor, p. 416; Hallett p. 128.

29. Dauben, chapter 6, and pp. 236–39; Hallett, pp. 9–11, 35–36.
30. Hill, *Word and Object in Husserl, Frege and Russell*, p. 14.
31. Dauben, pp. 6, 49, 120, 121, 127, 147, 150, 154, 159; Hallett, pp. xi, 67, 49.
32. H. Spiegelberg, *The Context of the Phenomenological Movement* (The Hague: M. Nijhoff, 1981), pp. 7–63; Bell, pp. 3–28.
33. Husserl, *Logical Investigations,* p. 242.
34. Ibid., p. 42.
35. Husserl, *Introduction to the Logical Investigations,* pp. 34–35.
36. Edmund Husserl, "Persönliche Aufzeichnungen," *Philosophy and Phenomenological Research* 16 (1956): 294; Miller, chapters 1, 3, and 5; Kusch, pp. 12–55.
37. Husserl, *Introduction to the Logical Investigations,* pp. 34–35.
38. Husserl, *Logical Investigations,* p. 185.
39. G. Frege, "*Begriffsschrift,*" *From Frege to Gödel,* ed. J. van Heijenoort (Cambridge, MA: Harvard University Press, 1967), p. 8.
40. B. Russell, *My Philosophical Development* (London: Allen & Unwin, 1985 [1959]), p. 100; Russell links his views on presentation with those of Brentano's school most explicitly in the articles he published in *Mind* 8, 13–16 (1899–1907) on Brentano's student A. Meinong, reprinted in *Essays on Analysis* (New York: Braziller, 1973). He discusses his views on presentation at length in "The Nature of Acquaintance," *Logic and Knowledge,* pp. 127–74; "Knowledge by Acquaintance and Knowledge by Description," chapter 10 of *Mysticism and Logic* (London: Allen and Unwin, 1986 [1917]); chapter 5 of *Problems of Philosophy* (Oxford: Oxford University Press, 1967 [1912]). See also A. Garciadiego, *Bertrand Russell and the Origins of the Set-theoretic Paradoxes* (Basel: Birkhäuser, 1992); Grattan-Guinness, "Georg Cantor's Influence on Bertrand Russell" as cited in note 7 above.
41. B. Russell and A. N. Whitehead, *Principia Mathematica to *56* (Cambridge, UK: Cambridge University Press, 1964 [1927, 2nd ed. rev.]), p. viii.
42. Hill, *Word and Object in Husserl, Frege and Russell,* pp. 61–66, 134–35.

7

Claire Ortiz Hill

ABSTRACTION AND IDEALIZATION IN EDMUND HUSSERL AND GEORG CANTOR PRIOR TO 1895

After two years of training in philosophy under Franz Brentano, Edmund Husserl arrived in Halle, Germany in 1886 with a plan to help provide radical foundations for mathematics by engaging in psychological analyses of the concept of number. To this end, he espoused a theory of abstraction of the concept of number akin to the one his colleague Georg Cantor was propounding in conjunction with the Platonistic theories about sets he was hard at work developing and defending.

However, Husserl rather quickly grew profoundly dissatisfied with the psychological foundations he had begun laying for mathematics in his "On the Concept of Number" (Husserl 1887) and then in the *Philosophy of Arithmetic* (Husserl 1891). He even confessed to having experienced doubts about his approach right from the very beginning (Husserl 1913, 34–35). The first clues to the answers to the many questions his efforts to provide more secure foundations for arithmetic raised would ultimately be found in Hermann Lotze's theory of Platonic ideas (Husserl 1913, 36–42, 44–47). And during years of hard, solitary work in Halle Husserl developed the original interpretation of Platonic idealism which went into the making of the phenomenological method (Husserl 1900/01; Husserl 1913).

Very little is known of Husserl's encounter with the Georg Cantor's theories about Platonic idealism and the abstraction of number concepts. Much, however, can be gleaned about this from a close study and comparison of Cantor's and Husserl's writings during those crucial years in Husserl's development. So in the following pages I try to shed light on that dark period in Husserl's development by studying the evolution his ideas underwent as this relates to Cantor's philosophizing about abstraction, Platonic idealism, and the concept of number. I focus on the important changes which took place in Husserl's ideas during his first ten years in Halle.

1. Creating Numbers in the *Mannigfaltigkeitslehre*

During Husserl's fifteen-year sojourn in Halle, Georg Cantor was hard at work laying the foundations of set theory and reconnoitering, conquering, colonizing, and defending the new world of transfinite numbers. Any further progress of his work on set theory, he had explained in the beginning of his *Grundlagen einer allgemeinen Mannigfaltigkeitslehre* (Cantor 1883), was absolutely dependent upon the expansion of the concept of real whole numbers beyond the present boundaries and in a direction which, as far as he knew, no one had yet searched. He had, he claimed, burst the confines of the conceptual formation of real whole numbers and broken through into a new realm of transfinite numbers. As strange and daring as his ideas might now seem, he was convinced that they would one day be deemed completely simple, appropriate, and natural (p. 165). Much of the work he would do in the coming years would be aimed at showing they were.

Upon discovering the transfinite numbers, Cantor tells readers of the *Mannigfaltigkeitslehre*, he had not been clearly conscious of the fact that these new numbers possessed the same concrete reality the whole numbers did. He was, however, now persuaded that they did (p. 166) and intent upon proving that "after the finite there is a *transfinite* . . . which by its very nature is not finite, but infinite, which, however, can be determined by *numbers* which are definite, well-defined, and distinct from one another just as the finite can" (p. 176).

Acting on a conviction, spelled out in a 1884 letter to Gösta Mittag-Leffler, that the only correct way to proceed was "to go from what is most simple to that which is composite, to go from what already exists and is well-founded to what is more general and new by continually proceeding by way of transparent considerations, step by step without making any leaps" (Cantor 1991, 208), Cantor began devising a strategy as to how to provide his "strange" new numbers with secure foundations by demonstrating precisely how the transfinite number system might be built from the bottom up.

Through the combined action of two principles, he argued in the *Mannigfaltigkeitslehre*, one might break through any barrier in the conceptual formation of the real, whole numbers (pp. 166–67) and with the greatest confidence and self-evidence arrive at ever new number classes and numbers having the same concrete definiteness and reality as objects as the previous ones (p. 199). In §§1, 11, and 12 of the work he showed how his principles might lead to the definitions of the new numbers and produce number classes. By the first principle, the first series of positive real whole numbers $1, 2, 3, \ldots, v$ would have their origin and basis in the repeated positing and adding of underlying units considered to be identical, the number v being the expression both for a specific finite number of units posited in this way and for the joining of the posited units into a whole. Numbers thus manufactured would be members of the first number class which was infinite and for which there was no greatest number (p. 195). By the second principle, whenever there was any definite succession of defined whole numbers for which no greatest one existed, a new number could be created considered as the boundary of those numbers, i.e., defined as the next greater number to them all (pp. 195–96). By a third principle, the second number class defined would not only receive a higher power than the first, but precisely the *next higher* one, hence the *second power* (p. 167; 197). By "power" Cantor meant cardinal number.

In a note to the *Mannigfaltigkeitslehre* Cantor also provided a rather inchoate account of his idea of the procedure for the correct formation of concepts:

> One lays down a propertyless thing, which is at first nothing other than a name or a sign *A*, and systematically gives it different, even infinitely many distinct predicates whose meaning is generally known through already existing ideas and which may not contradict each other; through this, *A*'s relations to already existent concepts, and in particular to related concepts, will be determined. When one has completely finished with this, then all the conditions for awakening the concept *A* which was slumbering in us are present and it comes into existence. . . . (p. 207 nn. 7, 8)

2. Cantor's Theory of Abstraction

One of the many questions Cantor's *Mannigfaltigkeitslehre* account of concept formation raises is that of how exactly one manages to lay hold of the propertyless objects by which number concepts are to be manufactured. The answer, though, was forthcoming. For since 1883, Cantor had begun making his thereafter oft repeated claims that his transfinite numbers could be "produced through abstraction from reality *with the same necessity* as the ordinary finite whole numbers by which alone all other mathematical conceptual formations thus far have been produced" (Cantor 1991, 136; Kreiser 1979).

Cantor's earliest attempt to publish his theory of abstraction, however, met with failure (Cantor 1884; Cantor 1991, 226–30, 240–44). In the posthumously published "Principien" of 1884 he had written:

> The *power* of a set *M* is hereupon defined as the *presentation* of what is *common* to *all* of the sets *M* of *equivalent* sets and *only those* and hence also of the set *M* itself; it is the *representatio generalis . . .* for all sets *of the same class as M*. It therefore seems to me to be the most *primitive, psychologically*, as well as *methodologically simplest root concept*, arisen through *abstraction*, from all *particular characteristics*

which a set of a *specific class* may display, both with respect to the *nature* of its *elements*, as well as with regard to the *relations* and *order* in which the *elements are to each other* or can stand to *things lying outside the set*. The concept of *power* originates in reflecting only upon *what is in common to all of one and the same class of member sets*. (Cantor 1884, 86)

Accounts of this very process appear in Cantor's correspondence of the time (ex. Cantor 1991, 178–80; Eccarius 1985). And when the theory was first published in the "Mitteilungen zur Lehre vom Transfiniten" (Cantor 1887/88), Cantor would stress that he had advocated and repeatedly taught it in his courses as much as four years earlier. Integrated directly into the "Mitteilungen" are parts of letters he had written on the subject during the years prior to the publication of the work (pp. 378–79, 387 n., 411 n.).

When the abstraction theory did find its way into print, it formed an integral part of Cantor's endeavor to provide solid foundations for transfinite arithmetic in the "Mitteilungen" (Cantor 1887/88), where Cantor particularly sought to show how his theorems about transfinite numbers were firmly secured "through the logical power of proofs" which, proceeding from his definitions, which were "neither arbitrary nor artificial, but originate naturally out of abstraction, have, with the help of syllogisms, attained their goal" (p. 418). Much of the "Mitteilungen" is, in fact, devoted to explaining exactly how numbers are to be procured from reality by abstraction and, in particular, how the actual infinite number concept is to be formed through appropriate natural abstractions in the way the finite number concepts are won through abstraction from finite sets (p. 411; Cantor 1991, 329, 330).

For Cantor, cardinal numbers were, as he explained to Giulio Vivanti in 1888, the general concepts assigned to sets which one may obtain by abstracting both from the properties of the elements and from the order in which they are given (Cantor 1991, 302). In the "Mitteilungen" he repeatedly gave essentially the same recipe for extracting cardinal numbers from reality through abstraction:

> In abstracting from a given set M, composed of determinate, completely distinct things or abstract concepts called the elements of the set and thought of as a thing for itself, both the properties of the elements and the order in which they are given, is produced in us a determinate general concept . . . which I call the *power* of M or the *cardinal number* corresponding to the set M. (p. 411; see also pp. 379, 387)

There he also provided this concrete illustration of the procedure he was prescribing:

> For the formation of the general concept "five" one needs *only a* set (for example all the fingers of my right hand) which corresponds to this cardinal number; the

act of abstraction with respect to both the properties and the order in which I encounter these wholly distinct things, produces or rather awakens the concept "five" in my mind. (p. 418 n. 1)

Cantor believed his theory of abstraction to be the distinctive feature of his number theory and that it represented an entirely different method for providing the foundations of the finite numbers than was to be had in the theories of his contemporaries. As he wrote to Giuseppe Peano:

> my view of "numbers" is *fundamentally* toto coelo different from those we find in Grassmann, Dedekind, Helmholtz, Weierstrass, Kronecker, etc.; and from this results, as you see from my work, a completely different method for the grounding of "finite number theory." With those conceptions one would have never have come upon the transfinite numbers whose grounding is only possible in the way which I have carved out. (Cantor 1991, 365; 363)

With his theory as to how number concepts might be abstracted from reality, Cantor explained in the "Mitteilungen," he was laying bare the roots from which the organism of transfinite numbers develop with logical necessity (Cantor 1887/88, 380). He said that it was the thought process involved in forming the actual infinite number concept through abstraction like that by which finite number concepts were won from finite sets that had led him to the theory of transfinite numbers he had begun to outline in the *Mannigfaltigkeitslehre* (Cantor 1887/88, 411).

The definition by abstraction would eventually make its way into the *Contributions* where Cantor maintained that the principles he had "laid down, and on which later the theory of the actually infinite or transfinite cardinal numbers will be built, afford also the most natural, shortest, and most rigorous foundation of the theory of finite numbers" (Cantor 1895/97, §5).

Since Cantor was propounding this theory of abstraction at the very time Husserl was writing "On the Concept of Number" (1887) and the *Philosophy of Arithmetic* (1891) where a quite similar theory was advocated, it is important to note here that in the latter work, Husserl approvingly noted that Cantor's definition of number in the "Mitteilungen" is profoundly different from that of the *Mannigfaltigkeitslehre* (Husserl 1891, 126 n.). Husserl specifically points to two passages of the "Mitteilungen." In the first one, Cantor had written:

> By the *power* or *cardinal number* of a set (*Menge*) M (which is made up of distinct, conceptually separate elements m, m', . . . and is to this extent determined and limited), I understand the general concept or species concept (*Gattungsbegriff*) (universal) which one obtains by abstracting from the properties of the elements of the set, as well as from all the relations which the elements may have, whether themselves or to other things, but especially from the order reigning

among the elements and only reflect upon what is common to all sets *equivalent* to *M*. (Cantor 1887/88, 387; also Cantor 1991, 178)

Calling Cantor a mathematician of genius and referring to the above cited passage concerning the formation of the general concept "five," Husserl further commends Cantor for having written with a great deal of precision in the "Mitteilungen" that for "the formation of the general concept "five" one needs *only a* set . . . which corresponds to this cardinal number" (Husserl 1891, 126 n.).

3. A Detour by Way of Gottlob Frege and Louis Couturat

Two of Cantor's and Husserl's contemporaries, Gottlob Frege and Louis Couturat, offered contrasting interpretations of the abstraction theory the Halle men favored, the judgments they passed affording valuable insight into the nature of the philosophical issues involved in Cantor's and Husserl's attempts to win numbers from reality through abstraction.

In his cranky reviews of Cantor's "Mitteilungen" (Frege 1892; Frege 1979) and Husserl's *Philosophy of Arithmetic* (Frege 1894), Frege charged both men with attempting to achieve magical effects by using abstraction to destroy the properties things have (Frege 1979, 69–70; Frege 1894, 205). Abstraction, Frege said, would endow mathematicians with the miraculous, supernatural ability (Frege 1979, 69, 71) to change things (Frege 1979, 70; Frege 1894, 197–98, 204–5) in "the wash-tub of the mind" (Frege 1894, 205; chapter 6 of the present book).

Frege himself held that the "properties which serve to distinguish things from one another are, when we are considering their Number, immaterial and beside the point" and that "we want to keep them out of it" (Frege 1884, §34); according to him one was to "follow pure logic" by "disregarding the particular characteristics of objects" (Frege 1879, 5). He furthermore believed that the propositions of pure logic could not "come to consciousness in a human mind without any activity of the senses" since "without sensory experience no mental development is possible in the beings known to us" (Frege 1879, 5). But it was the "psychological and hence empirical turn" (Frege 1892, 180, 181) he believed Cantor and Husserl had given the matter that particularly irked him (Frege 1894, 197, 208–9).

Couturat interpreted Cantor's and Husserl's efforts otherwise. In his criticism of empirico-psychologistic theories of number in *De l'infini mathématique* (1896), he argued that, despite claims by Leopold Kronecker and

Hermann von Helmholtz (as well as Richard Dedekind and Carl Friedrich Gauss) that their theories of number were "pure of any worldly blemish and any commerce with the world of the senses" (p. 319), their theories of number were in reality empiricist (pp. 318–31) because they just described the psychological process of counting without inquiring into the ideal conditions which make it possible and intelligible (p. 331). According to their theories, Couturat explains, the concept of whole cardinal number would originate in experience, in the counting of a concrete set of external objects given in perception (p. 319; p. 329). The two words "psychologistic" and "positivistic," he says, sum up and display the inadequacy and vice of such an empirical approach (pp. 330–31). And, interestingly, he praises Husserl (p. 331 n.) for having come to that very conclusion in the *Philosophy of Arithmetic.*

The elements of a collection cannot, Couturat objects, be counted as concrete objects, but only, as Cantor does, as a set of abstract, propertyless units obtained when by "abstracting from the particular nature of the objects given and from their distinctive properties, one considers each one of them as *one*, meaning one reduces it to a unit, and embraces all those abstract units in a single mental act. . . . " (pp. 325–26).

So for Couturat, Cantor's theory of number is only "empiricist in appearance," but is "rationalist in reality" (p. 332; p. 335) because it is based on the "genuinely *a priori*" (p. 332) "metaphysical idea of unit" (p. 341), which could not be the residue of any abstraction performed on sensory data. For a unit is neither a perception nor anything given in perception (p. 340), but is "a pure rational form which the mind imposes *a priori* on all its objects just by the fact that it thinks them" (p. 341).

Such an interpretation lends insight into the nature of Cantor's theories and into the development of Husserl's ideas in the same area. It is in fact easy to find passages in Cantor's work where any appearance of empiricism or psychologism is belied by an underlying rationalism. For, however empirical or psychological Cantor's mysterious references to inner intuition (ex. Cantor 1883, 168, 170, 201), or to experiences helping produce concepts in his mind (Cantor 1887/88, 418 n. 1) may appear, he believed abstraction would liberate mathematicians to engage in strictly arithmetical forms of concept formation by freeing them from psychologism, empiricism, Kantianism, and insidious appeals to intuitions of space and time (ex. Cantor 1883, 191–92; Cantor 1885; Cantor 1887/88, 381 n. 1; Eccarius 1985, 19–20).

In speaking out in the *Mannigfaltigkeitslehre* against the new empiricism, sensualism, skepticism, and Kantianism which, he argued, mistakenly located the sources of knowledge and certainty in the senses or in the "supposedly pure forms of intuition of the world of presentation," Cantor had affirmed that sure knowledge could "only be obtained through concepts and ideas

which, at most stimulated by external experience, are on the whole formed
through inner induction and deduction as something which in a way already
lay within us and was only awakened and brought to consciousness" (p. 207,
n. 6).

4. Cantor and Platonic Ideas

Cantor surely made no secret of his intention to supply his numbers with
adequate philosophical and metaphysical foundations. The *Mannig-
faltigskeitslehre* (1883) came replete with epistemological and metaphysical
reflections aimed at explaining and justifying his novel ideas to a readership
chary of such talk (ex. Cantor 1991, 100, 113, 118, 178, 199, 227). In the
beginning of the "Principien," expressing the high regard he had for metaphysics and his belief in a close alliance between metaphysics and mathematics, he thanked Jules Tannery for having paid him the honor of according
philosophical, and even metaphysical, worth to his writings (Cantor 1884,
83–84). In 1885, Mittag-Leffler even felt the need to warn Cantor that his
new terminology and philosophical way of expressing himself might be so
frightening to mathematicians as to seriously damage his reputation among
them (Cantor 1991, 241). The "Mitteilungen" (1887/88) that Cantor published in a philosophical journal largely represented an attempt on his part to
provide philosophical justification for his new numbers, a concern which fills
the pages of his letter books as well (Cantor 1991).

Cantor was also most explicit about the precise nature of the metaphysical underpinnings he was proposing for his numbers. He believed that the
transfinite "presented a rich, ever growing field of ideal research" (Cantor
1887/88, 406) and saw abstraction as showing the way to that new, abstract
realm of ideal mathematical objects which could not be directly perceived or
intuited. His talk of awakening and bringing to consciousness the knowledge, concepts, and numbers slumbering in us (Cantor 1883, 207 n. 6, 7, 8;
Cantor 1887/88, 418 n. 1) is an unmistakeable allusion to Plato's theory of
recollection and Socratic theories of concept formation (ex. the *Meno*
81C–86C; *Phaedo* 72E, 75E–76A). He considered his transfinite numbers to
be but a special form of Plato's *arithmoi noetoi* or *eidetikoi*, which he thought
probably even fully coincided with the whole real numbers (Cantor 1884,
84; Cantor 1887/88, 420). To Giuseppe Peano he wrote: "I conceive of
numbers as 'forms' or 'species' (general concepts) of sets. In essentials this is
the conception of the ancient geometry of Plato, Aristotle, Euclid etc."
(Cantor 1991, 365). By manifold or a set, he wrote in the *Mannigfaltigkeitslehre*, he was defining something related to the Platonic *eidos* or *idea*, as also
to what Plato called a *mikton* in the *Philebus* (Cantor 1883, 204 n. 1). "My

idealism," he wrote to Paul Tannery, "is related to the Aristotelian-Platonic kind, which as you know is at the same time a form of *realism*. I am just as much a *realist* as an *idealist*" (Cantor 1991, 323). In the *Mannigfaltigkeit-slehre* he had emphasized that the "certainly realist, at the same time, however, no less than idealist foundations" of his reflections were essentially in agreement with the basic principles of Platonism according to which only conceptual knowledge in Plato's sense afforded true knowledge, but that the nearer our presentations came to the truth, the nearer their objects must come to being real and vice versa (Cantor 1883, 181, 206 n. 6).

However, despite his explicit appeals to Platonism and his multiple references to philosophical writings, Cantor did not explicitly refer readers to any of Plato's writings other than the *Philebus* (Cantor 1883, 204 note 1). He did, though, refer them (ex. Cantor 1883, 206 note 6; 205 n. 2) to Eduard Zeller's books on Greek philosophy (Zeller 1839; Zeller 1875; Zeller 1879), which Cantor's "frightening" new terminology and philosophical way of expressing himself so echoes as to afford insight into the precise nature of Cantor's understanding of the Platonic Ideas he was espousing. So examining the sixty pages of Zeller's synthesis of Plato's doctrines (Zeller 1875, 541–602) that Cantor directly cited in the *Mannigfaltigkeitslehre* (Cantor 1883, 206 n. 6)—much of which is a standard interpretation of Plato's writings—adds to what we can harvest about Cantor's Platonism from the spare remarks he scattered throughout his writings.

For one thing, Zeller's account of Platonic Ideas sheds some light on how Cantor might have come to think it reasonable to marry a theory of abstraction with a theory of sets as Platonic *eidos* or *idea* and numbers as Platonic *eidetikoi*. For the formation of concepts through abstraction is a process generally associated with Aristotelian and empiricist philosophy and is generally considered to be incompatible with the basic principles underlying a strictly Platonic theory of forms or Ideas known through recollection (Weinberg 1968, 1). Aristotle himself made no secret of his opposition to Plato's Ideas and *eidetikoi* (*Metaphysics* Books M–N). Given these facts, any attempt to marry abstraction and Platonic Ideas might be viewed as crude and uncouth.

Yet we find Cantor maintaining that his technique for abstracting numbers from reality provides the only possible foundations (Cantor 1991, 365, 363; Cantor 1887/88, 380, 411) for his Platonic conception of numbers. Knowledge, he maintained, could "only be obtained through concepts and ideas which, at most stimulated by external experience, are on the whole formed through inner induction and deduction as something which in a way already lay within us and was only awakened and brought to consciousness" (Cantor 1883, 207 n. 6). By engaging in the process of concept formation prescribed in the *Mannigfaltigkeitslehre*, Cantor maintained one fulfilled the conditions for "awakening" a concept which "slumbering in us" "comes into existence" (Cantor 1883, 207 nn. 7, 8). And it was his theory of abstraction that was going to yield those concepts and ideas.

Zeller, however, actually attributed a theory of abstraction to Plato, giving some clue as to how Cantor might have justified his own interpretation. "The Ideas," Zeller wrote in pages Cantor cited (Cantor 1883, 206 n. 6), "arisen out of Socratic concepts, are in fact, like these, abstracted from experience, however little Plato is willing to have this word; they thus first present a particular, and only step by step move up from this particular to the general, from lower to higher concepts" (Zeller 1875, 584). Plato, Zeller contended, was "not aiming for a pure a priori construction, but only for a complete logical ordering of the Ideas, which he himself found through induction, or if we prefer: through an increasing recollection of what is sensory" (Zeller 1875, 584).

In so interpreting Plato, Zeller was certainly thinking of passages from the *Dialogues* in which there is talk of passing from a plurality of perceptions to a unity gathered together by reasoning (ex. *Phaedrus* 249B–C), or in which Plato wrote of using of our senses in connection with objects to recover or recollect previously acquired knowledge (ex. *Phaedo* 75E). The soul, Socrates would demonstrate in the *Meno*, has learned everything and nothing prevents someone from discovering everything because searching and learning are recollection (*Meno* 81C–D,) so that those who do not know have within themselves true opinions about the things they do not know, and these opinions can be stirred up like a dream in such a way that in the end their knowledge about these things would be as accurate as anyone's (*Meno* 85C).

Cantor directly appealed to Zeller's interpretation of Plato's theories in a note (Cantor 1883, 206 n. 6) to the important section of the *Mannigfaltigkeitslehre* in which the Halle mathematician explains and justifies his conviction that "mathematics is entirely free in her development and only bound to the obvious consideration that her concepts both be not self-contradictory and stand in ordered relationships fixed through definitions to previously formed, already existing and proven concepts" (Cantor 1883, 182).

This freedom, Cantor maintained, was a consequence (Cantor 1883, 182) of the connection between the two kinds of reality or existence which both finite and infinite whole numbers enjoy and which, strictly speaking, they share with any concepts or Ideas whatsoever. First, Cantor maintained, these numbers have intrasubjective or immanent reality "inasmuch as on the basis of definitions they occupy a fully determinate place in our understanding, are as distinct as possible from all other components of our thought, stand in determinate relations to them and consequently modify the substance of our mind in determinate ways" (Cantor 1883, 181). A second kind of reality may, Cantor believed, also be ascribed to these numbers. They may have transsubjective or transient reality inasmuch as "they must be regarded as an expression or image of processes of the external world lying outside of the intellect, as further the different number classes . . . are representatives of powers which are actually present in corporeal and intellectual nature" (Cantor 1883, 181).

The "thoroughly realistic, but no less idealistic, foundation" of these reflections, Cantor explains, leaves no doubt in his mind "that these two kinds of reality come together constantly in the sense that a concept existing in the first sense always also possesses transient reality in certain, even infinitely many respects . . ." (Cantor 1883, 181). "Only conceptual knowledge," Cantor now cites Zeller, "is said (according to Plato) to afford true knowledge. The nearer, however, our presentations come to the truth . . . the nearer their objects must come to being real and vice versa. What is knowable, is; what is not knowable, is not, and to the same extent something is, it is also knowable" (Cantor 1883, 206–7 n. 6; Zeller 1875, 541–42).

As a consequence, Cantor concludes, mathematics has "*purely* and *simply* to take into consideration the immanent reality of her concepts and hence no obligation whatsoever to investigate their transient reality." "In particular," he continues, "with the introduction of new numbers she is only dutybound to give definitions of them that bestow upon them such determinacy and, if need be, such a relationship to the older numbers, that in given cases they may be distinctly differentiated from one another. As soon as a number meets these conditions, it can and must be considered as existent and real in mathematics." This is the reason, Cantor maintains, "why one is to regard the rational, irrational, and complex numbers as altogether just as existent as the finite positive whole numbers." The actual basis of the connection between these two kinds of reality, he tells us, lies "in the *unity* of the *universe, to which we ourselves belong*" (Cantor 1883, 182).

Unity was in fact a major theme of Cantor's philosophy of arithmetic. Though he maintained that to be considered existent and real in mathematics numbers must be distinctly differentiated from one another and as distinct as possible from all other components of our thought, he stressed over and over that these independent numbers in and for themselves organically coalesce into a unified whole in special ways (Cantor 1887/88, 379, 380, 381 n. 1).

As an example of what he meant, Cantor once gave the equation $5 = 2 + 3$. Two and three, he reasoned, are not contained in the concept five. Were they, he asked, what would that mean of one and four? 1, 2, 3, 4 are, however, virtual components of 5, he explains, and the equation indicates a specific ideal connection of the three cardinal numbers two, three and five for themselves. It is thus, he concludes, five in and for itself independent from four or three and from any other number. Each number is by essence a *simple* concept in which a manifold of ones are combined together into an organic whole in *special ways* (Cantor 1887/88, 418 n. 1).

A good measure of the freedom Cantor felt he enjoyed in fact came from his adoption of the Platonic *eidetikoi* or ideal numbers alluded to in the above paragraph. And in affirming that the whole real numbers were "related to the *arithmoi noetoi* or *eidetikoi* of Plato with which they probably even fully coincide" (Cantor 1884, 84) and that his transfinite numbers were but a special form of these *eidetikoi* (Cantor 1887/88, 420; Cantor 1884, 84), Cantor himself provided an important clue as to how exactly he thought one might understand the seemingly paradoxical union of the One and the Many.

According to Zeller, Plato expressed the combining of the One and the Many by referring to Ideas as numbers, thus distinguishing between an empirical treatment of numbers and pure, ideal *arithmoi eidetikoi* which by their very nature are detached from things perceptible by the senses and which unlike the other, mathematical, numbers stand in a before and after relationship to one another. The essential thing for Plato, Zeller maintains, was only the thought, which underlies his number theory, that "in what is real the One and the Many must be organically combined" (Zeller 1875, 574). In such a theory, he explains, numbers become the connecting link between Ideas and appearance. It is by this combining of the One and Many that Plato was able to "put the concreteness of Socrates's concepts in the place of the abstract, Eleatic One, to link the concepts dialectically, and place them in not merely a negative, but also a positive relationship to appearance, that the Many of appearance is borne by and included in the unitary concept." And this, then, gave him the right to set forth a multiplicity (*Vielheit*) of logically interrelated Ideas, a world of Ideas (Zeller 1875, 583–84; 567–70; Zeller 1839, 239–41; see also Aristotle *Metaphysics* Books M–N).

According to Cantor's theories, the numbers obtained through his abstraction processes, both the whole numbers (cardinal numbers, powers) and the order types (the *Anzahlen*, or numbers of a well-ordered set) were simple conceptual formations, each of which was a genuine unity (*monas*) in which a plurality and manifold of ones became bound together in a uniform fashion. Abstracting from both the characteristics of the elements of the set and the order in which they are given, we obtain the cardinal numbers or powers; abstracting only from the characteristics of the elements and leaving their order intact, we obtain the ideal numbers or *eidetikoi* (1887/88, 379–80; 1883, 180–81).

The elements of a set, he explains, are to be thought of as separate. In the intellectual image (*intellektualen Abbild*), which he calls the order type or ideal number, the ones (*Einsen*) are, however, united into a single organism. In a certain sense, he explains, each ideal number can be looked upon as something composed of *matter* and *form*. The conceptually distinct ones contained therein supply the *matter*, while the order subsisting among them corresponds to the *form* (Cantor 1887/88, 380).

In finite sets, Cantor tells us, these two kinds of numbers coincide. The differences, however, come most distinctly to the fore in infinite sets (1887/88, 379–80). In the *Mannigfaltigkeitslehre*, he describes his real delight in seeing how when we proceed up into the infinite, the concept of whole number, which in the finite only serves as a backdrop for ideal numbers, splits into two concepts. And how in proceeding back down from the infinite to the finite he sees how beautifully and clearly the two concepts become one again and flow together into the concept of finite whole number. Without these two concepts, Cantor believed one could not progress in his theory of manifolds and he lamented Bernard Bolzano's failure to resort to this distinction (Cantor 1883, 180–81).

In 1890 Cantor wrote to Giuseppe Veronese that contradictions he had found in Cantor's theories were but apparent and that one must distinguish between the numbers which we are able to grasp in our limited ways and "numbers as they are *in and for themselves, and in and for the Absolute intelligence*," each of which "is a simple concept and a unity, just as much a unity as one itself. Taken absolutely," he explained, "the smaller numbers are only *virtually* contained in the bigger ones. They are, taken absolutely, all independent one from the other, all equally good and all equally necessary metaphysically" (Cantor 1991, 326). In 1895 he wrote to Charles Hermite that "the reality and absolute uniformity of the whole numbers seems to be *much stronger* than that of the world of sense. That this is so has a single and quite simple ground, namely the whole numbers both separately and in their actual infinite totality exist in that highest kind of reality as eternal ideas in the Divine Intellect" (cited Hallett 1984, 149).

Another clue as to how Cantor might realize a Platonic union of the One and the Many lies in Cantor's pronouncement that by a manifold or set he generally meant "any Many which can be thought of as a One, any totality of determinate objects which can be united by a law into a whole." He thus believed he was defining, he wrote, something related to the Platonic *eidos* or *idea* and to what Plato called a *mikton* in the *Philebus*, an ordered mixture of the *peras* (limit) and the *apeiron* (limitlessness, indeterminacy), what Cantor himself called the improper or potential infinite (Cantor, 1883, 204 n. 1).

It was, in fact, in the *Philebus*, Zeller tells us, that the solution to the problem as to how unitary concepts might traffic in the multiplicity of appearance was said to lie in the principle that what is real (*das Wirkliche*) unites unity and plurality, limit and limitlessness (Zeller 1875, 567 n. 1). It was there, Zeller explains, that Socrates showed that the One is Many and the Many is One, and that this was as true, not just of what is changing and transient, but of pure concepts which are also composed of the One and the Many, and have within them limit and limitlessness (Zeller 1875, 565).

In the dialogue Socrates confesses to having been perplexed by the assertion that the Many are One and the One is Many (14C–15D). The best way of avoiding chaos he had found, he says, lies in appealing to a method which had often eluded him and left him alone and confused, but by which every matter appropriate for scientific consideration had been brought to light and of which he had always been enamored. He said that superior people of old closer to the gods had received this method as a divine gift to mankind (16A–C).

According to those people of old, things said to exist consist of one and many and also have limit and limitlessness inherently within themselves. Being so composed it must be assumed that they have a single concept in every case. This must be looked for and, being present, will be found. Grasping it, one must find out whether in the next stage there are two or three or some other number. And each of these units must be treated in the same way until it is seen both that the original unity is One and Many and an unlimited number, and just how many it is. Only when the whole number between the unlimited and the single unit has been grasped can the concept of limitlessness be applied to the plurality, and only then can each of the units be dismissed and released into the indeterminate. Misguided people make the unit limitless right away and fail to demarcate what is intermediate (16C–17A).

By doing our best to assemble fragments and segments and recognize the mark of a single nature, everything found becoming more and less is to be classified as unlimited and put into a single class (24E–25A), which is then to be mixed with the class of limit, the class of whatever keeps opposites from being at odds and makes them proportionate and harmonious by implanting number (25D). The mixing of these two classes, Socrates maintains, gives rise to a third class made possible by the measures which are produced with the limit, and from this, all fine things come (26B–D). As an example Socrates gave vocal sound which, he explains, is both a single phenomenon and quantitatively unlimited, but when one knows the nature and boundaries of the intervals with their height and depth of sound and systems that they form, the scales, tempos and measures does one become really knowlegeable. The other alternative, that of limitless plurality, he says, condemns one to ignorance (17B–D). (Compare with Cantor's examples of sets: Cantor 1887/88, 421–22; 412).

5. Husserl on Abstraction and the Concept of Number

Taught by Franz Brentano to despise metaphysical idealism, Edmund Husserl came within close range of Cantor's ideas upon arriving at the University of Halle in 1886 to prepare his *Habilitationsschrift* entitled "On the Concept of Number" (Husserl 1887). In this short work, a revised and much enlarged version of which was later published as the *Philosophy of Arithmetic* (Husserl 1891), Husserl set out to lay bare the roots out of which arithmetic develops by logical necessity. Cantor served on Husserl's *Habilitationskommittee* and approved the mathematical portion of the work (Gerlach and Sepp 1994). The two men became close friends (M. Husserl 1988, 114; Cantor 1991).

"On the Concept of Number" bore the subtitle "Psychological Analyses" and in it Brentano's disciple set out to anchor arithmetical concepts in direct experience by analyzing the actual psychological processes to which, he believed, the concept of number owed its genesis. In view of his entire training, he later said, it had been obvious to him when he started the work that "what mattered most for a philosophy of mathematics was a radical analysis of the 'psychological origin' of the basic mathematical concepts" (Husserl 1913, 33). Psychology, he maintained in the introduction to the work, was the indispensable tool for analyzing the concept of number, and the analysis of elementary concepts one of the more essential tasks of the psychology of the time. "In truth," Husserl even declared there, "not only is psychology indispensable for the analysis of the concept of number, but rather this analysis even *belongs within* psychology" (Husserl 1887, 94–95).

Brentano tried to inculcate in his students a model of philosophy based on the natural sciences and taught them to abhor the very kind of metaphysical idealism that pervaded Cantor's work so that any such considerations are markedly absent from both "On the Concept of Number" and *Philosophy of Arithmetic*. Yet, in spite of that major difference, many of the basic convictions underlying what Husserl hoped to accomplish in "On the Concept of Number" and the *Philosophy of Arithmetic* were perfectly compatible with the embattled creator of set theory's efforts to put the new numbers he was inventing on sound foundations. "With respect to the starting point and the germinal core of our developments toward the construction of a general arithmetic," Husserl wrote in about 1891, "we are in agreement with mathematicians that are among the most important and progressive ones of our times: above all with *Weierstrass*, but not less with *Dedekind*, Georg *Cantor* and many others" (Husserl 1994a, 1).

For one, it was Husserl's goal to analyze the concepts and relations which are in themselves simpler and logically prior, and then move on to analyze the more complicated and more derivative ones. The definitive removal of the real and imaginary difficulties in problems on the borderline between mathematics and philosophy, he then believed, would only come about in this way (Husserl 1887, 94–95).

And, under the influence of Weierstrass, Husserl still trusted that "a rigorous and thoroughgoing development of higher analysis . . . would have to emanate from elementary arithmetic alone, in which analysis is grounded," that the sole foundation of elementary arithmetic was "in that never-ending series of concepts which mathematicians call 'positive whole numbers'," and that "all the more complicated and artificial forms which are likewise called numbers—the fractional and irrational, and negative and complex numbers—have their origin and basis in the elementary number concepts and their interrelations" (Husserl 1887, 95).

It was also Husserl's early conviction that set theory lay at the basis of mathematics. The most primitive concepts, Husserl informed his readers, are the *general* concepts of set and of number which are grounded in the *concrete* sets of specific objects of any kind whatsoever and to which particular numbers are assigned. Just as cardinal numbers relate to sets, he affirmed in the introduction to the *Philosophy of Arithmetic*, so ordinals relate to series which are themselves ordered sets (Husserl 1891, 4).

Citing Euclid's classical definition of the concept of number as "a multiplicity of units," *eine Vielheit von Einheiten*, Husserl began the analyses of the *Philosophy of Arithmetic* by asserting that "the analysis of the concept of number presupposes the concept of multiplicity (*Vielheit*)" (p. 8). He then comments that in place of the word '*Vielheit*' the practically synonymous terms '*Mehrheit*', '*Inbegriff*', '*Aggregat*', '*Sammlung*', '*Menge*', etc. have been used, and he informs his readers that in order to neutralize any differences in meaning among the terms, he will not restrict himself to the exclusive use of any one of them (p. 8 and note). (In the parts of the *Philosophy of Arithmetic* under study here, he uses '*Inbegriff*', '*Menge*', and '*Vielheit*' interchangeably. Since Cantor also used the terms '*Menge*', '*Mannigfaltigkeit*', and '*Inbegriff*' interchangeably, I have chosen to adopt the common term 'set' ('*Menge*'), the choice which I consider best limns the issues in any comparative study of Husserl's and Cantor's theories in the late twentieth century).

Husserl began his analysis of the concept of number affirming that there could be no doubt that multiplicities or sets of determinate objects were the concrete phenomena which formed the basis for the abstraction of the concepts in question. And he furthermore maintained that everyone knew what the terms 'multiplicity' and 'set' meant, that the concept itself was well-defined and that there was no doubt as to its extension which, though we may not as yet be entirely clear as concerns the essence and formation of the corresponding concept, might be taken as given (Husserl 1887, 96–97, 111; Husserl 1891, 9–10, 13).

Initially limiting himself to analyses of sets of individually given, collectively grouped objects and numbers as known through direct experience (like Cantor's finger example) Husserl turned to the experience of concrete sets to consider how it is from them that both the more indeterminate universal concept of set and the determinate number concepts are to be abstracted (Husserl 1891, 10, 13).

And just as Cantor was trying to show that the transfinite numbers were "produced through abstraction from reality *with the same necessity* as the ordinary finite whole numbers by which alone all other mathematical conceptual formations thus far [have] been produced" (Cantor 1991, 136), a theory of abstraction (which Husserl considered as psychological process) would be granted a significant role in "On the Concept of Number" (Husserl 1887, 97–98, 115–16), and even more so in *Philosophy of Arithmetic* (Husserl 1891, 10–16, 82–96, 165–66).

The concrete phenomena upon which abstraction is performed, Husserl told his readers, were sets of determinate objects of any kind whatsoever. Any objects "whether physical or mental, abstract or concrete, whether given in experience or fantasized," Husserl emphasized, can be grouped together in a set and be counted (Husserl 1891, 10). However, if the nature of the particular contents makes no difference at all in forming the general concept of set and definite number, then the question arises, Husserl realized, as to what sort of abstraction process might enable one to obtain the general concepts of set and numbers from such concrete groups of objects. For according to the traditional theory of concept formation through abstraction, one is to disregard the properties which distinguish the objects while focussing on the properties they have in common which then go into the making of the general concept. But, as Husserl explains:

> It is right away obvious that a comparison of the individual contents we come across in the sets given would not yield the concept of set or number for us, and it was (even if this occurred) absurd to expect it.... The sets are merely composed of individual contents. So how are any common properties of the whole to come to the fore when the constituent parts may be so utterly heterogeneous? (Husserl 1891, 13; also Husserl 1887, 97, 112)

Such an objection may be overcome, Husserl explains, by observing that the sets are not merely composed of individual contents. Any talk of sets or multiplicities necessarily involves the combination of the individual elements into a whole, a unity containing the individual objects as parts. And though the combination involved may be very loose, there *is* a particular sort of unification there which would also have to have been noticed as such since the concept of set could never have arisen otherwise. "So," Husserl concludes, "if our view is correct, the concept of set arises through reflection on the particular . . . way in which the contents are united together . . . in a way analogous to the manner in which the concept of any other kind of whole arises through reflection upon the mode of combination peculiar to it." Husserl called the mode of combination characteristic of sets collective combination (Husserl 1891, 14–15; Husserl 1887, 97–98).

Husserl is now ready to characterize the distinctive abstraction process which produces the concept of set. Having before us any determinate, individual contents combined into a collection, he informs readers, we abstract the general concept by totally abstracting from the characteristics of the individual contents collected. Abstracting from something, he explains, simply means not paying any particular attention to it, and that absolutely does not cause the contents and their interconnections to disappear from our consciousness. Our interest is mainly focussed on the collective combination while the contents are only considered and attended to as any content whatsoever, anything whatsoever, any unit whatsoever (Husserl 1891, 84–85).

Number, the author of the *Philosophy of Arithmetic* explained, is the general form of a set under which the set of objects a, b, c falls. To obtain the concept of number of a concrete set of like objects, for example A, A, and A, one abstracts from the particular characteristics of the individual contents collected, only considering and retaining each one of them insofar as it is a something or a one, and thus regarding their collective combination, one obtains the general form of the set belonging to the set in question: one and one, etc. and . . . and one, to which a number name is assigned (Husserl 1891, 88, 165–66; Husserl 1887, 116–17).

Owing to confusions about the precise nature of the abstraction process Cantor and Husserl advocated, it is important to note that both men considered the abstraction process which yields the concept of set to be of a distinctive kind (Cantor 1991, 363, 365; Cantor 1887/88, 380, 411; Husserl 1891, 84). In the *Philosophy of Arithmetic*, Husserl underscored this conviction with criticisms aimed at dissociating the theory he adopted from the better known abstraction theories of Locke and Aristotle, something it is important to stress because philosophers have been all too wont to assimilate the process Husserl advocated to theories more familiar to them (Hill 1991, 15, 68). The analyses of the *Philosophy of Arithmetic*, Husserl declared, lead to the important, securely substantiated discovery that one cannot elucidate the formation of the concept of number in the same way one elucidates the concepts of color, form, etc. and that Aristotle and Locke were wrong to try to do so (Husserl 1891, 91–92). Mill was no more successful in winning Husserl's favor (Husserl 1891, 167).

In the late 1880s that is how Husserl believed that that most elementary of all arithmetical concepts, the concept of whole number could be produced through abstraction from reality to provide a sound basis for deriving more complicated and artificial forms of numbers in a strictly logical way. Change, however, was on the horizon and hints of the changes to come are already to be found in the *Philosophy of Arithmetic*. For conspicuously absent from that work is the enthusiastic espousal of psychologism of the opening pages of "On the Concept of Number." And conspicuously apparent in the opening pages of the 1891 work is Husserl's tergiversation regarding his earlier conviction that all the more complicated and artificial forms of numbers had their origin and basis in the concept of positive whole numbers and their interrelations and could be derived from them in a strictly logical way. Weierstrass's thesis is never embraced in the *Philosophy of Arithmetic* in the enthusiastic way it was in "On the Concept of Number." Instead we find Husserl initially writing that he will use it as his point of departure, but that it may prove false, and then finally in the introduction written just before the work was published, a statement that it is false (Hill 1991, 81–88).

6. Husserl Reasons his Way into the Realm of the Ideal

Completely under Brentano's influence in the beginning, Husserl had initially viewed idealistic systems with a jaundiced eye (Husserl 1919, 345). There were, however, ways in which Brentano's psychologism never came to satisfy him, and as he analyzed the basic concepts of mathematics he grew increasingly troubled by doubts of principle as to how to reconcile the objectivity of mathematics with psychological foundations for logic. His inability to silence his doubts undermined his confidence in psychologism, and he felt increasingly pushed to probe more deeply into the essence of logic, and especially to try to resolve "the profound difficulties which are tied up with the opposition between the subjectivity of the act of knowledge and the objectivity of the content and object of knowledge (or of truth and being)" (Husserl 1994a, 250; Husserl 1900/01, 42).

Husserl repeatedly said that the immediate cause of his intellectual crisis lay in his inability to answer questions about "imaginary" numbers (ex. negative square roots, negative, irrational, complex, transfinite numbers) that arose while trying to complete the *Philosophy of Arithmetic* (Husserl 1913, 33; Husserl 1891, viii, 5–6; Husserl 1970, 430–47; Husserl 1994a, 15–16; chapter 9 of the present book). By 1891 we find him already protesting that "a utilization of symbols for scientific purposes, and with scientific success, is still not therefore a *logical* utilization" (Husserl 1994a, 48), and he lamented the mental energy wasted in "the endless controversies over negative and imaginary numbers, over the infinitely small and the infinitely large, over the paradoxes of divergent series, and so on" (Husserl 1994a, 49). How much quicker and more secure the progress of arithmetic would have been, he believed, "if already upon the development of its methods there had been clarity concerning their logical character" (Husserl 1994a, 49). He thought one might "search logical works in vain for light on what really makes such mechanical operations, with mere written characters or word signs, capable of vastly expanding our actual knowledge concerning the number concepts" (Husserl 1994a, 50). He knew of "no logic that would even do justice to the very possibility of a genuine calculational technique" (Husserl 1994a, 17).

Husserl has also told of how his mathematical investigations left him facing a host of burning questions about the incredibly strange realms of actual consciousness and of pure logic (Husserl 1994a, 490–91), a category comprising "all of the pure 'analytical' doctrines of mathematics (arithmetic, number theory, algebra, etc.) and the entire area of formal theories . . . the theory of manifolds in the broadest sense" (Husserl 1913, 28), "the traditional syllogistic . . . the pure theory of cardinal numbers, the pure theory of ordinal numbers, of *Cantorian* sets . . . the pure mathematical theory of probability" which it would be absurd to classify under psychology (Husserl 1994a, 250). He had no idea of how to bring the two worlds together. Yet he believed they had to interrelate and form an intrinsic unity (Husserl 1994a, 490–91). He wanted to know how symbolic thinking was possible, how objective, mathematical, and logical relations constituted themselves in subjectivity, how insight into that was to be understood, and how the mathematical in itself, as given in the medium of the psychical, could be valid (Husserl 1913, 35).

All this ambiguity, Husserl said, "found ever new nourishment in the expanded philosophical-arithmetic studies, which extended to the broadest field of modern analysis and theory of manifolds and simultaneously to mathematical logic and to the entire sphere of logical in general" (Husserl 1913, 35). Logical research into formal arithmetic and the theory of manifolds presented him with particular difficulties (Husserl 1900/01, 41).

He confessed that he had been disturbed, and even tormented, by doubts about sets right from the very beginning. The concept of collection in Brentano's sense, Husserl explained, was to arise through reflection on the concept of collecting. Sets, he thus had reasoned, arose out of collective combination, in being conceived as one. This combining process involved when objects are brought together to make a whole only consists in that one thinks of them "together" and was obviously not grounded in the content of the disparate items collected into the set. It could not be physical, so it must be psychological, a unique kind of mental act connecting the contents of a whole. But then, he asked, echoing Cantor's concern to maintain the distinction between counting and numbers, was "the concept of number is not something basically different from the concept of collecting which is all that can result from the reflection on acts?" (Husserl 1913, 34–35; re. Husserl 1887, 97–112, 115 and Husserl 1891, chapter 3).

Husserl's questions were ones Cantor's philosophy of arithmetic raises. And if it seems exaggerated and unwarranted to suggest that someone's logical assumptions could be shaken by an encounter with Cantor's work, it is wise to remember that Husserl was on hand as Cantor began discovering the antinomies of set theory (Dauben 1979, 240–70), which played such a role in rocking the ground upon which many others had hoped to derive arithmetic. Indeed it was in studying Cantor's 1891 proof by diagonal argument that there is no greatest cardinal number that Russell discovered the contradiction of the set of all sets that are not members of themselves (Grattan-Guinness 1978; Grattan-Guinness 1980; Hill 1991, 1), which David Hilbert described as having had "a downright catastrophic effect in the world of mathematics," having led Dedekind and Frege to abandon their theories and "quit the field." Hilbert characterized the reaction to Cantor's theory of transfinite numbers as having been "dramatic" and "violent" (Hilbert 1925, 375).

Cast intellectually adrift as his earliest convictions gave way, Husserl set off on his own to answer his questions (Husserl 1913, 17; Husserl, 1900/01, 41–43), ultimately reasoning his way into the realm of ideal as he began to accord idealist systems the highest value, seeing them as shedding light on totally new, radical dimensions of philosophical problems. The ultimate and highest goals of philosophy, Husserl came to believe, are only opened up when the philosophical method these particular systems call for is clarified and developed (Husserl 1919, 345).

Although his experience of Cantor's work may have acted to pry Husserl away from psychologism and to steer him in the direction of idealism, Husserl said it was Hermann Lotze's work which was responsible for the fully conscious and radical turn from psychologism and the Platonism that came with it. In his *Logic* Lotze had unequivocally proclaimed that psychology's achievements, "do not reach those obscure regions of enquiry, the illumination of which might open new paths to Logic," and that he would "like to make clear . . . that Logic would have to renounce for a long time yet any profounder understanding of the operations of thought if she had to look for it in the psychological analysis of their origin" (Lotze 1888, §333).

Husserl said that he gained his first major insight in studying Lotze's interpretation of Plato's doctrine of Ideas, which exercised a profound effect on him and became a determining factor in all his further studies. Lotze, Husserl realized, was already writing about truths in themselves, and so the idea came to him to transfer all of the mathematical and a major part of the traditionally logical into the realm of the ideal. His own concepts of "Ideal" significations, and "Ideal" contents of presentations and judgments originally derived from Lotze (Husserl 1913, 36; Husserl 1994a, 201; Lotze 1888, chapter II).

Only by reflecting on Lotze's ideas did Husserl find the key to what he called "the curious conceptions of Bolzano" that had originally seemed naive and unintelligible to him (Husserl 1994a, 201). Husserl had known Bolzano's writings since he was a student of Weierstrass, had studied the 'paradoxes of infinite' under Brentano, and knew Bolzano's work through Cantor (Husserl 1913, 37), but had been disinclined to traffic in any mystico-metaphysical exploitation of Ideas or ideal possibilities and such (Husserl 1994b, 39). Though references to numbers in themselves and things in themselves appeared in Husserl's earliest writings (Hill 1991, 80–88), Husserl had originally thought of "propositions in themselves" as mythical entities, suspended between being and non-being (Husserl 1913, 37–38; Husserl 1994a, 201–2).

Viewed from the vantage point of Lotze's theory of Platonic Ideas, however, Bolzano's ideas came to have a powerful impact on Husserl. Plato, Lotze had insisted, never asserted the existence of the Ideas apart from things (Lotze 1888, §317), but intended to teach what Lotze called their validity (*die Geltung*), "a form of reality not including Being or Existence." In relegating the Ideas to a home which is not in space, Lotze argued, Plato was not trying to hypostasize their mere "validity" into any kind of real existence, but was plainly seeking to guard altogether against any such attempt (Lotze 1888, §§314, 317, 318, 319).

It suddenly became clear to Husserl that Bolzano had not hypostasized presentations and propositions in themselves (Husserl 1994b, 39), which Husserl now saw

> were what were ordinarily called, the 'senses' of statements, what is said to be one and the same when, for example, different persons are said to have asserted the same thing, or what scientists simply call a theorem (for example the theorem about the sum of the angles in a triangle) which no one would think of as being someone's experience of judging and that this identical sense could be none other than the universal, the species belonging to a certain Moment present in all actual assertions having the same sense which makes possible the identification in question, even when the descriptive content of individual experiences of asserting varies considerably in other respects. (Husserl 1994a, 201)

"Viewed this way," Husserl echoed Lotze, "Bolzano's theory that propositions are objects which nonetheless have no 'existence' is quite intelligible. They enjoy the ideal existence or validity characteristic of universals (Husserl 1994a, 201–02).

The first two volumes of the *Wissenschaftslehre* on presentations and propositions in themselves, Husserl concluded, were to be seen as an initial attempt to provide a unified presentation of the domain of pure ideal doctrines and already provided a complete plan of a pure logic, an insight which proved immensely helpful enabling him to use Bolzano's account step by step to verify the "Platonic" interpretation (Husserl 1913, 37). Lying hidden in Bolzano's *Wissenschaftslehre* was something that seemed to Husserl to be "one of the most momentous logical insights" that the "core content of any normative and practical logic consists in propositions that do *not* deal with acts of thought, but rather with those Ideas instanced in certain of their Moments" (Husserl 1994a, 209), a key thesis of Lotze (Lotze 1888, chapter II).

However, neither Bolzano nor Lotze provided the full answers Husserl needed, for there was no trace in the work of either of any thought of the pure phenomenological elucidation of knowledge that Husserl ultimately concluded was needed to solve the puzzles surrounding the being in itself of the ideal sphere and its relationship to consciousness (Husserl 1994b, 39; Husserl 1913, 38–40, 45–49; Husserl 1994a, 201–3, 209. Husserl 1900/01, *Prolegomena*, §61, Appendix). These could only be solved through his own phenomenological investigations for which logic had to be more than a formal, mathematical, theory, but required "phenomenological and epistemological elucidations in virtue of which we not merely are completely certain of the validity of its concepts and theories, but also truly understand them" (Husserl 1994a, 215). These phenomenological investigations became the stuff of the *Logical Investigations* (Husserl 1900/01) that Husserl wrote during his last decade in Halle and in which he has said that all possible efforts had been "taken to dispose the reader to the recognition of this ideal sphere of being and knowledge . . . to side with 'the ideal in this truly Platonistic sense,' 'to declare oneself for idealism' with the author" (Husserl 1913, 20).

7. Concluding Remark

By describing the crucial shift in Husserl's views on psychologism and metaphysical idealism as they relate to Georg Cantor's theories of abstraction and Platonic Ideas, I have tried to fit these two major themes of Cantor's philosophy of arithmetic into a now rather standard version of the development of Husserl's thought. I have used textual analysis to achieve my goals. Plumbing Husserl's and Cantor's writings of the 1880s and 1890s, I have described something of the relationship between the two men's efforts to provide solid foundations for the positive whole numbers and the more complicated and artificial forms of numbers, thereby establishing connections between their ideas which have until now gone virtually unsuspected. In so doing, I hope to have contributed to filling a gaping hole in the understanding of the development of the ideas which went into the making of phenomenology, as well as to a better understanding of Cantor's philosophy of arithmetic by examining the philosophical, metaphysical theories within which, to the chagrin of his contemporaries, he chose to frame his ideas.

It is my conviction that further research into the relationship between Husserl's and Cantor's ideas will show that no complete grasp of Husserl's phenomenology, and his philosophy of logic and mathematics in particular, is possible without a clear understanding of the role Cantor's philosophy of arithmetic played in the development of Husserl's thought. And I would like to take this opportunity to suggest that a thorough understanding of the impact of Cantor's ideas on all periods of Husserl's work will prove fruitful and constructive in ways which, given then the current state of research into the subject, could only seem surrealistic at the present time. Cantor's work embodied many of the very problems Husserl found so distressing, and it is certain that in searching for solutions he felt compelled to turn elsewhere. However, further research will surely show that phenomenology, while not being Cantorian, distinctly bears Cantor's imprint.

ACKNOWLEDGMENTS

I wish to express my gratitude to Ivor Grattan-Guinness for all the advice and documents he has given me, to Emiko Ima for her hard work, and to William Gallagher for help transliterating the Greek terms.

REFERENCES

Aristotle. *Works.* 1983. Cambridge, MA: Harvard University Press.
Bolzano, B. 1831. *Paradoxes of the Infinite.* London: Routledge and Kegan Paul, 1950. Translation of his *Paradoxien des Unendlichen* of 1831.
———. 1837. *Theory of Science.* Dordrecht: Reidel, 1973. Partial translation of his *Wissenschaftslehre* of 1837.
Cantor, G. 1883. *Grundlagen einer allgemeinen Mannigfaltigkeitslehre. Ein mathematisch-philosophischer Versuch in der Lehre des Unendlichen.* Leipzig: Teubner, 1883. Cited as appears in Cantor 1932, 165–246.
———.1884. "Principien einer Theorie der Ordnungstypen" (dated November 6, 1884).First published as "An unpublished paper by Georg Cantor" by I. Grattan-Guinness in *Acta Mathematica* 124 (1970): 65–107.
———. 1885. "Rezension von Freges *Grundlagen.*" *Deutsche Literaturzeitung* 6: 728–29, in Cantor 1932, 440–41.
———. 1887/88. "Mitteilungen zur Lehre vom Transfiniten." *Zeitschrift für Philosophie und philosophische Kritik* 91 (1887): 81–125; 92 (1888): 240–65, also published as *Gesammelte Abhandlungen zur Lehre vom Transfiniten,* Halle: Pfeffer, 1890. Cited as appears in Cantor 1932, 378–439.
———. 1895/97. *Contributions to the Founding of the Theory of Transfinite Numbers.* New York: Dover, 1955 (1915). Translation of his "Beiträge zur Begründung der transfiniten Mengenlehre," 1895/97.
———. 1932. *Gesammelte Abhandlungen.* Ed. E. Zermelo. Berlin: Springer.
———. 1991. *Georg Cantor Briefe.* Ed. H. Meschkowski and W. Nilson. New York: Springer.
Cavaillès, J. 1962. *Philosophie mathématique.* Paris: Hermann.
———. 1937. *Méthode axiomatique et formalisme.* Paris: Hermann, 1981.
Charraud, N. 1994. *Infini et inconscient, essai sur Georg Cantor.* Paris: Anthropos.
Couturat, L. 1896. *De l'infini mathématique.* Paris: Blanchard, 1973.

Dauben, J. 1979. *Georg Cantor, His Mathematics and Philosophy of the Infinite.* Princeton: Princeton University Press.

Desanti, J. 1968. *Les idéalités mathématiques.* Paris: Seuil.

Dugac, P. 1976. *Richard Dedekind et les fondements des mathématiques.* Paris: Vrin.

Eccarius, W. 1985. "Georg Cantor und Kurd Lasswitz: Briefe zur Philosophie des Unendlichen." *NTM Schriftenr. Gesch. Naturwiss., Technik, Med.* 22: 7–28.

Frege, G. 1879. "*Begriffsschrift*, a formula language, modeled upon that of arithmetic, for pure thought." In van Heijenoort 1967, 5–82.

———. 1892. "Review of Georg Cantor, *Zur Lehre vom Transfiniten: Gesammelte Abhandlungen.*" In Frege 1984, 178–81.

———. 1894. "Review of E. G. Husserl, *Philosophie der Arithmetik I.*" In Frege 1984, 195–209.

———. 1979. "Draft Towards a Review of Cantor's *Gesammelte Abhandlungen zur Lehre vom Transfiniten.*" *Posthumous Writings.* Oxford: Blackwell, 1979, 68–71.

———. 1984. *Collected Papers on Mathematics, Logic and Philosophy.* Oxford: Blackwell.

Gerlach, H. and H. Sepp, eds. 1994. *Husserl in Halle.* Bern: Peter Lang.

Gilson, L. 1955. *Méthode et métaphysique selon Franz Brentano.* Paris: Vrin.

Grattan-Guinness, I. 1978. "How Russell Discovered His Paradox." *Historia Mathematica* 5 (1978): 127–37.

———. 1980. "Georg Cantor's Influence on Bertrand Russell." *History and Philosophy of Logic* 1: 61–93.

———. 1982. "Psychology in the Foundations of Logic and Mathematics: The Cases of Boole, Cantor and Brouwer." *History and Philosophy of Logic* 3: 33–53.

———. 1996. "Numbers, Magnitudes, Ratios and Proportions in Euclid's *Elements*: How Did He Handle Them?" *Historia Mathematica* 23: 355–75.

Hallett, M. 1984. *Cantorian Set Theory and Limitation of Size.* Oxford: Clarendon.

Hilbert, D. 1925. "On the Infinite." In van Heijenoort 1967, 369–92.

Hill, C. O. 1991. *Word and Object in Husserl, Frege and Russell: The Roots of Twentieth Century Philosophy.* Athens, OH: Ohio University Press.

———. 1994. "Frege's Attacks on Husserl and Cantor." *The Monist* 77, no. 3 (July): 347–57, chapter 6 of the present book.

———. 1995. "Husserl and Hilbert on Completeness." *From Dedekind to Gödel, Essays on the Development of the Foundations of Mathematics.* Ed. J. Hintikka. Dordrecht: Kluwer, 143–63, chapter 10 of the present book.

———. 1997. "Did Georg Cantor Influence Edmund Husserl?" *Synthese* 113: 145–70, chapter 8 of the present book.

———. 1998. "From Empirical Psychology to Phenomenology, Husserl on the Brentano Puzzle." *The Brentano Puzzle.* Ed. R. Poli. Ashgate: Aldershot, 151–67.

Husserl, E. 1887. "On the Concept of Number." In *Husserl: Shorter Works.* Ed. P. McCormick and F. Elliston. Notre Dame: University of Notre Dame Press, 1981, 92–120. Translation of his 1887 "Über den Begriff der Zahl," in Husserl 1970, 289–339.

———. 1891. *Philosophie der Arithmetik.* Halle: Pfeffer, 1891. In Husserl 1970.

———. 1900/01. *Logical Investigations.* New York: Humanities Press, 1970.

———. 1913. *Introduction to the Logical Investigations.* The Hague: M. Nijhoff, 1975.

———. 1919. "Recollections of Franz Brentano." *Husserl: Shorter Works.* Ed. P. McCormick and F. Elliston. Notre Dame: University of Notre Dame Press, 1981, 342–49. Also translated by L. McAlister in her *The Philosophy of Brentano*, London: Duckworth, 1976, 47–55.

———. 1970. *Philosophie der Arithmetik, mit ergänzenden Texten (1890-1901).* Husserliana, vol. XII. The Hague: M. Nijhoff, 1970.

———. 1983. *Studien zur Arithmetik und Geometrie.* Husserliana, vol. XXI. The Hague: M. Nijhoff.

———. 1994a. *Early Writings in the Philosophy of Logic and Mathematics.* Dordrecht: Kluwer.

———. 1994b. *Briefwechsel, Die Brentanoschule.* Vol. 1. Dordrecht: Kluwer.

Husserl, M. 1988. "Skizze eines Lebensbildes von E. Husserl." *Husserl Studies* 5: 105–25.

Illemann, W. 1932. *Husserls vorphänomenologische Philosophie.* Leipzig: Hirzel.

Jourdain, P. 1908–14. "The Development of the Theory of Transfinite Numbers." *Archiv der Mathematik und Physik* 14 (1908/09): 289–311; 16 (1910): 21–43; 22 (1913/14): 1–21. Reprinted in Jourdain's *Selected Essays on the History of Set Theory and Logics,* ed. I. Grattan-Guinness. Bologna: CLUEB, 1991, 33–99.

Kerry, B. 1885. "Über Georg Cantors Mannichfaltigkeitsuntersuchungen." *Vierteljahrsschrift für wissentschaftliche Philosophie* 9: 191–232.

Kreiser, L. 1979. "W. Wundts Auffassung der Mathematik. Briefe von G. Cantor an W. Wundt." *Wissenschaftliche Zeitschrift, Karl-Marx Universitat, Ges. u. Sprachwissenschaft* 28: 197–206.

Kusch, M. 1989. *Language as Calculus vs. Language as Universal Medium.* Dordrecht: Kluwer.

Lotze, H. 1879. *Metaphysik.* Leipzig: Hirzel.

———. 1888. *Logic.* New York: Garland, 1980. Reprint of B. Bosanquet's translation of his *Logik.*

Mahnke, D. 1966. "From Hilbert to Husserl: First Introduction to Phenomenology, especially that of formal mathematics." *Studies in the History and Philosophy of Science* 8 (1966): 71–84. Translation of "Von Hilbert zu Husserl: Erste Einführung in die Phänomenologie, besonders der formalen Mathematik."

Meschkowski, H. 1965. "Aus den Briefbüchern Georg Cantors." *Archive for the History of the Exact Sciences* 2, no. 6: 503–19.

———. 1967. *Probleme des Unendlichen. Werk und Leben Georg Cantors.* Braunschweig: Vieweg.

Miller, J. 1982. *Numbers in Presence and Absence.* The Hague: M. Nijhoff.

Neemann, U. 1977. "Husserl und Bolzano." *Allgemeine Zeitschrift für Philosophie* 2: 52–66.

Panza, M. and J-M. Salanskis. 1995. *L'objectivité mathématique, platonismes et structures formelles.* Paris: Masson.

Picker, B. 1962. "Die Bedeutung der Mathematik für die Philosophie Edmund Husserls." *Philosophia Naturalis* 7: 266–355.

Plato. *Dialogues.* Cambridge, MA: Harvard University Press, 1926–1946.

Purkert, W., and H. Ilgauds. 1991. *Georg Cantor 1845–1918.* Basel: Birkhäuser.

Rosado Haddock, G.E. 1973. "Edmund Husserls Philosophie der Logik und Mathematik im Lichte der gegenwärtigen Logik und Grundlagenforschung." Ph.D. dissertation, Rheinische Friedrich-Wilhelms-Universität, Bonn.

Russell, B. 1973. *Essays in Analysis.* London: Allen & Unwin.

Schmit, R. 1981. *Husserls Philosophie der Mathematik: platonische und konstruktivische Momente in Husserls Mathematik Begriff.* Bonn: Bouvier.

Schuhmann, K. 1977. *Husserl Chronik.* The Hague: M. Nijhoff.

———. 1990/91. "Husserls doppelter Vorstellugsbegriff: die Texte von 1893." *Brentano Studien* 3: 119–36.

Sebestik, J. 1992. *Logique et mathématique chez Bernard Bolzano.* Paris: Vrin.

Smith, B. 1994. *Austrian Philosophy: The Legacy of Franz Brentano.* La Salle, IL: Open Court.

Spiegelberg, H. 1982. "Franz Brentano (1838–1917): Forerunner of the Phenomenological Movement." *The Phenomenological Movement.* The Hague: M. Nijhoff, 27–48.

Tiles, M. 1989. *The Philosophy of Set Theory, An Historical Introduction to Cantor's Paradise.* Oxford: Blackwell.

van Heijenoort, J., ed. 1967. *From Frege to Gödel.* Cambridge, MA: Harvard University Press.

Weinberg, J. 1968. "Abstraction in the Formation of Concepts." *Dictionary of the History of Ideas.* Ed. P. Wiener. New York: Charles Scribner's Sons, 1–9 .

Willard, D. 1980. "Husserl on a Logic That Failed." *The Philosophical Review* 89, no. 1 (January): 46–64.

———. 1984. *Logic and the Objectivity of Knowledge.* Athens, OH: Ohio University Press.

Zeller, E. 1839. *Platonische Studien.* Tübingen: Osiander.

———. 1875. *Die Philosophie der Griechen (Sokrates und die Sokratiker. Plato und die alte Akademie).* 3rd ed. Leipzig: Fues's Verlag.

———. 1879. *Die Philosophie der Griechen (Aristoteles und die alten Peripatetiker).* 3rd ed. Leipzig: Fues's Verlag.

8

Claire Ortiz Hill

DID GEORG CANTOR INFLUENCE EDMUND HUSSERL?*

Few have entertained the idea that the creator of set theory may have influenced the founder of the phenomenological movement,[1] and not a few would consider the mere suggestion utterly preposterous, if not downright blasphemous. Yet Georg Cantor and Edmund Husserl were close friends at the University of Halle[2] during the last fourteen years of the nineteenth century, when Cantor was at the height of his creative powers and Husserl in the throes of an intellectual struggle during which his ideas were particularly malleable and changed considerably and definitively (Husserl 1913). It was there and then that Husserl wrote "On the Concept of Number" (Husserl 1887), the *Philosophy of Arithmetic* (Husserl 1891), and the groundbreaking *Logical Investigations* (Husserl 1900/01), where he lay the foundations of his phenomenology that went on to shape the course of twentieth-century philosophy in Continental Europe.

Although the relationship between Husserl's and Cantor's ideas has gone all but unnoticed, there is much information to be gleaned from a close examination of their writings. So in the following pages I turn to that source to show how Husserl's and Cantor's ideas overlapped and crisscrossed during those years in the areas of philosophy and mathematics, arithmetization, abstraction, consciousness and pure logic, psychologism, metaphysical idealism, new numbers, and sets and manifolds. In so doing I hope to shed some needed light on the evolution of Husserl's thought during that crucial time in Halle.

Before turning to the examination of these questions, however, it is important to bear in mind that influence may be negative or positive, temporary or permanent, and it may show through in different areas in different ways. Moreover, in the case of an original thinker of Husserl's stature it could not be a matter of the simple imitation of someone else's ideas. For example, few question that Franz Brentano influenced Husserl, but Husserl actually eradicated the most distinctive features of Brentano's teaching from phenomenology and embraced metaphysical and epistemological views which Brentano considered odious and despicable (Hill 1998).

1. Marrying Philosophy and Mathematics

Recently converted to philosophy by Franz Brentano (Husserl 1919), Husserl took a front row seat at the creation of set theory and the transfinite number system when he arrived in Halle in 1886 to prepare his *Habilitationsschrift*, "On the Concept of Number," of which his *Philosophy of Arithmetic* would be a revised and much enlarged version. Cantor served on Husserl's *Habilitationskommittee* and approved the mathematical portion of Husserl's work (Gerlach and Sepp 1994). The two became close friends.

In the late 1880s Husserl and Cantor figured among the small number of their contemporaries intent upon wedding mathematics and philosophy (Frege being, of course, another). Over the years Cantor's thoughts had been increasingly turning to philosophy and in 1883 he published the *Grundlagen einer allgemeinen Mannigfaltigkeitslehre* (Cantor 1883), a work, according to its 1882 foreword, "written with two groups of reader in mind—philosophers who have followed the developments in mathematics up to the present time, and mathematicians who are familiar with the most important older and newer publications in philosophy." [3]

As earnest as Cantor's forays into philosophy were, they hardly met with universal acclaim. In 1883 Gösta Mittag-Leffler deemed it fit to warn Cantor that his work would be much more easily appreciated in the mathematical world "without the philosophical and historical explanations" (Cantor 1991, 118). And in 1885 he warned him that he risked shocking most mathematicians and damaging his reputation with his new terminology and philosophical way of expressing himself in the "Principien einer Theorie der Ordnungstypen" (Cantor 1991, 244), a posthumously published work which opened with a strong statement in favor of an admixture of philosophy, metaphysics, and mathematics (Cantor 1884, 83–84).

By the time Husserl arrived in Halle, Cantor was practically ready to abandon mathematics for philosophy. He was much less in a position than Husserl was to change course in mid-stream, but this did not keep him from trying to teach philosophy (Cantor 1991, 210, 218) and from seasoning his writings with philosophical reflections and references to philosophers like Democritus, Plato, Aristotle, Augustine, Boethius, Aquinas, Descartes, Nicolas von Cusa, Spinoza, Leibniz, Kant, Comte, Bacon, Locke, and so on. In 1894 Cantor wrote to the French mathematician Hermite that "in the realm of the spirit" mathematics had no longer been "the essential love of his soul" for more than twenty years. Metaphysics and theology, he "openly confessed," had so taken possession of his soul as to leave him relatively little time for his "first flame." It might have been otherwise, he said, had his wishes for a position in a university where his teaching might have had a greater effect been fulfilled fifteen, or even eight years earlier, but he was now serving God better than, "owing to his apparently meager mathematical talents," he might have done through exclusively pursuing mathematics (Cantor 1991, 350).

During the late 1880s, the embattled creator of set theory was hard at work trying to put the new numbers he was inventing on solid foundations and philosophically justifying the claims he was making about them (Dauben 1979, chapter 6). Those attempts found expression in his correspondence (Cantor 1991) and most particularly in the "Mitteilungen zur Lehre vom Transfiniten" (Cantor 1887/88), which he said he published in the *Zeitschrift für Philosophie und philosophische Kritik* because he had grown disgusted with the mathematical journals (Dauben 1979, 139, 336 n. 29).

2. Arithmetization

Much of the initial intellectual kinship between Husserl and Cantor can be explained by the mighty influence the great mathematician Karl Weierstrass exercised on both of them. "Mathematicians under the influence of Weierstrass," Bertrand Russell once wrote, "have shown in modern times a care for accuracy, and an aversion to slipshod reasoning, such as had not been known among them previously since the time of the Greeks." Analytical Geometry and the Infinitesimal Calculus, he went on to explain, had produced so many fruitful results since their development in the seventeenth century that mathematicians had neither taken the time, nor been inclined to examine their foundations. Philosophers who might have taken up the task were lacking in mathematical ability. Mathematicians were only finally awakened from their "dogmatic slumbers" by Weierstrass and his followers in the latter half of the nineteenth century (Russell 1917, 94).

Cantor[4] and Husserl had both fallen under the influence of Weierstrass and Husserl had even been his assistant.[5] It was Weierstrass, Husserl has said, who awakened his interest in seeking radical foundations for mathematics. Weierstrass's aim, Husserl once recalled, was to expose the original roots of analysis, its elementary concepts and axioms on the basis of which the whole system of analysis might be deduced in a completely rigorous, perspicuous way. Those efforts made a lasting impression on Husserl who would say that it was from Weierstrass that he had acquired the ethos of his intellectual endeavors (Schuhmann 1977, 7). Late in his philosophical career Husserl would say that he had sought to do for philosophy what Weierstrass had done for mathematics (Becker 1930, 40–42; Schuhmann 1977, 345).

With "respect to the starting point and the germinal core of our developments toward the construction of a general arithmetic," Husserl wrote as he was carrying out his plan to help supply radical foundations for mathematics, "we are in agreement with mathematicians that are among the most important and progressive ones of our times: above all with *Weierstrass,* but not less with *Dedekind,* Georg *Cantor* and many others" (Husserl 1994, 1).

Husserl began "On the Concept of Number" writing of the exhilaratingly creative period mathematics had known over the previous hundred years, during which new and very far-reaching instruments of investigation had been found and an almost boundless profusion of important pieces of knowledge won. Mathematicians, however, he explained, had neglected to examine the logic of the concepts and methods they were introducing and using and there was now a need for logical clarification, precise analyses, and a rigorous deduction of all of mathematics from the least number of self-evident principles. The definitive removal of the real and imaginary difficulties on the borderline between mathematics and philosophy, he deemed, would only come about by first analyzing the concepts and relations which are in themselves simpler and logically prior, and then analyzing the more complicated and more derivative ones (Husserl 1887, 92–95).

The natural and necessary starting point of any philosophy of mathematics, Husserl considered, was the analysis of the concept of whole number (Husserl 1887, 94–95). For the still faithful disciple of Weierstrass then believed the "domain of 'positive whole numbers' to be the first and most underivative domain, the sole foundation of all remaining domains of numbers" (Husserl 1994, 2). "Today there is a general persuasion," he wrote in "On the Concept of Number,"

> that a rigorous and thoroughgoing development of higher analysis . . . would have to emanate from elementary arithmetic alone in which analysis is grounded. But this elementary arithmetic has . . . its sole foundation . . . in that never-ending series of concepts which mathematicians call 'positive whole numbers'. All of the more complicated and artificial forms which are likewise called numbers the fractional and irrational, and negative and complex numbers have their origin and basis in the elementary number concepts and their interrelations. (Husserl 1887, 95)

So in "On the Concept of Number" and *Philosophy of Arithmetic* Husserl set out to carry Weierstrass's work to arithmetize analysis a step further by submitting the concept of number itself to closer scrutiny than Weierstrass had.

As for Cantor, after working for several years to clear up theoretical obscurities regarding infinity and the real number system raised but not fully answered by Weierstrass's lectures on analytic functions, the late 1880s found him hard at work laying the foundations of his set theory and reconnoitering, conquering, colonizing and defending the world of transfinite numbers (Jourdain 1908–1914; Hilbert 1925; Hallett 1984; Dauben 1979).

Like Husserl, Cantor was immersed in a project aimed at demonstrating that the positive whole numbers formed the basis of all other mathematical conceptual formations inspired by Weierstrass's famous theory to that effect. Any further progress of the work on set theory, Cantor explained in the beginning of his *Mannigfaltigkeitslehre* was absolutely dependent upon the expansion of the concept of real whole numbers beyond the present boundaries and in a direction which, as far as he knew, no one had yet searched. He had, he claimed, burst the confines of the conceptual formation of real whole numbers and broken through into a new realm of transfinite numbers. Initially he had not been clearly conscious of the fact that these new numbers possessed the same concrete reality the whole numbers did. He had, however, become persuaded that they did. As strange and daring as his ideas might now seem, he was convinced that they would one day be deemed completely simple, appropriate, and natural (Cantor 1883, 165–66). Much of the work he was doing was aimed at showing they were.

Acting on a conviction, spelled out in a 1884 letter to Gösta Mittag-Leffler, that the only correct way to proceed was "to go from what is most simple to that which is composite, to go from what already exists and is well-founded to what is more general and new by continually proceeding by way of transparent considerations, step by step without making any leaps" (Cantor 1991, 208), Cantor began devising a strategy as to how to provide his "strange" new transfinite numbers with secure foundations by demonstrating precisely how the transfinite number system might be built from the bottom up. He was in possession of principles by which, he claimed, one might break through any barrier in the conceptual formation of the real, whole numbers and with the greatest confidence and self-evidence arrive at ever new number classes and numbers having the same concrete definiteness and reality as objects as the previous ones (Cantor 1883, 166–67, 199).

3. Abstraction

Cantor developed a theory of abstraction which he believed was the distinctive feature of his number theory and represented an entirely different method for providing the foundations of the finite numbers than was to be found in the theories of his contemporaries (Cantor 1991, 365, 363). He accorded his theory a key role in his efforts to provide solid foundations for transfinite arithmetic, believing that with it he was laying bare the roots from which the organism of transfinite numbers develop with logical necessity (Cantor 1887/88, 380).

For Cantor, cardinal numbers were the general concepts assigned to sets which one may obtain by abstracting both from the properties of the elements and from the order in which they are given (Cantor 1991, 302) and the transfinite numbers could be "produced through abstraction from reality with *the same necessity* as the ordinary finite whole numbers by which alone all other mathematical conceptual formations thus far have been produced" (Cantor 1991, 136; Kreiser 1979).

The earliest accounts he gave of his abstraction process appear in his correspondence (Cantor 1991, 178–80) and in the posthumously published "Principien einer Theorie der Ordnungstypen" (Cantor 1884). The theory did not, however, make its way into print until it appeared in the "Mitteilungen" (Cantor 1887/88), into which he directly incorporated letters he had written and where he stressed that he had advocated and repeatedly taught the theory in his courses as many as four years earlier (Cantor 1887/8, 378–79, 387 n., 411 n.).

In the "Mitteilungen" Cantor was particularly intent upon proving that his theorems about transfinite numbers were firmly secured "through the logical power of proofs" which, proceeding from his definitions which were "neither arbitrary nor artificial, but originate naturally out of abstraction, have, with the help of syllogisms, attained their goal" (Cantor 1887/88, 418). And much of the work is devoted to explaining precisely how one might procure numbers by abstraction from reality and, in particular, how the actual infinite number concept might be formed through appropriate natural abstractions in the way the finite number concepts are won through abstraction from finite sets (Cantor 1887/88, 411; also Cantor 1991, 329, 330).

In the work Cantor repeatedly gave essentially the same recipe for extracting cardinal numbers from reality through abstraction (Cantor 1887/88, 379, 387, 411, 418 n. 1). In a text, which Husserl approvingly cites in the *Philosophy of Arithmetic* (Husserl 1891, 126 n.) Cantor explained that for "the formation of the general concept 'five' one needs only a set (for example all the fingers of my right hand) which corresponds to this cardinal number; the act of abstraction with respect to both the properties and the order in which I encounter these wholly distinct things, produces or rather awakens the concept 'five' in my mind" (Cantor 1887/88, 418 n. 1).

"By the *power* or *cardinal number* of a set M (which is made up of distinct, conceptually separate elements $m, m', \ldots$ and is to this extent determined and limited)," Cantor said he understood

> the general concept or species concept (universal) which one obtains by abstracting from the properties of the elements of the set, as well as from all the relations which the elements may have, whether themselves or to other things, but especially from the order reigning among the elements and only reflects upon what is common to all sets *equivalent* to M. (Cantor 1887/88, 387; also Cantor 1991, 178)

In the "Principien" Cantor had written that the cardinal number of a set seemed to him "to be the most primitive, *psychologically*, as well as methodologically simplest root concept, arisen through abstraction, from all particular characteristics which a set of a specific class may display, both with respect to the nature of its elements, as well as with regard to the *relations* and *order* in which the *elements are to each other* or can stand to *things lying outside the set*" (Cantor 1884, 86).

Cantor was propounding this theory of abstraction just as Husserl was writing "On the Concept of Number" and the *Philosophy of Arithmetic*, where a similar theory is espoused. Husserl actually had offprints of the "Mitteilungen" and marked and underlined precisely those passages (and almost exclusively those passages) in which Cantor explained the abstraction process.[6] In "On the Concept of Number," Husserl wrote:

> It is easy to characterize the abstraction which must be exercised upon a concretely given multiplicity in order to attain to the number concepts under which it falls. One considers each of the particular objects merely insofar as it is a something or a one herewith fixing the collective combination; and in this manner there is obtained the corresponding general form of multiplicity, one and one and . . . and one, with which a number name is associated. In this process there is total abstraction from the specific characteristics of the particular objects. . . . To abstract from something merely means to pay no special attention to it. Thus in our case at hand, no special interest is directed upon the particularities of content in the separate individuals. (Husserl 1887, 116–17)[7]

In the *Philosophy of Arithmetic* Husserl studied this same process in greater detail (chapter 4 especially), particularly underscoring the uniqueness of the abstraction process which yields the number concept. According to the traditional theory of concept formation through abstraction, he pointed out there, one is to disregard the properties which distinguish the objects while focussing on the properties they have in common and out of which the general concept will be made. But Husserl thought it absurd to expect that the concept of set or number might result from any comparison of the individual contents making up the sets which may be utterly heterogeneous. He emphasized that his analyses had led to the important, securely substantiated finding that it is impossible to elucidate the formation of the concept of number in the same way one elucidates the concepts of color, form, etc. And he pointedly disassociated his theory from the better known abstraction theories of Locke and Aristotle, something it is important to call attention to because philosophers have been all too wont to assimilate the process Husserl advocated to theories more familiar to them (Husserl 1891, 12–16, 50, 84–93, 182–86).

4. Psychologism

As similar as Husserl's and Cantor's theories of abstraction appear to have been in the late 1880s, there was, nonetheless, a major difference between them which reflected deep contradictions within both men's philosophies of arithmetic and soon brought Husserl to rethink and profoundly modify his entire approach to philosophy.

This major difference between them lie in the fact that Husserl began writing "On the Concept of Number" as a committed empirical psychologist à la Brentano, whose philosophical ideal was most nearly realized in the exact natural sciences (Husserl 1919, 344–45). "In view of my entire training," Husserl wrote of his early work on the foundations of arithmetic, "it was obvious to me when I started that what mattered most for a philosophy of mathematics was a radical analysis of the 'psychological origin' of the basic mathematical concepts" (Husserl 1913, 33; see Husserl 1891, 16). In "On the Concept of Number" and the *Philosophy of Arithmetic* he set out to anchor arithmetical concepts in direct experience by analyzing the actual psychological processes to which, he believed, the concept of number owed its genesis. He considered the analysis of elementary concepts to be one of the more essential tasks of the psychology. "Not only is psychology indispensable for the analysis of the concept of number," he declared in "On the Concept of Number," "but rather this even *belongs* within psychology" (Husserl 1887, 95).

Given his early training, it is easy to see how Husserl might have considered Cantor's philosophy of arithmetic ripe for the kind of analyses Brentano trained his students to undertake—how Cantor's talk of intuitions and presentations might grow in sophistication through analysis in accordance with Brentano's teachings about presentation and intentionality. Cantor's appeals to inner intuition and talk of things like the fingers of his right hand helping produce or awaken concepts in his mind do sound unabashedly empiricistic or psychologistic. Frege certainly saw them as such as he decried the "psychological and hence empirical turn" in the "Mitteilungen" (Frege 1892, 180, 181).

However, psychology from the empirical standpoint was strictly incompatible with the ideas of Cantor, who considered himself to be an adversary of psychologism, empiricism, positivism, naturalism, and related trends. However psychologistic his mysterious references to inner intuition (ex. Cantor 1883, 168, 170, 201) or to experiences helping produce concepts in his mind (Cantor 1887/88, 418 n. 1) may appear, he opposed the new empiricism, sensualism, skepticism, and Kantianism which, he argued, mistakenly located the sources of knowledge and certainty in the senses or in the "supposedly pure forms of intuition of the world of presentation." He maintained that certain knowledge could "only be obtained through concepts and ideas which, at most stimulated by external experience, are on the whole formed through inner induction and deduction as something which in a way already lay within us and was only awakened and brought to consciousness" (Cantor 1883, 207 n. 6). In his 1885 review of the *Foundations of Arith-*

metic, he praised Frege for demanding that all psychological factors and intuitions of space and time be banned from arithmetical concepts and principles because that was the only way their strict logical purity and validity might be secured (Cantor 1885, 440).

For Cantor, the transfinite "presented a rich, ever growing field of ideal research" (Cantor 1887/88, 406) and abstraction was to show the way to that new, abstract realm of ideal mathematical objects which could not be directly perceived or intuited. For him it was a way of producing purely abstract arithmetical definitions, a properly arithmetical process as opposed to a geometrical one with appeals to intuitions of space and time (Cantor 1883, 191–92). He envisioned it as a technique for focussing on pure, abstract arithmetical properties and concepts which divorced them from any sensory apprehension of the particular characteristics of the objects figuring in the sets, freed mathematics from psychologism, empiricism, Kantianism and insidious appeals to intuitions of space and time to engage in strictly arithmetical forms of concept formation (ex. Cantor 1883, 191–92; Cantor 1885; Cantor 1887/88, 381 n. 1; Eccarius 1985, 19–20; Couturat 1896, 325–41).

In this, Cantor was in step with Weierstrass's aim to arithmetize analysis (Jourdain 1908–14; Kline 1972). And herein lies a contradiction in Husserl's analyses. For Husserl too shared those particular aims and long sections of "On the Concept of Number" and the *Philosophy of Arithmetic* were devoted to discussions aimed at obtaining pure arithmetical concepts by detaching the concept of number from any spatio-temporal intuitions and so also from any taint of Kantianism in keeping with the goals Weierstrass had set (Husserl 1887, Section 2; Husserl 1891, chapter 2).

Husserl did not initially see Brentano's empirical psychology as empirical and psychological in a pernicious sense and continued to resort to psychological analyses in the *Philosophy of Arithmetic*. However, the confidence he expressed in "On the Concept of Number" eroded quickly and the enthusiastic espousal of psychologism of "On the Concept of Number" does not even appear in the later work—a sign that change was on the horizon.

There were respects, Husserl would confess in the foreword to the *Logical Investigations*, in which psychological foundations had never come to satisfy him. The psychological analyses of his earliest work on the foundations of arithmetic, he explained, left him deeply dissatisfied and he "became more and more disquieted by doubts of principle, as to how to reconcile the objectivity of mathematics, and of all science in general, with a psychological foundation for logic." His whole method by which he had hoped to illuminate mathematics through psychological analyses became shaken and he felt himself "more and more pushed towards general critical reflections on the essence of logic, and on the relationship, in particular, between the subjectivity of knowing and the objectivity of the content known" (Husserl 1900/01, 42).

5. Actual Consciousness and Pure Logic

In the foreword to the *Logical Investigations* Husserl also tells something of the evolution of his thought during his time in Halle. After working for many years to bring philosophical clarity to pure mathematics, he had found himself up against difficulties connected with developments in mathematics which defied his efforts at logical clarification. The problems were so compelling as to make him set aside his investigations into philosophico-mathematical matters until he had "succeeded in reaching a certain clearness on the basic questions of epistemology and in the critical understanding of logic as a science" (Husserl 1900/01, 41–43).

Husserl wrote of having felt "tormented by those incredibly strange realms: the world of the purely logical and the world of actual consciousness" which he saw opening up all around him while he was laboring to understand the logic of mathematical thought and calculation. He believed that the two spheres had "to interrelate and form an intrinsic unity," but had no idea as to how to bring them together (Husserl 1994, 490–91; Husserl 1913, 20–22).

Into the strange world of pure logic whose interaction with the world of consciousness puzzled him, Husserl placed "all of the pure 'analytical' doctrines of mathematics (arithmetic, number theory, algebra, etc.)" (Husserl 1913, 28), the pure theory of cardinal numbers, the pure theory of ordinal numbers, the traditional syllogistic and the pure mathematical theory of probability. He also put Cantorian sets, "the *Mannigfaltigkeitslehre* in the broadest sense" into the category of pure logic (Husserl 1994, 250; Husserl 1913, 28).

Now the new numbers and countless infinities the emancipated Cantor was producing proved counter-intuitive and paradoxical enough to challenge accepted logical assumptions (Hilbert 1925, 375). And Cantor himself was the first to admit that his theories were strange. He saw himself as exploring terra incognita and referred to his new number classes as "strange things," writing that he had been logically compelled to introduce them almost against his will, but did not see how he might proceed further with set theory and function theory without them (Cantor 1883, 165, 175; Cantor 1991, 95).

So the nature of the problems which beset Husserl becomes clearer when one takes into account that Cantor was busy exploring, mapping and inventing the strange world of transfinite sets at the very time Husserl broke out in crisis, making it easy to believe that the crisis was aggravated, if not actually induced by Cantor's bold experiments with mathematics and epistemology.

In this respect the section on infinite sets in the *Philosophy of Arithmetic* (Husserl 1891, 246–50) yields insight into the nature of the problems tormenting Husserl as he tried to complete the book. There was, Husserl noted there, a particularly odd way of extending the original concept of multiplicity or quantity, which by its very nature reaches beyond the necessary bounds of human cognition, and so wins for itself an essentially new content. These were the infinite sets, multiplicities or collections. "Infinite," he wrote, calling to mind some of Cantor's main preoccupations, "are the extensions of most general concepts. Infinite is the sequence of numbers extended by symbolic means, infinite is the set of points on a line, and in general that of the limits of a continuum" (Husserl 1891, 246–47).

Signs, symbolic presentation à la Brentano, Husserl explained, might aid the mind in reasoning in regions of thought beyond what could be known through direct cognitive processes like perception or intuition. The repeated application of operations permitting the collecting together of a multitude of objects one after the other into a set could take the place of the direct cognitive grasp of sets with hundreds, thousands or millions of members, and he was satisfied that this was a way of actually representing collections in an ideal sense and essentially unproblematic from a logical point of view (Husserl 1891, 246).

However, Husserl considered, this became impossible in the case of infinite totalities, multiplicities or collections since the very principle by which they are formed or symbolized itself immediately makes collecting of all their members together one by one a logical impossibility. By no extension of our cognitive faculties could we conceivably cognitively grasp or even successively collect such sets, he points out. So the logical problems connected with infinite sets were of a completely different order. With them we had reached the limits of idealization (Husserl 1891, 246–47).

With infinite sets, he reasoned, it is always a case of the symbolic presentation of a never-ending process of concept formation. We are in possession of a clear principle by which we can transform any already formed concept of a certain given species into a new concept which is plainly distinct from the previous one. And we can do this over and over in such a way as to be certain a priori of never coming back to the original concept and to previously generated concepts. Repeated application of this process yields successive presentations of continually expanding sets, and if the generating principle is really determinate, then it is determined a priori whether or not any given object can belong to the concept of the expanding set of concepts (Husserl 1891, 247).

In the case of the concept of the infinite set of numbers we begin with a direct presentation, he explained. The other natural numbers can be reached by repeated additions of one, and nothing prevents us from advancing indefinitely in this way to new cardinals; we have a method for adding one to a previously given number, an operation that necessarily generates one new number after another without return and limitation, each new number being determined by the process (Husserl 1891, 247–48).

It is easy, Husserl continues, to see why mathematicians have tried to transpose the concept of quantity to such constructions, which are, however, of an essentially different logical nature. In the usual cases, the process by which the sets were generated was finite, there was always a last stage, it was sometimes possible actually to bring the process to a halt, and also to construct the corresponding set. However, this is quite absurd in the case of infinite sets. The process used to generate them is nonterminating, and the idea of a last stage, of a last member of the set is meaningless. And this constitutes an essential logical difference (Husserl 1891, 248).

Nonetheless, Husserl points out, despite the absurdity of the idea, the analogies pertaining foster a tendency to transpose the idea of constructing a corresponding collection for infinite sets, thereby creating what he calls a kind of "imaginary" concept whose antilogical nature is harmless in everyday contexts precisely because its inherent contradictoriness is never obvious in life. This is, he explains, the case when "All S" is treated as a closed set (Husserl 1891, 249).

However, Husserl warns, the situation changes when this imaginary construct is actually carried over into reasoning and influences judgments. It is clear, he concludes, that from a strictly logical point of view we must not ascribe anything more to the concept of infinite sets than is actually logically permissible, and above all not the absurd idea of constructing the actual set (Husserl 1891, 249).

As for those other "strange worlds," the strange worlds of pure consciousness, the naive epistemological theorizing which Cantor was so earnestly engaging in while Husserl was grappling with analogous questions could easily have impressed upon him an urgent need to develop the more sophisticated logical and epistemological tools needed to attain a deeper, clearer understanding of how the human mind interacted with the world of numbers. For the "world" of consciousness and the "world" of pure mathematics mingle together in Cantor's work in confusing and frustrating ways which cry out for clarification.

For instance, one might fairly wonder in just what way the cardinal number belonging to a set is an abstract image in our intellect (Cantor 1887/88, 416), exactly how the act of abstraction awakened the number concepts in Cantor's mind (Cantor 1887/88, 418 n. 1), or how concepts and ideas are formed through inner induction like something already lying within us which is merely awakened and brought to consciousness (Cantor 1883, 207 n. 6). Frege was perfectly justified in qualifying Cantor's appeals to direct inner intuition (Cantor 1883, 168, 170, 201) as "rather mysterious" (Frege 1884, §86).

So given the particular nature of Cantor's experiments it is not surprising to find Husserl asking in those years how rational insight was possible in science (Husserl 1994, 167), how the mathematical in itself as given in the medium of the psychical could be valid (Husserl 1913, 35), how logicians penetrated an objective realm entirely different from themselves (Husserl 1913, 222), how objective, mathematical and logical relations constituted themselves in subjectivity (Husserl 1913, 35), how symbolic thinking was possible (Husserl 1913, 35), how abandoning oneself completely to thought that is merely symbolic and removed from intuition could lead to empirically true results (Husserl 1994,167), or how mechanical operations with mere written characters could vastly expand our actual knowledge concerning number concepts (Husserl 1994, 50). These are all questions Cantor's theories raise.

Facing only what he variously called riddles, tensions, puzzles, and mysteries about consciousness and pure logic, and seeing all around him only unclear, undeveloped, ambiguous, confused ideas, but no "full and truly satisfactory understanding of symbolic thought or of any logical process" (Husserl 1994, 169), Husserl set out on his own to solve the problems his investigations into the foundations of mathematics had raised, concluding after a "decade of solitary, arduous labor" that the puzzles surrounding the being in itself of the ideal sphere and its relationship to consciousness would only be solved through the pure phenomenological elucidation of knowledge he developed (Husserl 1994, 251; Husserl 1900/01, 42–43, 223–24; Husserl 1913).

6. Metaphysical Idealism

To the chagrin of his contemporaries, as we saw in the preceding chapter, (Cantor 1991, 110, 113, 118, 178, 227, 241) Cantor persisted in clothing his theories about numbers in a metaphysical garb. In the *Mannigfaltigkeitslehre* he emphasized that the idealist foundations of his reflections were essentially in agreement with the basic principles of Platonism according to which only conceptual knowledge in Plato's sense afforded true knowledge (Cantor 1883, 181, 206 n. 6). His own idealism being related to the Aristotelian-Platonic kind, he wrote in an 1888 letter, he was just as much a realist as an idealist (Cantor 1991, 323).

By "manifold" or a "set" he explained in the *Mannigfaltigkeitslehre*, he was defining something related to the Platonic *eidos* or *idea*, as also to what Plato called a *mikton* (Cantor 1883, 204 n. 1). "I conceive of numbers," he informed Giuseppe Peano, "as 'forms' or 'species' (general concepts) of sets. In essentials this is the conception of the ancient geometry of Plato, Aristotle, Euclid etc." (Cantor 1991, 365). To Charles Hermite he wrote that "the whole numbers both separately and in their actual infinite totality exist in that highest kind of reality as eternal ideas in the Divine Intellect" (cited Hallett 1984, 149).

For Cantor the realm of the transfinite was "a rich, ever growing field of ideal research" (Cantor 1887/88, 406). He considered his transfinite numbers to be but a special form of Plato's *arithmoi noetoi* or *eidetikoi*, which he thought probably even fully coincided with the whole real numbers (Cantor 1884, 84; Cantor 1887/88, 420). And he considered that his technique for abstracting numbers from reality provided the only possible foundations for that Platonic conception of numbers (Cantor 1991, 363, 365; Cantor 1887/88, 380, 411). Abstracting from both the characteristics of the elements of the set and the order in which they are given, one obtained the cardinal numbers; abstracting only from the characteristics of the elements and leaving their order intact, one obtained the ideal numbers or *eidetikoi* (Cantor 1887/88, 379–80; Cantor 1883, 180–81). His talk of awakening and bringing to consciousness the knowledge, concepts and numbers slumbering in us (Cantor 1883, 207 nn. 6, 7, 8; Cantor 1887/88, 418 n. 1) is, of course, an unmistakable allusion to Plato's theory of recollection and Socratic theories of concept formation (ex. the *Meno* 81C–86C; *Phaedo* 72E, 75E–76A).

A good measure of the freedom Cantor felt he possessed as a mathematician in fact derived from his distinguishing between an empirical treatment of numbers and Plato's pure, ideal *arithmoi eidetikoi* which by their very nature are detached from things perceptible by the senses. In 1890 he wrote to Veronese that contradictions he had found in Cantor's theories were but apparent and that one must distinguish between the numbers which we are able to grasp in our limited ways and "numbers as they are *in and for themselves, and in and for the Absolute intelligence*" (Cantor 1991, 326; see also 267, 282).

In sharp contrast to this, Husserl came to Halle free of Platonic idealism. For Franz Brentano inculcated in his students a model of philosophy based on the natural sciences and taught them to despise metaphysical idealism. So completely under Brentano's influence in the beginning, Husserl was, initially quite disinclined to traffic in the kind of idealism that so pervaded Cantor's writings (Husserl 1919, 344–45) and any such considerations are conspicuously absent from both "On the Concept of Number" and *Philosophy of Arithmetic*.

Notwithstanding, Husserl left Halle a committed Platonic idealist persuaded that pure mathematics was a strictly self-contained system of doctrines to be cultivated by using methods that are essentially different from those of natural science (Husserl 1913, 29). "The empirical sciences—natural sciences," he wrote to Brentano in 1905, "—are sciences of 'matters of fact'. . . . Pure Mathematics, the whole sphere of the genuine Apriori in general, is free of all matter of fact suppositions. . . . We stand not within the realm of nature, but within that of Ideas, not within the realm of empirical . . . generalities, but within that of the ideal, apodictic, general system of laws, not within the realm of causality, but within that of rationality. . . . Pure logical, mathematical laws are laws of essence. . . ." (Husserl 1905, 37).

Phenomenology would be an "eidetic" discipline. The "whole approach whereby the overcoming of psychologism is phenomenologically accomplished," Husserl explained, "shows that what . . . was given as analyses of immanent consciousness must be considered as a pure a *priori* analysis of essence" (Husserl 1913, 42). He came to consider idealistic systems to be of "the highest value," that entirely new and totally radical dimensions of philosophical problems were illuminated in them, and "the ultimate and highest goals of philosophy are opened up only when the philosophical method which these particular systems require is clarified and developed" (Husserl 1919, 345). Every possible effort, he wrote, had been made in the *Logical Investigations* "to dispose the reader to the recognition of this ideal sphere of being and knowledge . . . to side with 'the ideal in this truly Platonistic sense', 'to declare oneself for idealism' with the author" (Husserl, 1913, 20).

While Cantor's experiments very likely acted to pry Husserl away from psychologism and to steer him in the direction of Platonic idealism, Husserl always maintained that it was Hermann Lotze's work, we saw in the previous chapter, which was responsible for the fully conscious and radical turn from psychologism and the Platonism that came with it. Lotze's interpretation of Plato's doctrine of Ideas, Husserl said, provided him with his first major insight, became a determining factor in all his further studies, and gave him the idea to transfer all of the mathematical and a major part of the traditionally logical into the realm of the ideal. Through Lotze's work Husserl also found the key to understanding what he called Bolzano's "curious conceptions," which had originally seemed naive and unintelligible to him, but in which he was to discover a complete plan of a pure logic and an initial attempt to provide a unified presentation of the domain of pure ideal doctrines (Husserl 1994, 201–2; Husserl 1913, 36–38, 46–49).

7. Imaginary Numbers

In the foreword to the *Logical Investigations* Husserl specifically alluded to having been troubled by the theory of manifolds, the *Mannigfaltigkeitslehre*, with its expansion into special forms of numbers and extensions. The fact, he explained, that one could obviously generalize, produce variations of formal arithmetic which could lead outside the quantitative domain without essentially altering formal arithmetic's theoretical nature and calculational methods had brought him to realize that there was more to the mathematical or formal sciences, or the mathematical method of calculation than would ever be captured in purely quantitative analyses (Husserl 1900/01, 41–42; Husserl 1913, 35).

And he always said that it was particularly difficulties he experienced in trying to answer questions raised by "imaginary" numbers that arose while trying to complete the *Philosophy of Arithmetic* which had marked the turning point in his thinking. He used the term "imaginary" in a very broad sense to cover negative numbers, negative square roots, fractions and irrational numbers, and so on (Husserl 1994, 13–16; Husserl 1913, 33; Husserl 1970, 430–51; Husserl 1929, §31; Hill 1991, 81–86). As we saw above he called infinite sets "imaginary concepts" (Husserl 1891, 249). [8]

In "On the Concept of Number" Husserl had maintained that all the more complicated and artificial forms of numbers had their origin and basis in the concept of positive whole numbers and their interrelations and were derivable from them in a strictly logical way. However, hints of a change of mind were already to be found in the very first pages of the *Philosophy of Arithmetic*, where Weierstrass's thesis that cardinal numbers form the sole basis for arithmetic is never endorsed in the confident way it was in "On the Concept of Number" (Husserl 1887, 94–95).

Conspicuously apparent in the later work is Husserl's tergiversation regarding Weierstrass's teaching. In the *Philosophy of Arithmetic* we find Husserl explaining that he will provisionally use it as a springboard for his own analyses. However he warns readers that, although cardinal numbers in a certain way seem to be the basic numbers involved in arithmetic because the signs for them figure in expressions for positive, negative, rational, irrational, real, imaginary, alternative, ideal numbers, quaternions etc. the analyses of the second volume would perhaps show that thesis to be untenable (Husserl 1891, 5–6). In the April 1891 preface to the book he went further to state that the analyses of the second volume would actually show that in no way does a single kind of concept, whether that of cardinal or ordinal numbers, form the basis of general arithmetic (Husserl 1891, VIII).

Finally, before the *Philosophy of Arithmetic* had even made its way into print, Husserl wrote to Carl Stumpf that the opinion by which had been guided in writing "On the Concept of Number" that the concept of cardinal number formed the foundation of general arithmetic had soon proved to be false, that through no manner of cunning could negative, rational, irrational, and the various kinds of complex numbers be derived from the concept of the cardinal number (Husserl 1994, 13).

By 1891 Husserl had become keenly aware that "a utilization of symbols for scientific purposes, and with scientific success, is still not therefore a *logical* utilization" (Husserl 1994, 48). As he explained:

> General arithmetic, with its negative, irrational, and imaginary (impossible) numbers, was invented and applied for centuries before it was understood. Concerning the signification of these numbers the most contradictory and incredible theories have been held; but that has not hindered their use. One could quite

certainly convince oneself of the correctness of any sentence deduced by means of them through an easy verification. And after innumerable experiences of this sort, one naturally comes to trust in the unrestricted applicability of these modes of procedure, expanding and refining them more and more—all without the slightest insight into the *logic*. . . . (Husserl 1994, 48)

Alluding to "the endless controversies over negative and imaginary numbers, over the infinitely small and the infinitely large, over the paradoxes of divergent series, and so on," Husserl lamented the mental energy "wasted on this route, more governed by chance than logic" (Husserl 1994, 49). Arithmetic, he contended, would have made quicker and surer progress had there been clarity about the logic of its methods upon their development instead. He believed that insight into the logical character of arithmetic would have to play a decisive role in the future development of the field (Husserl 1994, 48–51).

A lecture Husserl gave before the Göttingen Mathematical Society in 1901 (Husserl 1970, 430–51) affords some insight into the steps in his reasoning which led to the discovery of the solution he would ultimately deem satisfactory. Questions regarding imaginary numbers, he explained to his listeners there, had come up in mathematical contexts in which formalization yielded constructions which arithmetically speaking were nonsense but which, astonishingly, could nevertheless be used in calculations. It became apparent that when formal reasoning was carried out mechanically as if these symbols had meaning, if the ordinary rules were observed, and the results did not contain any imaginary components, then these symbols might be legitimately used. And this could be empirically verified (Husserl 1970, 432).

However, this fact raised significant questions about the consistency of arithmetic and about how one was to account for the achievements of certain purely symbolic procedures of mathematics despite the use of apparently nonsensical combinations of symbols. Faced with these problems, Husserl said that his main questions were: (1) Under what conditions can one freely operate within a formally defined deductive system with concepts which according to the definition of the system are imaginary and have no real meaning? (2) When can one be sure of the validity of one's reasoning, that the conclusions arrived at have been correctly derived from the axioms one has, when one has appealed imaginary concepts? And (3) To what extent is it permissible to enlarge a well-defined deductive system to make a new one that contains the old one as a part? (Husserl 1970, 433; Husserl 1929, §31).

Again, the work on numbers Cantor was doing in the late 1880s makes such questioning on Husserl's part completely understandable. Those were years during which Cantor was particularly engaged in what Grattan-Guinness has called "rather strange work on theory of numbers" (Grattan-Guinness 1971, 369), producing what Dauben has called "dinosaurs of his mental creation, fantastic creatures whose design was interesting, over-

whelming, but impractical to the demands of mathematicians in general"
(Dauben 1979, 158–59).

Husserl found the answers he was looking for in his own theory of complete manifolds (Hill 1995), the *definite Mannigfaltigkeiten* of the "Prolegomena to Pure Logic" in the *Logical Investigations* (§§69–70), in which
believed he had discovered "the key to the only possible solution of the
problem" as to "how in the field of numbers impossible (essenceless) concepts can be methodically treated like real ones" (Husserl 1900/01, 242).

8. Sets and Manifolds

In "On the Concept of Number" and the *Philosophy of Arithmetic* Husserl
made set theory the basis of mathematics. Using the terms 'multiplicity' and
'set' interchangeably to neutralize any differences in meaning among the
terms, and citing Euclid's classical definition of the concept of number as "a
multiplicity of units," he began the analyses of the *Philosophy of Arithmetic* by
affirming that "the analysis of the concept of number presupposes the concept of multiplicity" (Husserl 1891, 8 and note; also Husserl 1887, 96).

The most primitive concepts involved, he explained in the work, are the
general concepts of set and number which are grounded in the concrete sets
of specific objects of any kind whatsoever and to which particular numbers are
assigned. There could be no doubt, he maintained, that multiplicities or sets
of determinate objects were the concrete phenomena which formed the basis
for the abstraction of the concepts in question. He supposed that everyone
knew what the terms 'multiplicity' and 'set' meant, that the concept itself was
well-defined and that there was no doubt as to its extension, which might be
taken as given (Husserl 1891, 9 10,13; Husserl 1887, 96–97, 111).

In the *Philosophy of Arithmetic*, number was the general form of a set
under which the set of objects *a*, *b*, *c* fell. To obtain the concept of number
of a concrete set of like objects, for example *A*, *A*, and *A*, one abstracted
from the particular characteristics of the individual contents collected, only
considering and retaining each one of them insofar as it was a something or a
one, and thus obtaining the general form of the set belonging to the set in
question: one and one, etc. and . . . and one, to which a number name was
assigned (Husserl 1891, 88, 165–66; also Husserl 1887, 116–17).

However, Husserl confessed to having been disturbed, and even tormented, by doubts about sets right from the very beginning (Husserl 1913,
35). In the *Logical Investigations*, we have seen, he specifically mentioned
having been troubled by the theory of manifolds (Husserl 1900/01, 41–42;
Husserl 1913, 35) and he specifically put Cantorian sets, "the *Mannigfaltigkeitslehre* in the broadest sense" into the category of pure logic which
was a source of torment to him (Husserl 1994, 250; Husserl 1913, 28).

The concept of collection in Brentano's sense, he explained in 1913, was to arise through reflection on the concept of collecting. Sets, he had reasoned, arose out of collective combination, in being conceived as one. This combining process involved when objects are brought together to make a whole only consisted in that one thought of them "together" and was obviously not grounded in the content of the disparate items collected into the set. It could not be physical, so it must be psychological, a unique kind of mental act connecting the contents of a whole. But then, he began asking, was "the concept of number not something basically different from the concept of collecting which is all that can result from the reflection on acts?" (Husserl 1913, 34–35; re. Husserl 1891, 14–15; Husserl 1887, 97–98).

Now it is not surprising to find that Husserl had doubts about sets when one considers that he was on hand as Cantor began discovering the antinomies of set theory (Cantor 1991, 387–465; Dauben 1979, 240–70). Others certainly had their logical assumptions shaken upon contact with transfinite set theory. David Hilbert has described the reaction to it as having been dramatic and violent (Hilbert 1925, 375). And it was in studying Cantor's proof that there can be no greatest cardinal number that Bertrand Russell was led to the famous contradiction of the set of all sets that are not members of themselves (Russell 1903, 100, 344, 500; Russell 1959, 58–61; Frege 1980, 133–34, 147; Grattan-Guinness 1978; 1980) which Hilbert has described as having had a "downright catastrophic effect in the world of mathematics," making Dedekind and Frege abandon "their standpoint and quit the field" (Hilbert 1925, 375).

That first-hand experience of inconsistent sets may have actually permanently inoculated Husserl against any recourse to sets or classes. In the early 1890s he was already expressing grave doubts about extensional logic, by which he meant a calculus of classes (Husserl 1994, ex. p. 121). His chief target then was Ernst Schröder (Husserl 1994, 52–91, 421–41). But his antipathy is evident in several articles of the period (ex. Husserl 1994, 92–114, 115–20, 121–30, 135–38, 443–51) in which he was intent upon laying bare "the follies of extensional logic" (Husserl 1994,199), which he would replace by a calculus of conceptual objects. In these texts he seeks to show "that the *total* formal basis upon which the class calculus rests is valid for the relationships between conceptual objects," and that one could solve logical problems without "the detour through classes" (Husserl 1994, 109), which he considered to be "totally superfluous" (Husserl 1994, 123). Late in his life Husserl was still denouncing extensional logic as naive, risky, doubtful and the source of many a contradiction requiring every kind of artful device to make it safe for use in reasoning (Husserl 1929, §§23, 26).

While Husserl used the various terms for set interchangeably in the late 1880s, he rarely resorted to use of the more Cantorian term 'manifold',

'*Mannigfaltigkeit*'. In the 1890s, though, he did began studying manifolds, but in the Riemannian, rather than the Cantorian sense (Husserl 1983, 92–106, 408–11; Husserl 1970, 475–78, 493–500). Those investigations culminated in the above-mentioned theory of complete *Mannigfaltigkeiten* expounded in the *Logical Investigations* (Husserl 1900/01, §70). He always considered that theory to represent the highest task of formal logic and the formulation of it in the *Prolegomena* to have been definitive (Husserl 1929, §28).

According to that theory, logic defying creations like those flowing from Cantor's pen could be shown to have redeeming scientific value when they were integrated into a whole which was greater than the sum of its parts, where the anti-logical character of imaginary constructions would become neutralized within the context of a consistent, complete deductive system (Husserl 1970, 441).

This solution was in keeping with Cantor's conviction that mathematical concepts need only be both non-self-contradictory and stand in systematically determined relations established through definition from the previously formed, proven concepts one already has on hand and that mathematicians are only obliged to provide definitions of the new numbers determinate in this way and, if need be, to establish this relationship to the older numbers (Cantor 1883, 182). But Husserl's *Mannigfaltigkeiten* were Riemannian more than they were Cantorian.

9. Conclusion

By showing how Husserl's and Cantor's ideas overlapped and crisscrossed during Husserl's time in Halle, I have tried to shed light on a complex period in the development of Husserl's thought. Four stages in that interaction are discernible from this study.

Initially, Husserl's ideas about mathematics and philosophy, sets, abstraction, and the arithmetization of analysis fit in with Cantor's ideas. That may be attributable to the influence of Weierstrass, to Cantor's position on Husserl's *Habilitationskommittee*, and to the fact that many of Cantor's theories might seem amenable to clarification through Brentano's teachings. For example: Brentano's collective unification might be what Cantor meant by the "special relationship" binding elements of a set; the objects of thought and intuition of Cantor's sets might be Brentano's intentional objects; Cantor's technique of extracting numbers from reality through abstraction might be a psychological process; and with Brentano's positivism Husserl might have been able to "banish all metaphysical fog and all mysticism" from mathematical investigations into numbers and manifolds like Cantor's (Husserl 1900/01, 242).

There was then a second stage during which Cantor's work must have been instrumental in unseating Husserl from his earliest convictions by raising hard questions about imaginary numbers, sets, consciousness and pure logic, idealism, etc., something not so astonishing when one considers that Husserl suffered the direct, early impact of theories about transfinite sets that played a role in rattling the logical assumptions of Bertrand Russell, Gottlob Frege, Richard Dedekind, and many others who worked on the foundations of arithmetic.

In a third stage Husserl drew near to some ideas Cantor was rather alone in espousing (metaphysical idealism and the renunciation of psychologism, empiricism, and naturalism) and turned away from other of his key ideas (his theory of sets, the arithmetization of analysis). Husserl then felt obliged to set out on his own, turning to Lotze's more sophisticated ideas about Platonic Ideas and to Bolzano's more sophisticated work on pure logic for guidance.

A fourth stage would consist of the assimilation of certain of Cantor's ideas into Husserl's phenomenology and extends far beyond the compass of this study. Here it would be a matter of studying the relationship between Cantor's theories and, for example, Husserl's *Mannigfaltigkeitslehre*, his theories about eidetic intuition, the phenomenological reductions, noemata, horizons, infinity, whole and part, formal logic, and of how Husserl's philosophy of logic and mathematics relates to Frege's, Russell's, or Gödel's ideas, in order to arrive at a fairer assessment of Husserl's place in the history of philosophy, logic, set theory, and the foundations of mathematics than has been possible, dreamt of, or even thought desirable up until now.

NOTES

* I wish to thank Emiko Ima and William Gallagher for their help. Emi is a particularly diligent worker. I have occasion to thank Ivor Grattan-Guinness in the notes, but that does not by any means account for all the friendly assistance he has provided. So I wish to thank him again here.

1. Exceptions: Fraenkel 1930, pp. 221, 253 n., 257; Schmit 1981, pp. 24, 40–48, 70–72, 77, 94, 124, 131–37; Picker 1962, pp. 266–73, 290, 302, 309–11, 328–29, 345–46; Rosado Haddock 1973, pp. 140–43; Hill 1991, pp. 2, 17, 20, 91; Grattan-Guinness, 1980, pp. 81–82, n. 56; Cavaillès 1962, p. 180; Illemann 1932, p. 50.
2. Primary sources are: M. Husserl 1988, p. 114 and letters of reference Cantor wrote for Husserl published in Cantor 1991, pp. 321, 373–74, 379–80, 423-24, and W. Purkert and H. Ilgauds 1991, pp. 206–7, which I must thank Ivor Grattan-Guinness for bringing to my attention. See also L. Eley's, "Editor's Introduction" to Husserl 1970, pp. XXIII–XXVIII; Husserl 1913, p. 37 and notes; Schuhmann 1977, pp. 19, 22.
3. I must thank Ivor Grattan-Guinness for sending this to me which he obtained from the Mittag-Leffler Institute in Sweden. The text is translated in Hallett 1984, pp. 6–7.

4. See Jourdain 1908–1914; Fraenkel 1930, pp. 193, 194, 199, 200, 208, 232–33, 236, 247, 251; Russell 1903, pp. 115, 166, 219, 285, 305, 314–15, 497, 519; Dauben 1979, pp. 151, 170–71, 176–77, 220–28, Hallett 1984, pp. 51–85, 119–64; Grattan-Guinness 1980, pp. 68–71.
5. Becker 1930, pp. 40–41; Schuhmann 1977, pp. 6–9, 11, 345. See also Picker 1962, pp. 266–70, 289–91, 302; Schmit 1981, pp. 24, 34–36; Osborn 1934, pp. 10–17, 37; Mahnke 1966, pp. 75–76. Miller 1982, pp. 1–13. Willard 1980, pp. 46–64, and 1984, pp. 3, 21, 110, 130; Kusch 1989, pp. 12–15.
6. Personally examined at the Husserl Archives and Library in Leuven, Belgium.
7. At this time in his career Husserl used the terms 'set', 'multiplicity', 'totality', 'manifold', etc. interchangeably. See Husserl 1891, p. 8 and note and Husserl 1887, p. 96.
8. It is worthwhile to note here that Cantor maintained that the transfinite numbers were "in a certain sense new irrationals," that both the transfinite numbers and the finite irrational numbers were definite, delineated forms of or modifications of the actual infinite (Cantor 1887/88, pp. 395–96; Cantor 1991, p. 182; Dauben 1979, p. 128).

REFERENCES

Becker, O. 1930. "The Philosophy of E. Husserl." *The Phenomenology of Husserl: Selected Critical Readings.* Ed. R. O. Elveton. Chicago: Quadrangle Books 1970, 40–72. Translation of "Die Philosophie Edmund Husserls."
Cantor, G. 1883. *Grundlagen einer allgemeinen Mannigfaltigkeitslehre. Ein mathematisch-philosophischer Versuch in der Lehre des Unendlichen.* Leipzig: Teubner. Cited as appears in Cantor 1932, 165–246.
———. 1884. "Principien einer theorie der Ordnungstypen" (dated November 6, 1884).
First published by I. Grattan-Guinness in *Acta Mathematica* 124 (1970): 65–107.
———. 1885. "Rezension von Freges *Grundlagen.*" *Deutsche Literaturzeitung* 6: 728–29. Cited as appears in Cantor 1932, 440–41.
———. 1887/88. "Mitteilungen zur Lehre vom Transfiniten." *Zeitschrift für Philosophie und philosophische Kritik* 91 (1887): 81–125; 92 (1888): 240–65, cited as appears in Cantor 1932, 378–439. Also published as Cantor 1890.
———. 1890. *Gesammelte Abhandlungen zur Lehre vom Transfiniten.* Halle: Pfeffer. See Cantor 1887/88.
———. 1895. *Contributions to the Founding of the Theory of Transfinite Numbers.* New York: Dover, 1955 (1915). Translation of "Beiträge zur Begründung der transfiniten Mengenlehre."
———. 1932. *Gesammelte Abhandlungen.* Ed. E. Zermelo, Berlin: Springer.
———. 1991. *Georg Cantor Briefe.* Ed. H. Meschkowski and W. Nilson. New York: Springer.
Cavaillès, J. 1962. *Philosophie Mathématique.* Paris: Hermann.
Couturat, L. 1896. *De l'infini mathématique.* Paris: Blanchard, 1973.
Dauben, J. 1979. *Georg Cantor, His Mathematics and Philosophy of the Infinite.* Princeton: Princeton University Press.
Eccarius, W. 1985. "Georg Cantor und Kurd Lasswitz: Briefe zur Philosophie des Unendlichen." *NTM Schriftenr. Gesch. Naturwiss., Technik, Med.* 22: 7–28.

Fraenkel, A. 1930. "Georg Cantor." *Jahresbericht der deutschen Mathematiker Vereinigung* 39: 189–286. Abridged in Cantor 1932, 452–83.

Frege, G. 1884. *Foundations of Arithmetic.* 2nd ed. rev. Oxford: Blackwell, 1986.

———. 1892. "Review of Georg Cantor, *Zur Lehre vom Transfiniten. Gesammelte Abhandlungen aus der Zeitschrift für Philosophie und philosophische Kritik.*" *Collected Papers on Mathematics, Logic and Philosophy.* Oxford: Blackwell, 1984, 178–81.

———. 1979. *Posthumous Writings.* Oxford: Blackwell.

———. 1980. *Philosophical and Mathematical Correspondence.* Oxford: Blackwell.

Gerlach, H., and H. Sepp, eds. 1994. *Husserl in Halle* Bern: P. Lang

Grattan-Guinness, I. 1971. "Towards a Biography of Georg Cantor." *Annals of Science* 27 (4): 345–91.

———. 1978. "How Russell Discovered His Paradox." *Historia Mathematica* 5: 127–37.

———. 1980. "Georg Cantor's Influence on Bertrand Russell." *History and Philosophy of Logic* 1 (1980): 61–93.

———. 1982. "Psychology in the Foundations of Logic and Mathematics: The Cases of Boole, Cantor and Brouwer." *History and Philosophy of Logic* 3: 33–53.

———. 1996. "Numbers, Magnitudes, Ratios, and Proportions in Euclid's *Elements*: How Did He Handle Them?" *Historia Mathematica* 23: 355–75.

Hallett, M. 1984. *Cantorian Set Theory and Limitation of Size.* Oxford: Clarendon.

Hilbert, D. 1925. "On the Infinite." In *From Frege to Gödel*, van Heijenoort, ed., 367–92.

Hill, C.O. 1991. *Word and Object in Husserl, Frege and Russell, the Roots of Twentieth Century Philosophy.* Athens, OH: Ohio University Press.

———. 1998. "From Empirical Psychology to Phenomenology: Husserl on the Brentano Puzzle." *The Brentano Puzzle.* Ed. R. Poli. Aldershot, UK: Ashgate, 151–67.

Husserl, E. 1887. "On the Concept of Number." *Husserl: Shorter Works.* Ed. P. McCormick and F. Elliston. Notre Dame: University of Notre Dame Press, 1981, 92–120. Translation of "Über den Begriff der Zahl," available in Husserl 1970, 289–339.

———. 1891. *Philosophie der Arithmetik.* Halle: Pfeffer. Also in Husserl 1970.

———. 1900/01. *Logical Investigations.* New York Humanities Press, 1970.

———. 1905. "Husserl an Brentano, 27.III.1905." *Briefwechsel, Die Brentanoschule.* Vol. I. Dordrecht: Kluwer, 1994.

———. 1913. *Introduction to the Logical Investigations.* The Hague: M. Nijhoff, 1975.

———. 1919. "Recollections of Franz Brentano." *Husserl: Shorter Works.* Ed. P. McCormick and F. Elliston. Notre Dame: University of Notre Dame Press, 1981. Translation of "Errinerungen an Franz Brentano."

———. 1929. *Formal and Transcendental Logic.* The Hague: M. Nijhoff, 1969.

———. 1970. *Philosophie der Arithmetik, mit ergänzenden Texten (1890–1901).* Husserliana, vol. XII. The Hague: M. Nijhoff. Introduction by L. Eley.

——— 1983. *Studien zur Arithmetik und Geometrie, Texte aus dem Nachlass (1886–1901).* Husserliana, vol. XXI. The Hague: M. Nijhoff.

———. 1994. *Early Writings in the Philosophy of Logic and Mathematics.* Dordrecht: Kluwer.

Husserl, M.1988. "Skizze eines Lebensbildes von E. Husserl." *Husserl Studies* 5: 105–25.

Illemann, W. 1932. *Husserls vorphänomenologische Philosophie.* Leipzig: Hirzel.

Jourdain, P. 1908–1914. "The Development of the Theory of Transfinite Numbers." *Archiv der Mathematik und Physik* 14 (1908/09): 289–311, 16 (1910): 21–43, 22 (1913–14): 1–21. Reprinted in Jourdain's *Selected Essays on the History of Set Theory and Logics*, ed. I. Grattan-Guinness, Bologna: CLUEB, 1991, 33–99.

Kline, M. 1972. *Mathematical Thought from Ancient to Modern Times*. Vol. 3. New York: Oxford University Press.

Kreiser, L. 1979. "W. Wundts Auffassung der Mathematik, Briefe von G. Cantor an W. Wundt." *Wissenschaftliche Zeitschrift, Karl-Marx Universität, Ges. u. Sprachwissenschaft* 28: 197–206.

Kusch, M. 1989. *Language as Calculus vs. Language as Universal Medium*. Dordrecht: Kluwer.

Lotze, H. 1880. *Logic*. Oxford: Clarendon Press. Translation of his *Logik* (1888).

Mahnke, D. 1966. "From Hilbert to Husserl: First Introduction to Phenomenology, especially that of formal mathematics." *Studies in the History and Philosophy of Science* 8: 71–84. Translation of his "Von Hilbert zu Husserl: Erste Einführung in die Phänomenologie, besonders der formalen Mathematik."

Miller, J. 1982. *Numbers in Presence and Absence*. The Hague: M. Nijhoff.

Osborn, A. 1934. *Edmund Husserl and his Logical Investigations*. New York: International Press.

Picker, B. 1962. "Die Bedeutung der Mathematik für die Philosophie Edmund Husserls." *Philosophia Naturalis* 7: 266–355.

Purkert, W., and H. Ilgauds. 1991. *Georg Cantor 1841–1918*. Basel: Birkhäuser.

Rosado Haddock, G. 1973. "Edmund Husserls Philosophie der Logik und Mathematik im Lichte der gegenwärtigen Logik und Grundlagenforschung." Ph.D. dissertation, Rheinische Friedrich-Wilhelms-Universität, Bonn.

Russell, B. 1903. *Principles of Mathematics*. London: Norton.

———. 1917 (1986). "Mathematics and Metaphysicians." *Mysticism and Logic*. London: Allen and Unwin, 75–95.

———. 1959 (1975). *My Philosophical Development*. London: Unwin.

Schmit, R. 1981. *Husserls Philosophie der Mathematik. platonische und konstruktivische Momente in Husserls Mathematik Begriff*. Bonn: Bouvier.

Schuhmann, K. 1977. *Husserl-Chronik*. The Hague: M. Nijhoff.

van Heijenoort, J., ed. 1967. *From Frege to Gödel, A Source Book in Mathematical Logic, 1879–1931*. Cambridge, MA: Harvard University Press.

Willard, D. 1980. "Husserl on a Logic That Failed." *The Philosophical Review* 89 (1): 46–64.

———. 1984. *Logic and the Objectivity of Knowledge*. Athens, OH: Ohio University Press.

9

Claire Ortiz Hill

HUSSERL'S
MANNIGFALTIGKEITSLEHRE

Kurt Gödel has been reported to have been especially interested in Edmund Husserl's ideas about axiomatization (ex. Wang 1996, 168, 334). It is, in fact, not well known that Husserl developed a theory of axiomatic systems and manifolds that, from the beginning of his career to its end, he maintained constituted the highest level of formal logic. This was his *Mannigfaltigkeitslehre*, which he began developing in the early 1890s, while in the throes of the intellectual struggle that caused him to abandon some of the most cherished ideas of his earliest mentors Karl Weierstrass, Franz Brentano, and Georg Cantor.

Husserl accorded the theory a preeminent role in the *Prolegomena to a Pure Logic* (Husserl 1900/01, *Prolegomena*, §§69–70), the first volume of his groundbreaking work the *Logical Investigations*. And as his academic career drew to a close in the late 1920s he reaffirmed, in *Formal and Transcendental Logic*, his continued espousal of that theory precisely as he had expounded it almost three decades earlier (Husserl 1929, §33). So with Husserl's *Mannigfaltigkeitslehre*, we find ourselves before a very significant, original theory that he propounded from one end of his philosophical career to the other.

However, of the few people who have inquired into Gödel's interest in Husserl's philosophy, I know of none who have sought any justification for Gödel's professed interest in this aspect of Husserl's thought. Hao Wang, the philosopher who reported Gödel's interest in Husserl's ideas on axiomatization, does not exhibit any particular knowledge or understanding of Husserl's ideas in this area (Wang 1986, 1987, 1996).

To shed light on the numerous dark areas still surrounding the relationship between Husserl's formal logic and his transcendental logic, and thus work to fill a gap in the literature on Husserl's philosophy, I propose here to take a closer look at how, by advocating a theory of deductive systems, or axiomatic systems as the highest task of pure logic Husserl sought to resolve certain problems that he encountered for the first time when writing the *Philosophy of Arithmetic*. I draw together what Husserl wrote about axiomatization and manifolds from the various sources now available in order to justify interest in this aspect of Husserl's thought. The lessons of the previous chapters of this book are necessarily intertwined into the discussions.

1. The Two Sides of Logic for Husserl

One important reason why Husserl's theory of axiomatic systems and manifolds has been overlooked and issues about his ideas about the mind, intuition, and intentionality stressed is that it has not been sufficiently emphasized that, for Husserl, logic had two sides. For him, logic turns towards what is subjective, towards what he called the deeply hidden subjective forms in which reason does its work (ex. Husserl 1929, §8). But for him, and this is what is less well appreciated, logic also turns towards the objective order, towards ideal objects, towards a world of concepts. There, truth is an analysis of essences or of concepts. Knowing subjects and the material world play no role.

Husserl always insisted on the primacy of the objective side of logic. It is knowledge of formal logic, he reminded readers in *Formal and Transcendental Logic*, that supplies the standards by which to measure the extent to which any presumed science meets the criteria of being a genuine science, the extent to which the particular findings of that science constitute genuine knowledge, the extent to which the methods it uses are genuine ones (Husserl 1929, §7).

The world constituted by transcendental subjectivity is a pre-given world, Husserl explained in *Experience and Judgement*. It is not a pure world of experience, but a world that is determined and determinable in itself with exactitude, a world within which any individual entity is given beforehand in an perfectly obvious way as in principle determinable in accordance with the methods of exact science and as being a world in itself in a sense originally deriving from the achievements of the physico-mathematical sciences of nature (Husserl 1939, §11).

These two sides of logic complement one another. Pure, objective, formal logic finds its necessary complement in subjective, transcendental logic. For example at its lowest level, that of traditional Aristotelian apophantic logic, formal logic deals with propositions composed of subjects and predicates. For Husserl, however, true philosophical logic requires a radical return to pre-predicative experience. It requires that one pierce through the logic of subject and predicates to the foundations of a underlying, hidden, logic in order to elucidate the origin of predicative judgments (Husserl 1939, §3).

One must not forget, Husserl reminds readers in *Experience and Judgement*, the importance of understanding the origins and particular legitimacy of the lower levels of logic in elucidating both the path one must take to attain evident knowledge at a higher level and the hidden presuppositions underlying this knowledge, presuppositions that determine and delimit its meaning (Husserl 1939, §§10, 11). This particular kind of analysis and understanding of the subjective foundations of traditional formal logic would be the job of the genealogy of transcendental logic that Husserl promulgated (Husserl 1939, §11; Husserl 1929, §40).

Husserl was perfectly aware of the extraordinary difficulties that this dual orientation of logic involved. Since, according to his theories, the ideal, objective, dimension of logic and the actively constituting, subjective, dimension interrelate and overlap, or exist side by side, logical phenomena thus seem to be suspended between subjectivity and objectivity in a confused way (Husserl 1929, §26c). In *Formal and Transcendental Logic*, he suggested that almost all that concerns the fundamental meaning of logic, the problems it deals with, its method, is laden with misunderstandings owing to the fact that objectivity arises out of subjective activity. He even considered that it was due to these difficulties that, after centuries and centuries, logic had not attained the secure path of rational development. Even the ideal objectivity of logical structures and a priori nature of logical doctrines especially pertaining to this objectivity, and the meaning of this a priori are afflicted with this lack of clarity, he maintained, since what is ideal appears as located in the subjective sphere and arises from it (Husserl 1929, §8).

2. The Problems that Led to the Development of the Theory

In *Formal and Transcendental Logic*, Husserl stressed that his *Mannigfaltigkeitslehre* dated back to the beginning of the 1890s, to a time in which he was trying to formulate a response to certain burning questions that he began asking during the writing of the aborted second volume of the *Philosophy of Arithmetic*. Unable to carry out the analyses of the concept of number that he had undertaken to perform there, he had, as we have seen in earlier chapters, found himself obliged to radically revise the methods of psychological analysis that he had learned from Franz Brentano, methods that he confessed he had never found to be entirely satisfactory (Husserl 1900/01, 42; Husserl 1975, 33–35). His conscience had tormented him, he would say, already at the time he published the *Philosophy of Arithmetic*, a book he considered he had gone beyond even as he published it (Husserl 1994, 490).

Husserl then found himself in the grips of an intellectual crisis. The *Logical Investigations*, he explained in the foreword to the book, had arisen out of reflections on certain unavoidable problems that had constantly hindered, and ultimately interrupted, the progress of many years of work devoted to achieving a philosophical clarification of pure mathematics (Husserl 1900/01, 41).

Husserl left dramatic descriptions of his intellectual crisis. It was a matter, he said, of ten years of hard, lonely work and of confused struggling. He aspired ardently after clarity, but only encountered confusion on all sides. He finally felt obliged to abandon ambiguous, undeveloped problems, obscure theories so as to begin somewhere on his own to assure greater clarity regarding fundamental epistemological questions and the critical understanding of logic as science (Husserl 1994, 497–98; Husserl 1975, 16–17; Husserl 1900/01, 42–43).

What were the problems that so tormented Husserl during those years?

He named five, and all five of them played a role in the genesis and development of his theory of manifolds. The first problem concerned pure logic and consciousness. While working on the logic of mathematical thought and mathematical calculation, Husserl has said, he was tormented by the "incredibly strange realms" of pure logic and actual consciousness (Husserl 1994, 490/91). For him, pure logic included "all of the pure 'analytical' doctrines of mathematics (arithmetic, theory of numbers, algebra, etc.) and the entire area of formal theories, or rather speaking in correlative terms, the theory of manifolds [*Mannigfaltigkeitslehre*] in the broadest sense" (Husserl 1975, 28), "the traditional syllogistic, but also the pure theory of cardinal numbers, the pure theory of ordinals, *Canto*rian sets, and so on" (Husserl 1994, 250).

Everything that is purely logical, he reasoned during his crisis, was an "in itself," something ideal that had nothing whatsoever to do with acts, subjects, or empirical persons belonging to actual reality. Yet this being so, he asked himself, how was symbolic thinking possible? How were objective, mathematical and logical relations constituted in subjectivity? How could the mathematical in itself, as given mentally, be valid? How does one move from mathematics to pure logic elucidated by a theory of knowledge? The "two worlds" should be interrelated and form a whole, but Husserl did not yet see how to join them (Husserl 1994, 490–91; Husserl 1975, 21–22, 35; Husserl 1900/01, 42–43).

Second Problem. When Husserl began the analyses of "On the Concept of Number" in 1886, it was obvious to him "that what mattered most for a philosophy of mathematics was a radical analysis of the psychological origin of the basic concepts of mathematics" (Husserl 1975, 33). "Not only is psychology indispensable for the analysis of the concept of number," he wrote there, "but rather this even *belongs within* psychology" (Husserl 1887, 95).

However, although the psychological analyses of "On the Concept of Number" were almost entirely incorporated into the first four chapters of the *Philosophy of Arithmetic*, the ardent defense of the psychologism of the former work is missing from the latter. Husserl would later confess that there had been "connections in which such a psychological foundation never came to satisfy" him, could bring "no true continuity and unity." He had grown "more and more disquieted by doubts of principle, as to how to reconcile the objectivity of mathematics, and of all of science in general, with a psychological foundation for logic" (Husserl 1900/01, 42; Husserl 1975, 34).

Third Problem. Husserl ran up against particularly disturbing problems while studying the logic of formal arithmetic and the theory of *Mannigfaltigkeiten*. These difficulties obliged him to engage in reflections of a very general nature which lead him beyond the confines of the mathematical realm towards a universal theory of formal deductive systems (Husserl 1900/01, 41).

Before looking further into the nature of those particular problems, it is important to recall the terminological obstacles one encounters when speaking of *Mannigfaltigkeiten* in this context. For the term '*Mannigfaltigkeit*', usually translated into English by 'multiplicity' or, better, 'manifold', is far from being univocal here.

Husserl himself was conscious of the terminological difficulties one faces in speaking of *Mannigfaltigkeiten*. In his earliest writings he acknowledged the ambiguity introduced into reasoning by the use of the terms '*Vielheit*', '*Mehrheit*', '*Inbegriff*', '*Aggregat*', '*Sammlung*', '*Menge*', etc. (which may be, and have been, variously translated by words like 'quantity', 'plurality', 'totality', 'aggregate', 'collection', 'set', 'multiplicity'). On the first page of the *Philosophy of Arithmetic*, Husserl states that, while recognizing the differences, he will abstain from using any single one of these terms exclusively. He hoped thus to be able to neutralize the differences (Husserl 1891, 8 and note). In so doing, he adopted the same strategy he had adopted in "On the Concept of Number" (Husserl 1887, 96). In these first two writings, he rarely employed the term '*Mannigfaltigkeit*'.

To better understand what is at stake here, it is also necessary to keep in mind the close professional and personal ties Husserl maintained with Georg Cantor, the creator of set theory and author of the *Mannigfaltigkeitslehre*, published just a few years earlier. Cantor was a member of the committee that approved "On the Concept of Number" (Gerlach and Sepp1994) and the theory of sets expounded there, a theory in many ways related to Cantor's. In *Formal and Transcendental Logic*, Husserl wrote of the *Philosophy of Arithmetic* as having represented an initial attempt on his part to achieve clarity regarding the true meaning of the concepts of set theory and the theory of cardinal numbers by going back to the spontaneous activities of collecting and counting in which the sets and the cardinal numbers are given (Husserl 1929, §27a).

Cantor himself used the terms '*Menge*', '*Mannigfaltigkeit*,' and '*Inbegriff*' without always distinguishing among them. Husserl only began using the term '*Mannigfaltigkeit*' more frequently in the 1890s when he studied Riemannian manifolds (ex. Husserl 1983, 92–103, 408–11; Husserl 1994, 488–89). Husserl's manifolds would finally bear little resemblance to Cantor's *Mannigfaltigkeiten*, except as concerns their Platonism (chapter 7 of the present book).

From the beginning of his investigations into the foundations of arithmetic, Husserl had been tormented by doubts about the psychological analysis of sets (*Mengen*). A collection could not be a physical entity composed of the items collected, he thought. So the idea of set, he had reasoned as a faithful disciple of Franz Brentano, must arise through psychological reflection upon the act of collecting (Husserl 1975, 34–35).

However, Husserl then asked, "isn't the concept of number not something basically different from the concept of collecting which is all that can result from the reflection on acts?" Doubts of this kind troubled him, tormented him, even from the very beginning. They eventually extended to all categorial concepts and to concepts of objectivities of any sort whatsoever, ultimately to include modern analysis and the theory of *Mannigfaltigkeiten* and simultaneously to mathematical logic and the entire field of logic in general (Husserl 1975, 34–35).

Husserl, we have seen, put Cantorian sets "the *Mannigfaltigkeitslehre* in the broadest sense" into the category of pure logic that raised thorny questions for him (Husserl 1994, 250). So, at first sight, one might think that this was merely a particular case of the already mentioned anguish regarding the relationship between consciousness and pure logic.

That would, however, be a mistake. Husserl, remember, frequented Cantor during the time that he was further developing the philosophically naïve, psychologistic theories of his famous *Mannigfaltigkeitslehre* (Cantor 1887/88; Cantor 1891) and as he began discovering the antinomies of set theory (Cantor 1991, 387–464; Dauben 1979, 240–70). It was in studying Cantor's theories that Bertrand Russell discovered the contradiction of the class of all the classes not belonging to themselves that put an end to the logical honeymoon that he was having when he began writing the *Principles of Mathematics* (ex. Russell 1903, §§100, 344, 500; Russell 1959, 58–61; Grattan-Guinness 1978, 1980).

Fourth Problem. During his years of crisis, Husserl reflected long and hard on the differences between Leibniz's *vérités de raison* and *vérités de fait* and Hume's relations of ideas and matters of fact, and between Kant's analytic and synthetic judgments (Husserl 1975, 36).

Husserl's very keen sense of the contrast between Hume's distinction and Kant's would play an important role in the formulation of the positions adopted in the years to come (Husserl 1975, 36). He would condemn Kant's logic as being utterly defective (Husserl 1900/01, *Prolegomena*, §58). Kant, Husserl insisted upon several occasions, had not understood the nature and role of formal mathematics. The way in which he defined the concept of analyticity was totally inadequate and even utterly wrong (Husserl 1906/07, §23). "Not only did he never see how little the laws of logic are all analytic propositions in the sense laid down by his own definition, but he failed to see how little his dragging in of an evident principle for analytic propositions really helped to clear up the achievements of analytic thinking" (Husserl 1900/01, Sixth Investigation, §66).

Fifth Problem. Husserl found himself up against questions raised by imaginary numbers. He used the term 'imaginary' in the broadest possible way to include negative, irrational numbers, fractions, negative square roots (Husserl 1983, 244–49; Husserl 1970, 432). He called infinite sets "imaginary concepts" (Husserl 1891, 249).

The most contradictory and incredible theories have been held about the meaning of negative, irrational, and imaginary ("impossible") numbers, Husserl noted in 1890. But that has not hindered their use. "One could quite certainly convince oneself of the correctness of any sentence deduced by means of them through an easy verification. And, after innumerable experiences of this sort one naturally comes to trust in the unrestricted applicability of these modes of procedure, expanding and refining them more and more—all without the slightest insight into the *logic* of the matter. . . ." Nonetheless, he stressed, the use of symbols to meet scientific ends and in a scientifically successful way is still not a logical use. He complained about all the mental energy wasted in this way and about how much quicker and more secure the progress of arithmetic would have been had there been more clarity and insight about the logic of these numbers (Husserl 1996, 48–49). He wished to know precisely what justified the use of such apparently meaningless signs in calculations, or in deductive thought (Husserl 1970, 433).

In "On the Concept of Number," Husserl had developed the thesis of his teacher Karl Weierstrass, according to which all the more complicated and more artificial constructions that are also called numbers, fractions and irrational numbers, negative and complex numbers have their origin and support in the elementary concepts of number and in the relations connecting them (Husserl 1887, 95). However, in a letter written to Carl Stumpf dating from 1890 or 1891, Husserl wrote that the theory that the concept of cardinal number forms the foundation of general arithmetic that he had tried to develop in "On the Concept of Number" had soon proved to be false. By no clever devices, he explained, "can one derive negative, rational, irrational, and the various sorts of complex numbers from the concept of cardinal number. The same is true of the ordinal concepts, of the concepts of magnitude, and so on. And these concepts themselves are not logical particularizations of the cardinal concept" (Husserl 1994, 13; also Husserl 1891, VIII, 5 and note).

3. Husserl's Theory of Manifolds

How then, might one characterize the theory of manifolds that Husserl began developing during those years of solitude and crisis?[1]

Husserl's reflections on logic, number theory, set theory, the theory of manifolds, the imaginary, etc. eventually led him to discern a certain natural order in formal logic and to broaden its domain to include two levels above traditional formal logic. He considered the detection of these three levels of formal logic to be of the greatest importance for the understanding of logic and philosophy. He explains this in a particularly clear way in *Einleitung in die Logik und Erkenntnistheorie*, §§18–19, published for the first time in 1984.

According to his new conception of formal logic, traditional, Aristotelian apophantic logic is to be found on the lowest level of formal logic and only constitutes a small area of pure logic in the broadest sense of the word. A logic of subject and predicate propositions and states of affairs, it deals with what might be stated about objects in general from a possible perspective.

The purely logical disciplines of the two levels of pure logic above apophantic logic still deal with individual things. However, these objects are no longer empirical or material entities. On the second level, one encounters objective constructions of a higher kind that are determined in purely formal terms and deal with objects in an indeterminate, general way. This is where one finds, for example, the theory of cardinal numbers, the theory of ordinals and set theory. It is here a matter of forms of judgments, and forms of their constituents, forms of deduction, forms of demonstration, sets and relationships between sets, combinations, orders, quantities, objects in general, etc. One investigates what is valid for these higher level objects.

On the second level, one may calculate, reason deductively, with concepts and propositions. Signs and rules of calculation suffice because each procedure is purely logical. One may manipulate signs like chess pieces for which rules having such and such a form are valid. One now only thinks of the signs that through the rules of calculation acquire their meaning within the game. One may proceed mechanically and the result will be accurate and justified. That proves to be an enormous help in reasoning. For is incomparably easier to think with signs. One is thereby freed from the equivocation and ambiguity that comes with words. And the process itself calls for the maximum of rigor.

Numbers and sets, then, play a different role in the second level than in the apophantic sphere. It is no longer a question of numbers as such, of numbers as objects about which one might predicate something. Numbers no longer function as independent entities, but are dependent structures. In the set theory of this level, it is not a matter of predicating something of the members, but of sets overall, having any members whatsoever.

Abstracting further, one reaches the third level, that of the theory of possible theories, the theory of manifolds. Manifolds are pure forms of possible theories which, like molds, remain totally undetermined as to their content, but to which thought must necessarily conform in order to be thought and known in a theoretical manner.

The general theory of the manifolds, or science of theory forms, is a field of free, creative investigation that is made possible once the form of the mathematical system is emancipated from its content. Indeed, once one discovers that deductions, series of deductions, continue to be meaningful and to remain valid when one assigns another meaning to the symbols, one is then free to liberate the mathematical system, which can henceforth be considered as being the mathematics of a domain in general, conceived in a general and indeterminate manner. Nothing more need be presupposed than the fact that the objects figuring in it are such that, for them, a certain connective supplies new objects and does so in such a way that the form determined is assuredly valid for them.

Here we have, Husserl maintained, a new discipline and a new method constituting a new kind of mathematics, the most universal one of all. Whereas on the second level, it was a matter of forms of propositions, forms of deduction, forms of demonstrations, etc., on this third level, formal logic deals with whole systems of propositions making up possible deductive theories. It is now a matter of theorizing about possible fields of knowledge conceived of in a general, undetermined way and purely and simply determined by the fact that they are in conformity with a theory having such a form, i.e., determined by the fact that its objects stand in certain relations that are themselves subject to certain fundamental laws of such and such determined form.

By using axioms of such and such a form, theories of such and such a form may be developed. These objects are exclusively determined by the form of the interconnections assigned them, meaning, neither directly inasmuch as individuals, nor indirectly by their kind or species. These interconnections themselves are just as little determined in terms of content as are the objects. Only their form determines them by virtue of the form of the elementary laws admitted as valid for these interconnections, laws that also determine the theory to be constructed, the form of the theories.

Husserl most frequently turned to Euclidean geometry to illustrate his ideas about manifolds. Moving from geometry to its theory form, he once suggested, this generalization results:

> We define an undetermined multiplicity of elements, or, what amounts to the same thing, a manifold using the following property: Two elements, called points, determine a figure called a straight line. Two straight lines have one and only one common point, etc. In other words, through axiom forms which are just formalizations of Euclidean axioms we define a multiplicity of any thing whatsoever in an indeterminate general way. The manifold is now no longer the manifold of points in the ordinary sense. It is by no means any longer a question of

space. Prior to the formalization one says: "Two points determine a straight line." That is a geometrical truth having a clear meaning. Everyone knows what points and straight lines signify from geometrical intuition. And it is the same with all other basic concepts and axioms of geometry. After formalization the words 'point,' 'straight line', 'angle', 'cut', etc. are completely empty signs, which only have the purely formal sense laid down for them by the axiom forms. One then speaks of a certain something, called a point, and a certain something called a straight line that must by definition in the defining manifold stand in a certain relationship to one another indicated by the words: Between any two points lies a straight line that "goes through them." Instead of space we have a concept defined by pure, categorial concepts that is purely set forth, namely defined, as a formal possibility. And the definition of a manifold as a Euclidean manifold says no more about existence than does the definition of a golden mountain as mountain of gold. . . . Only a form is defined, but whether axioms as truths have existence in any objective real or ideal spheres corresponding to the prescribed form remains open.

The Euclidean manifold defined stands in precisely the same relation to space as the number 2 does to any concrete two, to any concrete group of two things or two ideas, etc. . . .

The Euclidean manifold also in certain ways has "being," namely insofar as it is correctly defined, insofar as the axiom forms are ordered in such a way as to contain no formal contradictions, no violation of analytic principles. This being is thus only the being of formal analytical agreement. . . .

On the basis of the definition of this manifold . . . we can deduce conclusions . . . we can construct proofs, and it is then certain a priori that anything obtained in this way will correspond to something in the theory of space. We generally do not need to traffic in geometry any more. We can just traffic in the theory of Euclidean manifolds and obtain everything we could at any time obtain in spatial thought (*Raumdenken*). . . . (Husserl 1917/18, §57)

No longer restricted to operating in terms of a particular field of knowledge, we are free to reason completely on the level of pure forms. Operating within this sphere of pure forms, we quickly find that we can vary the systems in different ways. Instead of three dimensions, we may choose four or even n many dimensions and develop a theory of manifolds for all manifolds of n dimensions, which are still called Euclidean because, except for the number of dimensions, the axiom forms have not undergone any essential change. One finds ways of constructing an infinite number of forms of possible disciplines. And that is of inexhaustible practical interest.

Any individual theory is a particular instance of the theory form corresponding to it, just as all fields of theoretical knowledge are particular instances of manifolds. But not all the sciences, Husserl pointed out, are theoretical disciplines that, like mathematical physics, pure geometry or pure arithmetic, are characterized by the fact that their systemic principle is a purely analytical one. Theoretical disciplines have, Husserl explains, a systemic form that belongs to formal logic itself and that must be constructed a priori within formal logic itself and within its supreme discipline the theory of manifolds, as part of the overall system of forms of deductive systems that are possible a priori (Husserl 1929, §35a).

However, sciences like psychology, history, the critique of reason and, notably, phenomenology require that one go beyond the analytico-logical model. When they are formalized and one asks what it is that binds the propositional forms into a single system form, one finds oneself facing, Husserl maintained, nothing more than the empty general truth that there is an infinite number of propositions connected in objective ways that are compatible with one another in that they do not contradict each other analytically.

4. The Answers Husserl Found

How, then, did Husserl's theory of manifolds respond to the burning questions that had tormented him? First of all, Husserl's personal discovery in Halle of the strangeness of the worlds of pure logic and of consciousness led him to explore these two worlds and carefully inventory everything he found there. He devoted his life to this.

The solution he progressively found to the first question consisted of broadening the domain of traditional logic in such a way as to account for the progress made by modern mathematics and, most particularly, the progress represented by the theory of manifolds. He finally came to distinguish among three levels of pure logic, each one reflecting a higher degree of abstraction and each one further removed from subjectivity. As the supreme task at the highest level of this pure logic, the theory of manifolds would serve as Husserl's paradigm of logical reasoning purified of any trace of noxious psychologism.

Once liberated from things and psychologizing subjectivity, pure logic would find its necessary complement in a transcendental logic that would take into account the connections that philosophical logic inevitably maintains with the concrete world and knowing subjects. Once these connections with the concrete world and knowing subjects had been elucidated, pure logic would serve as a bulwark against the incursions of psychologizing subjectivity. The genealogy of logic that Husserl promulgated would detail how the logical relationships are constituted in subjectivity.

Husserl believed that he had thus brought needed clarity to the realm of pure logic and the relationships it maintains with the human consciousness. However, the solution he prescribed continues to inspire incomprehension, contempt, even anger among those who endeavor to elucidate the legitimate links existing between his formal logic and the philosophy of logic born of the best efforts of the most innovative and influential philosophers, logicians, and mathematicians of his time. Yet they shared his desire to discover secure, scientific foundations for mathematics and the theory of knowledge, his concern to reform logic, his intent to fight against psychologism, his desire to develop a theory of meaning, his questions concerning the role to be accorded to set theory in philosophy and mathematics, etc.

But all that cannot blind someone seeking to elucidate what Husserl meant by pure, formal logic to the blatant contradictions that seem evident when, for example, he declares that what is constituted by knowing subjects acquires the meaning of an ideal objectivity existing in itself, or when he writes that what is ideal appears as located in the subjective sphere and arises from it as a formation (Husserl 1929, §8).

For many, all the while insisting on the primacy and objectivity of pure logic, Husserl really outdid himself engaging in analyses of a subjective foundation of traditional formal logic through transcendental phenomenology. His works abound with analyses that seem to mix the two worlds of pure logic and actual consciousness, which in his writings diverge and become enmeshed anew in confusing ways that have left many philosophers unsatisfied with his response to the problems raised by the interaction between the subjective and objective orders that so disturbed him.

Let us turn to the second problem. When Husserl began working on the analyses of "On the Concept of Number," he believed that it was obvious that it was necessary to undertake a radical analysis of the psychological origin of the fundamental concepts of mathematics. In the *Philosophy of Arithmetic* he had harshly criticized what he called Frege's "ideal" of basing arithmetic on a series of formal definitions out of which all the theorems of this science might be derived in a purely syllogistic manner (Husserl 1891, 130). In the *Prolegomena*, Husserl retracted precisely those pages and only those pages upon which this particular pronouncement is to be found (Husserl 1900/01, *Prolegomena*, §45 n.). In the meantime, he had turned to developing his own ideas about grounding arithmetic on a series of formal definitions.

The value of his criticisms of logical psychologism, Husserl explained in *Formal and Transcendental Logic*, lie precisely in his drawing attention to a pure, analytic logic, distinct from any psychology, as being an independent field, like geometry or the natural sciences. Epistemological questions may well arise regarding this pure logic, but this must not interfere with its independent course, or involve delving into the concrete aspects of the logical life of the consciousness. For that would be psychology (Husserl 1929, §67).

With his theories about the three levels of formal logic, plus transcendental logic and the genealogy of logic, Husserl thought he had clearly determined the place of the propositions of arithmetic in relationship to the judgments of apophantic logic. Apophantic logic being found on the level lying in between actual numbers and abstract numbers, the theory of arithmetic was no longer to be seen as being in direct contact with acts of counting, ordering, combining, collecting, etc. There should, therefore, no longer be any radical analysis of the psychological origins of the fundamental concepts of mathematics *per se*.

Husserl's third problem concerned sets and manifolds. By assigning sets a place in the second level of pure logic, a step removed from the acts, subjects or empirical persons of actual reality, Husserl banished his doubts concerning the psychological analysis of sets. In addition, he distinguished sets from manifolds.

As someone who frequented Hilbert's school in Göttingen, and who upon several occasions pointed to the kinship existing between his own manifolds and Hilbert's axiomatic systems (Husserl 1913, §72, Husserl 1929, §31, and chapter 10 of the present book), one may imagine that Husserl considered, as so many mathematicians have, that properly carried out, the axiomatization of set theory might neutralize the contradictions found in Cantorian set theory. Defined and regulated by a complete axiomatic system, sets would thus be apt to play their fruitful role in mathematics.

With his theory of manifolds, Husserl still drew inspiration from the theories of mathematics, but he had come to use the word 'manifold' in his own way. He considered that the theory of manifolds of modern mathematics was already a realization of the idea of a science of possible deductive systems. He called that theory of manifolds "a fine flower of modern mathematics" (Husserl 1900/01, *Prolegomena*, §70). But, he had come to clearly distinguish his manifolds from Cantor's *Mannigfaltigkeiten* or sets, relegating the latter to a lower level of pure logic (Husserl 1906/07, §18). Husserl especially considered that in discoursing on manifolds unnamed mathematicians had often lacked clarity and that the theory of manifolds of modern mathematics, and all of modern formal analysis represented but a partial realization of his own ideal of a science of possible deductive systems (Husserl 1900/01, *Prolegomena*, §70).

The fourth problem concerned analyticity. Beginning in the early 1890s, Husserl would strive to develop the true concept of analyticity and to discover the basic philosophical line separating genuine analytical ontology from material (synthetic a priori) ontology which must be fundamentally distinct from it (Husserl 1975, 42–43).

By drawing the boundary line existing a priori between mathematics and natural sciences like psychology, Husserl believed that he was drawing the line of demarcation and expanding the domain of the analytical in keeping with the most recent discoveries in mathematics. He placed the fundamental concepts of mathematics on the second level of pure logic conceived as an expanded, completely developed, analytics. Husserl considered that one was proceeding in a purely formal manner there, since every single concept used was analytic (Husserl 1900/01, *Prolegomena*, §§69–70; Husserl 1917/18, §58).

Analytic logic, he explained in *Formal and Transcendental Logic,* is first of all valid as an absolute norm presupposed by any rational knowledge. His fight against psychologism, Husserl explained there "was in fact meant to serve no other end than the supremely important one of making the specific *province* of analytic logic visible in its purity and ideal particularity, freeing it from the psychologizing confusions and misinterpretations in which it had remained enmeshed with from the beginning" (Husserl 1929, §67).

The fifth problem concerned the questions raised by imaginary numbers and concepts. Husserl concluded that his theory of manifolds held the key to the only possible solution to the problem of how, in the domain of cardinal numbers, for example, concepts of impossible numbers might be methodically treated as real ones (Husserl 1900/01, *Prolegomena,* §70). He even once said that his chief purpose in developing the theory of manifolds had been to find a theoretical solution to the problem of imaginary quantities (Husserl 1913, §72 n.).

In the arithmetic of cardinal numbers, he finally concluded, the meaning of the axioms is so restrictive as to make subtracting 4 from 3 nonsense (*unsinnig*). There are no negative numbers in this arithmetic. Fractions are meaningless. So are irrational numbers, the $\sqrt{-1}$, and so on. Yet the remarkable thing, he considered, was that, in practice, all the calculations of the arithmetic of cardinal numbers can be carried out as if the rules governing the operations were unrestrictedly valid and meaningful (*sinnvoll*). One can disregard the limitations imposed in a narrower domain of deduction and act as if the axiom system were a more extended one (Husserl 1917/18, §56).

Understanding the nature of theory forms shows how such puzzling operations with imaginaries may be justified. One finds that there may be two valid discipline forms standing in relation to each other in such a way that the axiom system of one may be a formal limitation of that of the other. It then becomes plain that everything that can be deduced in the narrower axiom system is included in what can be deduced in the expanded system.

Now, all the theorems deducible in the expanded system must exclusively contain concepts that are either valid in terms of the narrower one, and thus not imaginary, or they must contain concepts that are imaginary. Thus it is that when one compares cardinal arithmetic and ordinal arithmetic (where the minuend may be greater than the subtrahend) and their respective discipline forms, besides theorems including only non-imaginary numbers having real meaning (*reale Bedeutung*) one finds formulas and theorems that also include negative numbers, which are imaginary in terms of the more restrictive axioms of the arithmetic of cardinal numbers (Husserl 1917/18, §56).

We cannot arbitrarily expand the concept of cardinal number, Husserl found. But we can abandon it and define a new, purely formal concept of positive whole number with the formal system of definitions and operations valid for cardinal numbers. And, as set out in our definition, this formal concept of positive numbers can be expanded by new definitions while remaining free of contradiction (Husserl 1970, 435).

Fractions do not acquire any genuine meaning through our holding onto the concept of cardinal number and assuming that units are divisible, Husserl theorized, but rather through our abandonment of the concept of cardinal number and our reliance on a new concept, that of divisible quantities. That leads to a system that partially coincides with that of cardinal numbers, but part of which is larger—meaning that it includes additional basic elements and axioms. And so in this way, with each new quantity (*Grösse*), one also changes arithmetics. The different arithmetics do not have parts in common. They have totally different domains, but have an analogous structure. They have forms of operation that are in part alike, but different concepts of operation (Husserl 1970, 436).

It was formal constraints requiring that one not resort to any meaningless expression, no meaningless imaginary concept that were restricting us in our theoretical, deductive work. But what is marvelous, Husserl believed, is that resorting to the infinity of pure forms and transformations of forms frees us from such conditions and at the same time explains to us why having used imaginaries, what is senseless, must yield, not senseless results, but to true ones (Husserl 1917/18, §57).

One may then, he concluded, operate freely within a multiplicity with imaginary concepts and be sure what one deduces is correct when the axiomatic system completely and unequivocally determines the body of all the configurations possible in a domain by a purely analytical procedure. A domain is complete, according to Husserl, when each grammatically constructed proposition exclusively using the language of this domain is, from the outset, determined to be true or false in virtue of the axioms, i.e., necessarily follows from the axioms (in which case it is true) or does not (in which case it is false). In this case, calculating with imaginary concepts can never lead to contradictions. It is this completeness of the axiomatic system that gives one the right to freely operate in this way (Husserl 1900/01, *Prolegomena*, §70; Husserl 1913, §72; Husserl 1917/18, §56; Husserl 1929, §31; Husserl 1970, 441).[2]

NOTES

1. Husserl did not always develop his theories as systematically as one may like. In this case, as elsewhere, the information needed to piece together an accurate picture of his theories is scattered throughout his life's work, obliging conscientious scholars to go out in pursuit of the facts necessary for elucidating texts that are often enigmatic. This description of his theory comes from Husserl's *Prolegomena to a Pure Logic*, vol. 1 of the *Logical Investigations*

§§69–70; *Ideas I*, §§71–72 and notes; *Formal and Transcendental Logic*, chapter 3; §§51–54 and notes; *Crisis in European Sciences and Transcendental Phenomenology* §9; "Arithmetic as an a priori science" in his *Early Writings*, pp. 7–11; "Mengen und Mannigfaltigkeiten" in his *Studien zur Arithmetik und Geometrie*, pp. 92–102; "Das Imaginäre in der Arithmetik," "Drei Studien zur Definitheit und Erweiterung eines Axiomensystems," "Das Gebiet eines Axiomensystems / Axiomensystem-Operationssystem," "Zur formalen Bestimmung einer Mannigfaltigkeit," published in the Husserliana edition of the *Philosophie der Arithmetik*. Most helpful of all was Husserl's very clear *Logik und allgemeine Wissenschaftstheorie* chapter 11, first published in 1996 and *Einleitung in die Logik und Erkenntnistheorie*, §§18–19, only available since 1984.

2. This chapter is based on a paper that I gave in French in Brussels in December 1998 at a Colloquium on the Genealogy of Logic organized by the Centre de Recherches Phénoménologiques of the Facultés Saint Louis. In writing the English version I benefited from the insights of Jairo da Silva of the mathematics department of the University of São Paulo.

REFERENCES

Cantor, G. 1883. *Grundlagen einer allgemeinen Mannigfaltigkeitslehre. Ein mathematisch-philosophischer Versuch in der Lehre des Unendlichen*. Leipzig: Teubner. Cited as appears in Cantor 1932, 165–246.

———. 1887/88. "Mitteilungen zur Lehre vom Transfiniten." *Zeitschrift für Philosophie und philosophische Kritik* 91: 81–125; 92: 240–65. Cited as appears in Cantor 1932, 378–439. Also published as Cantor 1890.

———. 1890. *Gesammelte Abhandlungen zur Lehre vom Transfiniten*. Halle: C.E.M. Pfeffer. See Cantor 1887/88.

———. 1932. *Gesammelte Abhandlungen*. Ed. E. Zermelo. Berlin: Springer.

Dauben, J. 1979. *Georg Cantor, His Mathematics and Philosophy of the Infinite*. Princeton: Princeton University Press.

Frege, G. 1885. "On Formal Theories of Arithmetic." *Collected Papers on Mathematics, Logic, and Philosophy*. Ed. B. McGuinness. Oxford: Blackwell, 1984, pp. 112–21.

———. 1891. "Function and Concept." *Translations from the Philosophical Writings*. 3rd ed. Oxford: Blackwell. 1980 (1952), pp. 21–41.

———. 1892. "On Sense and Meaning." *Translations from the Philosophical Writings*. 3rd ed. Oxford: Blackwell, 1980 (1952), pp. 56–78.

———. 1980. *Philosophical and Mathematical Correspondence*. Oxford: Blackwell.

Gerlach, H. and H. Sepp, eds. 1994. *Husserl in Halle*. Bern: P. Lang.

Grattan-Guinness, I. 1971. "Towards a Biography of Georg Cantor." *Annals of Science* 27, no. 4: 345–91.

———. 1978. "How Russell Discovered His Paradox." *Historia Mathematica* 5: 127–37.

———. 1980. "Georg Cantor's Influence on Bertrand Russell." *History and Philosophy of Logic* 1: 61–93.

———. 1996. "Numbers, Magnitudes, Ratios, and Proportions in Euclid's Elements: How Did He Handle Them?" *Historia Mathematica* 23: 355–75.

Hallett, M. 1984. *Cantorian Set Theory and Limitation of Size*. Clarendon: Oxford.

Hilbert, D. 1925. "On the Infinite." In van Heijenoort, 1967, 367–92.

Hill, C. O. 1991. *Word and Object in Husserl, Frege and Russell, the Roots of Twentieth Century Philosophy.* Athens: Ohio University Press.

———. 1995. "Husserl and Hilbert on Completeness." *From Dedekind to Gödel, Essays on the Development of the Foundations of Mathematics.* Ed. J. Hintikka. Dordrecht: Kluwer, 143–63 (chapter 10 of the present book).

———. 1998. "From Empirical Psychology to Phenomenology: Husserl on the Brentano Puzzle," *The Brentano Puzzle.* Ed. R. Poli. Aldershot, UK: Ashgate, 151–68.

———. 1997. "Did Georg Cantor Influence Edmund Husserl?" *Synthese* 113 (October): 145–70, chapter 8 of the present book.

———. 1999. "Abstraction and Idealization in Georg Cantor and Edmund Husserl." *Abstraction and Idealization. Historical and Systematic Studies, Poznan studies in the philosophy of the sciences and the humanities.* Ed. F. Coniglione, R. Poli, R. Rollinger. Amsterdam: Rodopi, chapter 7 of the present book.

Husserl, E. 1887. "On the Concept of Number." *Husserl: Shorter Works.* Ed. P. McCormick and F. Elliston. Notre Dame: University of Notre Dame Press, 1981, pp. 92–119.

———. 1891. *Philosophie der Arithmetik.* Halle: Pfeffer. Also Husserl 1970.

———. 1900/01. *Logical Investigations.* New York: Humanities Press, 1970.

———. 1906/07. *Einleitung in die Logik und Erkenntnistheorie.* Husserliana, vol. XXIV. Dordrecht: M. Nijhoff, 1984.

———. 1908. *Vorlesungen über Bedeutungslehre.* Husserliana, vol. XXVI. Dordrecht: M. Nijhoff, 1987.

———. 1913. *Ideas, General Introduction to Pure Phenomenology.* New York: Colliers, 1962.

———. 1917/18. *Logik und allgemeine Wissenschaftstheorie.* Husserliana, vol. XXX. Dordrecht: Kluwer, 1996.

———. 1929. *Formal and Transcendental Logic.* The Hague: M. Nijhoff, 1969.

———. 1939. *Experience and Judgement.* London: Routledge and Kegan Paul, 1973.

———. 1970. *Philosophie der Arithmetik, mit Ergänzenden Texten (1890–1901).* Husserliana, vol. XII. The Hague: M. Nijhoff, 1970. Introduction by L. Eley.

———. 1975. *Introduction to the Logical Investigations.* The Hague: M. Nijhoff, 1975.

———. 1983. *Studien zur Arithmetik und Geometrie, Texte aus dem Nachlass (1886–1901).* Husserliana, vol. XXI. The Hague: M. Nijhoff.

———. 1994. *Early Writings in the Philosophy of Logic and Mathematics.* Dordrecht: Kluwer.

Mancosu, P. 1998. *From Brouwer to Hilbert.* New York: Oxford University Press.

Peckhaus, V. 1990. *Hilbertsprogramm und kritische Philosophie, das Göttinger Modell interdisziplinärer Zusammenarbeit zwischen Mathematik und Philosophie.* Göttingen: Vandenhoeck & Ruprecht.

Russell, B. 1903. *Principles of Mathematics.* London: Norton.

———. 1959. *My Philosophical Development.* London: Unwin.

Scanlon, J. 1991. "'*Tertium Non Datur:*' Husserl's Conception of a Definite Multiplicity." *Phenomenology and the Formal Sciences.* Ed. T. Seebohm et al. Dordrecht: Kluwer, 139–47.

Tieszen, R. 1992. "Kurt Gödel and Phenomenology." *Philosophy of Science* 59: 176–94.

———. 1994. "Mathematical Realism and Gödel's Incompleteness Theorems." *Philosophia Mathematica* 3, no. 2: 177–201.

————. 1998a. "Gödel's Philosophical Remarks on Logic and Mathematics: Critical Notice of *Kurt Gödel: Collected Works, Vols. I, II, III.*" *Mind* 107: 219–32.

————. 1998b. "Gödel's Path from the Incompleteness Theorems (1931) to Phenomenology (1961)." *The Bulletin of Symbolic Logic* 4, no. 2: 181–203.

van Heijenoort, J., ed. 1967. *From Frege to Gödel: A Source Book in Mathematical Logic, 1879–1931.* Cambridge MA: Harvard University Press.

Wang, H. 1986. *Beyond Analytic Philosophy, Doing Justice to What We Know.* Cambridge MA: MIT Press.

————. 1987. *Reflections on Kurt Gödel.* Cambridge, MA: MIT Press.

————. 1996. A *Logical Journey, From Gödel to Philosophy.* Cambridge, MA: MIT Press.

Whitehead, A. 1911. *An Introduction to Mathematics.* Oxford: Oxford University Press, 1948

10

Claire Ortiz Hill

HUSSERL AND HILBERT ON COMPLETENESS*

In a 1900 paper entitled "On the Number Concept," the formalist mathematician David Hilbert proposed a set of axioms from which he hoped arithmetic might be derived. The last of these axioms was an "Axiom of Completeness" stipulating that: "It is not possible to adjoin to the system of numbers any collection of things so that in the combined collection the preceding axioms are satisfied; that is, briefly put, the numbers form a system of objects which cannot be enlarged with the preceding axioms continuing to hold."[1]

In his major works[2] the philosopher Edmund Husserl wrote that he had appealed to a concept of completeness closely related to the axiom of completeness Hilbert had introduced for the foundations of arithmetic. In these works Husserl is specifically referring to his theory of complete manifolds (*Mannigfaltigkeiten*) which, as he wrote in *Ideas* §72, have the "distinctive feature that a finite number of concepts and propositions—to be drawn as occasion requires from the essential nature of the domain under consideration—determines completely and unambiguously on the lines of pure logical necessity the totality of all possible formations in the domain, so that in principle, therefore, nothing further remains open within it." In such complete manifolds, Husserl maintained, "the concepts true and formal implication of the axioms are equivalent."

It is clear from Husserl's writings that he considers the fact of this kinship to be quite significant. And in §31 of his 1929 *Formal and Transcendental Logic,* Husserl even went so far as to say that the close study of his analyses would reveal that the underlying, though inexplicit, reasons which had led Hilbert to attempt to complete a system of axioms by adding a separate axiom of completeness were much the same as those which had played a determinant role in Husserl's own independent formulation of his concept of completeness.

The few people who have commented on Husserl's remarks have principally tried to determine whether or not Husserl's concept was in fact the same as Hilbert's, and a few have considered the relevance of Husserl's remarks to issues in mathematical logic.[3] Considering, howéver, what Gödel's theorems have brought to the subject, and given the radically different contexts into which Husserl and Hilbert integrated their ideas on completeness, I consider those particular questions to be rather academic. Since Husserl was ultimately inquiring into the foundations of all knowledge his notion played a role in investigations which were vastly broader in scope and essentially different in nature than Hilbert's inquiries into the foundations of mathematics were.

What intrigues me rather is Husserl's belief that, though inexplicit, Hilbert's deep underlying reasons for formulating his axiom of completeness were basically the same as those which had led Husserl to formulate his own concept of completeness. Unfortunately, however, Husserl was not himself very explicit about the steps in his own reasoning which had led to the formulation of his views on completeness, and he seemed to think the connection between his idea and Hilbert's was self-evident.[4]

Here I want to inquire further into the origins of Husserl's ideas on completeness, and then look at how Husserl thought he might provide more secure logical foundations for all knowledge by generalizing insights drawn from his investigations into the foundations of mathematics. Specifically, I argue that early in his philosophical career Husserl came to believe that having had recourse to the ideal elements he thought were necessary for the kind of foundations for knowledge he sought was justifiable when these elements admitted of formal definition within a complete deductive system. I draw Frege's ideas into the discussion because Frege tangled with the very same problems in the foundations of arithmetic that led Husserl to ally himself with formalism.

1. The Foundations of Mathematics and Imaginary Numbers

Husserl's interest in the foundations of mathematics has a noble ancestry, the full account of which has yet to be written and cannot be told here. Mathematically speaking, Husserl had the good fortune to be in the right place at the right time, and so was personally on hand to witness important developments in mathematics which Russell, Whitehead, Frege, and Wittgenstein mainly knew by description.

After studying under Karl Weierstrass and serving as his assistant,[5] Husserl studied for a time under Franz Brentano, the philosopher working to reform Aristotelian logic whose ideas on intentionality would later be so instrumental in freeing Bertrand Russell from the bonds of subjective idealism.[6] In 1886 Husserl moved to the University of Halle where for the next fourteen years he maintained close professional and personal ties with Georg Cantor.[7] Weierstrass, Brentano, and Cantor all three worked on the ideas of Bernard Bolzano,[8] one of the pioneers in the area of completeness.[9]

Husserl's interest in completeness and formalist foundations for mathematics is traceable back to those early years in Halle, and as he told his readers on several occasions, originally derived from problems regarding imaginary numbers which first came up while he was trying to complete his 1891 *Philosophy of Arithmetic*.[10]

The matter arose as he searched for answers to questions regarding the consistency of arithmetic, and it especially involved his attempts to account for the achievements of certain purely symbolic procedures of mathematics despite their appeal to apparently nonsensical combinations of symbols. Husserl sought for a way to explain, or explain away, the many expressions which appear in philosophical and mathematical reasoning but which do not and cannot designate objects.

Husserl's contemporary Alfred North Whitehead provides some insight into how mathematicians of the time viewed the problem which Husserl cited as the source of his views on completeness. Problems concerning imaginary numbers, Whitehead informed readers in 1911,[11] arose from the consideration of quadratic equations like $x^2 + 1 = 3$, $x^2 + 3 = 1$, and $x^2 + a = b$. The first equation, he explains, becomes $x^2 = 2$, and this has two alternative solutions, either $x = +\sqrt{2}$, or x equals $-\sqrt{2}$. And this is where the problem begins. For the equation $x^2 + 3 = 1$ yields $x^2 = -2$, but there is no positive number which when multiplied by itself will yield a negative square. So, if the symbols mean ordinary negative or positive numbers there is no solution to $x^2 = -2$, which must then be nonsense. One cannot therefore say that the symbols may mean numbers and "a host of limitations and restrictions" begin to accumulate. A new interpretation of such symbols is therefore required so that negative squares might have meaning. It was ultimately perceived, Whitehead writes, how convenient:

> it would be if an interpretation could be assigned to these nonsensical symbols. Formal reasoning with these symbols was gone through, merely assuming that they obeyed the ordinary algebraic laws of transformation; and it was seen that a whole world of interesting results could be attained if only these symbols might legitimately be used. Many mathematicians were not then very clear as to the logic of their procedure, and an idea gained ground that, in some mysterious way, symbols which mean nothing can by appropriate manipulation yield valid proofs of propositions.

Faced with this problem, Husserl said his main questions were: (1) Under what conditions can one freely operate within a formally defined deductive system with concepts which according to the definition of the system are imaginary and have no real meaning? (2) When can one be sure of the validity of one's reasoning, that the conclusions arrived at have been correctly derived from the axioms one has, when one has appealed to imaginary concepts? And (3) To what extent is it permissible to enlarge a well defined deductive system to make a new one that contains the old one as a part? He eventually concluded that if the system was complete, then calculating with imaginary concepts could never lead to contradictions.[12]

Such questions about the logical foundations of the real number system led Husserl to want to probe more deeply in order to clarify its structure. They were also, he tells us, instrumental in undermining his faith in Brentano's psychologism, which Husserl came to realize could not provide the real number system with the sound foundations he sought. For him these questions were also linked with deep frustrations he felt with regard to Kant's concept of analyticity which Husserl thought was too weakly formulated and in dire need of reform.[13] By early 1891 Husserl had in fact become convinced of the necessity of providing knowledge with a strong, formal, scaffolding in the Leibnizian sense of a *mathesis universalis* which would act as a guarantor of objectivity and be a safe bulwark against the incursions of psychologism or subjective idealism.[14]

It was in 1891, the year in which Husserl published the *Philosophy of Arithmetic*, that we first find clues that he was looking to formalist theories of arithmetic as a way of avoiding onerous problems with imaginary numbers and of establishing consistency in arithmetic. During that year he was working on the never to be published second volume of the *Philosophy of Arithmetic*.[15] In it he planned to deal with the fractions and negative and irrational numbers that he included under the heading of the imaginary.[16] Reading his 1890 drafts one is witness to his mounting frustration, and even anger, concerning the lack of any clear, logical understanding of the way in which mathematicians use such numbers, and then to the new confidence he finally won in 1891 in declaring arithmetic to be an analytic a priori discipline.[17]

That year Husserl wrote a letter to Carl Stumpf explaining his disillusionment with Brentano's methods and his new faith in the *arithmetica universalis*, which Husserl now thought of as a part of formal logic defined as a technique of signs making up a special, important chapter in logic as theory of knowledge. His new views, he wrote, would require important reforms in logic and he knew of no logic that would even give an account of the very possibility of a genuine calculational technique.[18]

So Husserl's earliest ideas on completeness were tied in not only with his inquiries into the logical foundations of the real number system, but also tied in with a more specifically philosophical quest to clarify the sense of the analytic a priori and develop a pure analytic logic free of any taint of psychologism. They reach deep into his reasons for abjuring Brentano's psychologism.

2. In Conflict with Frege

Husserl's earliest ideas on completeness also reach deep into his reasons for shunning Frege's logic, and so a good look at the connections between Husserl's ideas and formalist theories of mathematics actually provides insight into the nature of Husserl's clash with Frege, and sets into perspective the charge frequently leveled at Husserl that he lapsed back into psychologism not long after having repudiated it.

Husserl thoroughly studied Frege's *Foundations of Arithmetic* in the first volume of *Philosophy of Arithmetic*, sharply attacking the theory of extensionality and identity Frege espoused there.[19] Husserl never retracted those particular criticisms, but ten years later in the *Logical Investigations* he retracted the three pages of his criticism of Frege[20] in which he had denied that one could provide sound foundations for arithmetic by deriving theorems from a series of formal definitions in a purely logical fashion.[21]

In an 1891 letter, Husserl wrote to Frege that he had only a rough idea of how Frege would justify the imaginary in arithmetic since the path Husserl himself had found after much searching had been rejected by Frege in his 1885 article "On Formal Theories of Arithmetic." In the passage of the article cited in Husserl's letter,[22] Frege characterizes the unacceptable theory in question as one which but sets down rules by which one passes from the equations given to new ones in the way one moves chess pieces. Unless an equation contains only positive numbers, it no more has a meaning than the position of chess pieces expresses a truth. Now in virtue of these rules, Frege continues his criticism, an equation of positive whole numbers may actually appear. And if the rules are such that true equations can never lead to false conclusions, then only two results are possible: either the final equation is meaningless, or it has a content about which we can pass judgment. The latter will always be the case if it contains only positive whole numbers, and then it must be true, for it cannot be false. If the rules contain no contradictions among themselves, and do not contradict the laws of positive whole numbers, then no matter how often they are applied, no contradiction can ever enter in. Consequently, if the final equation has any meaning at all, it must be noncontradictory, and hence be true. This is a mistake, Frege concludes, for a proposition may very well be noncontradictory without being true.

The theory Frege has just opposed is apparently the one Husserl had come to favor. In his 1891 letter Husserl was not explicit about why he believed Frege's logic could not satisfactorily justify the imaginary in arithmetic. However, Husserl's reasons surely involved Frege's well-known thesis that in a logically perfect language, expressions that do not denote objects are unfit for scientific use. On his personal copy of Frege's article, Husserl particularly underlined the passage which reads: "The situation radically changes when one takes these figures to be signs of contents; in that case, the equation states that both signs have the same content. But if no content is present, the equation has no sense." In the margin Husserl wrote NB next to this passage.[23]

The use of signs or combinations of signs without reference was at the heart of Frege's dispute with formalists who, he believed, only manipulated signs without any regard for what those signs might stand for. In formal theories of arithmetic, Frege maintained, there is only talk of signs that neither have nor are meant to refer to objects. But, Frege insisted, "logic is not concerned with how thoughts, regardless of truth value, follow from thoughts" but "the laws of logic are first and foremost laws in the realm of reference . . . we have to throw aside proper names that do not designate or name an object. . . ."[24] Languages, Frege wrote in "On Sense and Reference":

> have the fault of containing expressions which fail to designate an object (although their grammatical form seems to qualify them for that purpose). . . . This arises from an imperfection of language from which even the symbolic language of mathematical analysis is not altogether free; even there combinations of symbols can occur that seem to mean something but (at least so far) do not mean anything, e.g. divergent infinite series. . . . A logically perfect language (*Begriffsschrift*) should satisfy the conditions, that every expression grammatically well constructed as a proper name out of signs already introduced shall in fact designate an object, and that no new sign shall be introduced as a proper name without being secured a meaning. . . . The history of mathematics supplies errors which have arisen in this way. . . . It is therefore by no means unimportant to eliminate the source of these mistakes.[25]

Frege explicitly allied himself with "extensionalist logicians." In my *Grundlagen* and the paper "Über formale Theorien der Arithmetik," he wrote,

> I showed that for certain proofs it is far from being a matter of indifference whether a combination of signs—e.g. $\sqrt{-1}$ has a reference or not, that, on the contrary, the whole cogency of the proof stands or falls with this. The reference is thus shown at every point to be the essential thing for science . . . the extensionalist logicians come closer to the truth in so far as they are presenting—in the extension—a reference as the essential thing.[26]

Husserl, however, sided with Hilbert who wrote of how:

> The method of ideal elements, that creation of genius, then allowed us to find an escape. . . . Just as i = $\sqrt{-1}$ was introduced so that the laws of algebra, those, for example, concerning the existence and number of the roots of an equation, could be preserved in their simplest form, just as ideal factors were introduced so that the simple laws of divisibility could be maintained even for algebraic integers (for example, we introduce an ideal common divisor for the numbers 2 and 1 +

$\sqrt{-5}$, while an actual one does not exist) so we must here adjoin the ideal propositions to the finitary ones. . . .[27]

Hilbert only stipulated one condition to which the method of ideal elements would be subject: "the proof of consistency; for extension by the addition of ideals is legitimate only if no contradiction is thereby brought in the old, narrower domain, that is, if the relations that result for the old objects whenever the ideal objects are eliminated are valid in the old domain."[28]

Over the next several years Husserl worked hard to sort through the problems raised as he tried to finish his book on the philosophy of arithmetic. The result was his *Logical Investigations*, the first volume of which is the antipsychologistic treatise, the *Prolegomena to Pure Logic*, in which Husserl formulated his views on completeness (§§69–70) in the way he would come to judge to be definitive (see *Formal and Transcendental Logic*, §28). The *Prolegomena* was surely written in the late 1890s.[29]

In 1900 Husserl was appointed to Göttingen where he would stay for the next sixteen years. There he was warmly welcomed into Hilbert's circle.[30] Husserl and Hilbert had much in common. Hilbert was just then lecturing on the calculus of variations (the subject of Husserl's doctoral thesis).[31] Hilbert had also just posited his axiom of completeness for arithmetic. He was an admirer of Weierstrass and Cantor.[32] Moreover, Hilbert was also just then corresponding with Frege on the very matter of truth and logical consistency at issue in the Frege passage I summarized above. As long as he had been thinking, writing, and lecturing on these things, Hilbert wrote of himself to Frege in late 1899, he had been saying that if the arbitrarily given axioms did not contradict one another, then they were true. This conception, he wrote, was the key to understanding his own recent work on the axioms of arithmetic and geometry, and the talk of completeness to be found there. Husserl had access to the Frege-Hilbert correspondence and partial copies of it, along with notes Husserl made on it which were found in his *Nachlass*.[33]

3. Completeness and the Imaginary

Invited by Hilbert and Felix Klein, Husserl addressed the Göttingen Mathematical Society in 1901 on the subject of completeness and the imaginary.[34] At the highest level, he told his audience, mathematics is the science of deductive systems in general. By appealing to a set of formal axioms which are consistent, independent, and purely logical in that they obey the principle of noncontradiction, it yields the set of propositions belonging to the theory defined. However, methodological questions arise when one tries to apply these formal techniques to the real number system and to specific domains of knowledge. These questions are a matter of serious concern to philosophers because their understanding of the general nature of the deductive sciences, and of theories in general, depends on their being able to resolve them. The development of the sciences, Husserl warned his listeners, had constantly shown that lack of clarity in the foundations ultimately wreaks its vengeance, that if certain levels of progress are reached, further progress is fettered by errors due to obscure methodological ideas (Husserl 1901, 431–32).

Questions regarding imaginary numbers, he continued, had come up in mathematical contexts in which formalization yielded constructions which arithmetically speaking were nonsense but which, astonishingly, could nonetheless be used in calculations. It became apparent that when formal reasoning was carried out mechanically as if these symbols had meaning, if the ordinary rules were obeyed, the results did not contain any imaginary components, then these symbols might be legitimately used. And this could be empirically verified (p. 432).

Husserl did not believe that general logic could shed light on the mystery because of the importance logicians accord to working with clear, precise, unambiguous concepts so that contradictions do not sneak in unnoticed. Logicians would ban contradiction, he said. For them contradictions only serve to show that a concept does not have an object, and contradictory concepts but yield contradictory consequences to which no object will correspond. But with the imaginary in mathematics that is plainly not the issue (p. 433).

Imaginary numbers may be countenanced, Husserl concluded, when they admit of formal definition within an enlarged consistent deductive system, and when the original formalized domain of deduction has the property that any proposition within the domain is either true on the basis of the axioms of that domain, or false, meaning in contradiction with the axioms. In addition, Husserl maintained, one needs to be clear about what is meant by a proposition's being in the domain. This, he argued, can only be determined if one can tell beforehand whether propositions deduced from the larger domain are situated in this sense in the more restricted domain and this can only be known if one knows from the outset that the proposition falls under this axiom in this sense. This is possible insofar as the axioms determine the domain completely, insofar as no other axiom may be added (p. 441).

Once in Hilbert's company Husserl did not slavishly copy Hilbert's views as is evinced by these remarks Husserl made regarding axioms of completeness. A domain, Husserl told the Göttingen Mathematical Society, could conceivably be defined as complete or incomplete by axioms. Namely, if one has all the basic principles from which all possible propositions of the domain are derivable, then one has the complete theory of the domain. Formalizing these basic principles one obtains a formal deductive system in which each proposition has a corresponding formal proposition. Among the axioms already defining the domain, however, is an axiom of closure which stipulates that the domain is determined by certain axioms and no others. Where this axiom is not added the domain remains open insofar as further axioms can perhaps be added and the objects of the domain receive different formal interpretations. This, Husserl warned, is not legitimate completeness, not something specifically characteristic of axiom systems, because we can make any axiom system, any deductive system quasi-complete by appealing to an axiom of that kind. So that kind of completeness can be of no use whatsoever to us. In extending an axiom system one obviously gives such an axiom up. A system of axioms with that kind of axiom cannot be extended. The concept of extending presupposes that no such axiom is involved. Moreover, Husserl continues, it is of course true that an axiom system closed from the outside in that spurious way already has the property sought, namely that one can tell whether or not a given proposition follows from the axioms or not. It need only contain the relations, forms of combination, in short the concepts, formally defined by the axiom system. If it has a meaning in terms of these definitions, it is either true in virtue of the definitions, or in contradiction with them (pp. 441–42).

Incidentally, Husserl considered the completeness of arithmetic to be self-evident because, as he explained to his Göttingen audience, any arithmetic, no matter whether it is defined to include just all positive numbers, all real numbers, positive rational numbers or rational numbers in general, etc., is defined by a system of axioms on the basis of which we can prove that any proposition derived exclusively from concepts established as being valid by the axioms either follows from the axioms or is in contradiction with them (p. 443).

4. Formal Logic and Complete Deductive Systems

Husserl's earliest struggles to provide the real number system with sound logical foundations and to establish the consistency of arithmetic soon evolved into a quest to secure sound logical foundations for all scientific knowledge. In 1929 he published *Formal and Transcendental Logic* (Husserl 1929) which was the product of decades of reflection upon the relationship between logic and mathematics, between mathematical logic and philosophical logic, between logic and psychology, and between psychologism and his own transcendental phenomenology. One of the stated goals of the book was to redraw the boundary line between logic and mathematics in light of the new investigations into the foundations of mathematics. A second goal was to examine the logical and epistemological issues such developments have raised (pp. 10–17, §11). The kinship with Hilbert's ideas is palpable, and the work surely benefited from years of direct participation in the events that had taken place in the mathematical world during his lifetime which would have had to condition any informed response to the questions Husserl was asking.

Now I want to look at some of the steps in Husserl's reasoning that led him from his questions about the consistency of the real number system to the actual development of a theory of formal logic with a theory of complete deductive systems as its highest task, and a transcendental logic as its complement. First, I need to look at how Husserl came to conceive the relationship between formal mathematics and formal logic.

Husserl believed that the formalization of large tracts of mathematics in the nineteenth century had laid bare the deep significant connections obtaining between formal mathematics and formal logic, and so had raised profound new questions about the deep underlying connections existing between the two fields. Logic and mathematics, he believed, had originally developed as separate fields because it had taken so long to elevate any particular branch of mathematics to the status of a purely formal discipline free of any reference to particular objects. Until that had been accomplished the important internal connections obtaining between logic and mathematics were destined to remain hidden. Once large tracts of mathematics had been formalized, however, parallels existing between its structures and those of logic became apparent, and the abstract, ideal, objective dimension of logic could be properly recognized, as it traditionally had been in mathematics.

In particular, developments in formalization had unmasked close relationships between the propositions of logic and number statements making it possible for logicians to develop a genuine logical calculus which would enable them to calculate with propositions in the way mathematicians did with numbers, quantities and the like (Husserl 1929, 23–27).

Extensionalist logicians, Husserl says, had also worked on the logical foundations of mathematics, and had come to some of the same conclusions he had. But in *Formal and Transcendental Logic*, Husserl condemns their work as fundamentally misguided and unclear. He qualifies extensional logic as naive, risky, and doubtful. He complains that it has been the source of many a contradiction requiring every kind of artful device to make it safe for use in mathematical reasoning. However, Husserl credits extensionalist logicians with having managed to raise some highly interesting philosophical questions, and with having succeeded in imparting a genuine sense of the common ground existing between mathematics and logic to mathematicians, whose work is relatively unhindered by the particular lack of clarity involved (Husserl 1929, 74, 76, 83).

Husserl turned to Bolzano for a theory of meaning and analyticity appropriate to the true logical calculus he now envisaged. Husserl's early work on the philosophy of arithmetic, I have argued, had convinced him that arithmetic was an analytic a priori discipline. Initially, Husserl had believed that Bolzano's theories regarding ideal propositions in themselves and truths in themselves involved an appeal to abstruse metaphysical entities, but in the 1890s it all of a sudden became clear to Husserl that Bolzano had actually been talking about something fundamentally completely understandable, namely the meaning of an assertion, what was declared to be one and the same thing when one says of different people that they affirm the same thing. This realization demystified meaning for Husserl.[35]

Husserl was persuaded of the inadequacy of Kant's analytic-synthetic distinction, and he came to believe that Bolzano's more Leibnizian approach to analyticity and meaning harbored the insights logicians needed to prove their propositions by purely logical means, meaning analytically in virtue of the meaning of their terms in a way analogous to mathematical reasoning. However, in Husserl's opinion, Bolzano never saw the internal equivalence between the analytic nature of both formal logic and formal mathematics made possible by developments in the field of mathematics that had only taken place after his death.[36]

At the same time, Husserl's familiarity with deductive techniques employed by his contemporaries drew him to see the advantages of deductive reasoning patterned after the formal reasoning advocated in formalist theories of mathematics. He believed this could be applied to formal logic as a technique for deriving propositions from propositions in a purely logical, analytic way. Hilbert's stringent requirements regarding consistency and completeness were much the same as Husserl's own ideas as to how propositions were to be derived from propositions.

Husserl believed the highest task of formal logic to be the theory of complete deductive systems and the complete manifolds which were their counterparts in the objective order. In *Formal and Transcendental Logic* §28, Husserl cites the characterization of complete deductive systems he had given in the 1900 *Prolegomena* §70 saying that he could not improve upon it. "The objective correlate of the concept of a possible theory determined only in its form," he had written there, "is the concept of any possible domain of knowledge that would be governed by a theory having such a form, what mathematicians call a manifold." It is then, according to his theory, a domain determined solely by the circumstance that it comes under a theory having such a form, that among the objects belonging to the domain certain connections are possible. In respect of their matter the objects remain completely indeterminate, are exclusively determined by the form of the combinations ascribed to them, combinations which are themselves only formally determined by the elementary laws assumed to hold good for them.

For Husserl, the great advance of mathematics as developed by Riemann and his successors consisted not just in appealing to the form of deductive systems, but also in having gone on to view such systems of forms as mathematical objects, to be altered freely, universalized mathematically.

Husserl wrote of what he called the hidden origin of the concept of a complete manifold that: "If the Euclidean ideal were actualized, then the whole infinite system of space-geometry could be derived from the irreducible finite system of axioms by purely syllogistic deduction according to the principles of lower level logic, and thus the a priori essence of space could become fully disclosed in a theory . . . the transition to form then yields the form idea of any manifold that conceived as subject to an axiom system by formalization could be completely explained nomologically in a deductive theory that would be 'equiform' with geometry." If, he continued, a manifold is conceived as defined and determined exclusively by such a system of forms of axioms belonging to the theorems and component theories, then ultimately the whole science forms necessarily valid for such a manifold can be derived by pure deduction.

And then Husserl asked just what it was that purely formally characterized a self-contained system of axioms as "complete," as a system by which actually a "manifold" would be defined. Every formally defined system of axioms, he noted, has an infinity of deducible consequences, but a manifold governable by an explanatory nomology includes the idea that there is no truth about such a domain that is not deducibly included in the "fundamental laws" of the corresponding nomological science; it is not defined by just any formal axiom system, but by a complete one. Such an axiom system is characterized by the fact "that any proposition . . . that can be constructed in accordance with the grammar of pure logic out of concepts . . . occurring in that system is either true, that is to say: an analytic (purely deducible) consequence of the axioms, —or false, —that is to say: an analytic contradiction—; *tertium non datur.*" For Husserl such axiom systems represented the highest level of formal logic (Husserl 1929, 28–36).

5. Formal and Transcendental Logic

However, Husserl's logicians cannot leave matters there. For them formal logic alone cannot suffice in Husserl's sense of logic as a theory of science, an enlarged analytics. As theoreticians of science in general, they are also obliged to contend with the question of basic truths about a universe of objects existing outside of formal systems; they are called upon to seek solutions to the problems that come up when scientific discourse steps outside the purely formal domain and makes reference to specific objects or domains of objects. They are not free to sever their ties with nature and science, to accept a logic that tears itself entirely away from the idea of any possible application and becomes a mere ingenious playing with thoughts, or symbols that mere rules or conventions have invested with meaning. They must step out of the abstract world of pure analytic logic with its ideal, abstract entities, and confront those more tangible objects that make up the material world of things. In addition, they are obliged to step back and investigate the theory of formal languages and systems themselves, and their interpretations, to engage in what we call metatheory (Husserl 1929, 109–10, 52).

Even in playing a game one actually judges, collects, counts, and draws conclusions, Husserl points out. On the purely formal level complete manifolds could be viewed as deductive games with symbols. But one is not dealing with an actual theory of manifolds, he maintains, until one regards the game symbols as being signs for actual abstract objects, units, sets, manifolds, etc., and until the rules of the game acquire the status of laws applying to these manifolds (Husserl 1929, 99).

Husserl's logicians must also see that the logical sense of the formal sciences also includes a sphere of cognitive functioning and a sphere of possible applications, and they must also set about trying to answer the difficult questions regarding the way they themselves interact with both the objective structures of the abstract realm of formal logic and mathematics, and those of the material order. So once armed with the objective structures of formal logic, Husserl's logicians are still obliged to go further and to come to terms with really hard ontological questions about the objects involved and equally hard epistemological questions about subjectivity. Philosophical logicians cannot ignore these problems. Formal logic requires a complement in the form of what Husserl called a transcendental logic (Husserl 1929, 109, 111).

In the introduction to *Formal and Transcendental Logic* Husserl wrote that the problem guiding him originally lay in determining the sense of, and isolating, a pure analytic logic of noncontradiction was one of evidence: the problem of the evidence of the sciences making up formal mathematics. He was, he says, particularly struck by the fact that the evidence making up the truths of formal mathematics and formal logic is of an entirely different order than that of other a priori truths in that the former do not involve any intuition of objects or states of affairs whatsoever (Husserl 1929, 12). The formalness of these disciplines lies precisely in their relationship to "anything whatever," with a most empty universality which makes no reference to any actual material interpretation, to any material particularly characterizing the objects or domain of objects (Husserl 1929, 87).

Mathematics has its own purity and legitimacy. Mathematicians are free to create arbitrary structures. They need not be concerned with questions regarding the actual existence of their formal constructs, nor with any application or relationship their constructs might have to possible experience, or to any transcendent reality. They are free to do ingenious things with thoughts or symbols that receive their meaning merely from the way in which they are combined, to pursue the necessary consequences of arbitrary axioms about meaningless things, restricted only by the need to be noncontradictory and coordinated to concepts previously introduced by precise definition (Husserl 1929, 138).

And the same, Husserl contends, is true for formal logic when it is actually developed with the radical purity which is necessary for its philosophical usefulness and gives it the highest philosophical importance. Severed from the physical world, it lacks everything that makes possible a differentiation of truths or, correlatively of evidences. However, real philosophical logic, in Husserl's sense of a theory of science, a *Wissenschaftslehre*, can only develop in connection with transcendental phenomenology by which logicians penetrate an objective realm which is entirely different from them (Husserl 1929, p. 8, §23).

Husserl's logicians are not brain-dead machines. Far from it. His logicians can only submit to a logic which they have thought through and thought through with insight, a fact which he believed cried out for epistemological investigations into the subjective and intersubjective cognitive processes that inevitably interact with the objective order.

For Husserl, as for Bertrand Russell and Jean-Paul Sartre, it was Brentano's theory of the intentionality of mental acts that indicated the way out of the mind to things. According to Brentano's famous thesis of intentionality every mental phenomenon is characterized, by reference to a content, direction toward an object. For Husserl intentionality acted as the philosophical logician's bridge from the mind to the objective order, and was the key that unlocked the way to transcendental analyses which chart the mind's path to things (Husserl 1929, §§42–44, 210, 245). Husserl was keenly aware of the pitfalls of subjectivity and as a part of his project to overcome them he undertook exhaustive studies of the knowing process, finally prescribing a demanding series of mental exercises designed to instill rigor into epistemology by training philosophers to reason (Husserl 1929, 274) in ways which Husserl hoped would make the psychological chaff fly, and the transcendental grain lie sheer and clear (Husserl 1929, 13, 237, §61).

Hilbert too believed that his formal logic alone could not suffice. "If scientific knowledge is possible," he maintained, "certain intuitive conceptions and insights are indispensable."[37] "No more than any other science," he wrote,

> can mathematics be founded by logic alone; rather as a condition for the use of logical inferences and the performance of logical operations something must already be given in our faculty of representation, certain extralogical concrete objects that are intuitively present as immediate experience prior to all thought. If logical inference is to be reliable, it must be possible to survey these objects completely in all their parts, and the fact that they occur, that they differ from one another, and that they follow each other, or are concatenated, is immediately given intuitively, together with the objects, as something that neither can be reduced to anything else nor requires reduction. This is the basic philosophic position that I regard as requisite for mathematics and, in general, for all scientific thinking, understanding and communication.[38]

In many respects Husserl may be viewed as one who endeavored to provide the philosophical complement to Hilbert's views.

Before concluding, I would like to draw attention to an additional difference between Husserl and Frege. In *Formal and Transcendental Logic*, Husserl also uses the example of identity statements to point out differences between the formal order and the material order which are relevant to philosophers who have followed the discussions on sense and reference in this century. This is in reference to identity statements of the form '*a* is *b*', and '*a* is *c*'. Husserl argues that the objects designated in these statements are non-self-sufficient under all circumstances. They are what they are within the context of the whole, and different wholes can have components which are equal, but not the same in all ways. If we say '*a* is *b*' and '*a* is *c*', the *a* in the first statement is not identical to the *a* of the second, he maintains. The same object *a* is meant in both cases, but in a different how (Husserl 1929, 295–96).

Mathematicians, Husserl acknowledges, are not in the least interested in the different ways objects may be given. For them objects are the same which have been correlated together in some self-evident manner. However, Husserl warns, logicians who do not bewail the lack of clarity involved here, or who claim that the differences do not matter are not philosophers, since here it is a matter of insights into the fundamental nature of formal logic, and without a clear grasp of the fundamental nature of formal logic, one is obviously cut off from the great questions that must be asked about logic and its role in philosophy (Husserl 1929, 147–48).

Here we have a fundamental and abiding difference between Husserl and Frege who always insisted that in his logic, as in mathematics, there was no difference between identity and equality.[39] Husserl had taken issue with Frege on this very matter in the *Philosophy of Arithmetic*,[40] and as can be imagined from what has just been said, Husserl's own views on the question played a determinant role in all his philosophical and logical investigations.

6. Conclusions

The formalistically inclined thinker that I have described as hard at work rigorizing philosophical thought is a far cry from being the man whom both detractors and disciples have so often depicted as wantonly engaging in an orgy of subjectivity. Husserl did have a distressing propensity to deal with everything that had to do with subjectivity, so those portrayals are not utterly without foundation. But he did insist that the subjective order could not be properly examined until the objective order had been, and until the objectivity of the structures girding scientific knowledge had been established and demonstrated. He maintained that pure logic with its abstract ideal structures would have to be clearly seen and definitely apprehended as dealing with ideal objects before transcendental questions about them could be asked (Husserl 1929, §§8, 9, 11, 26, 42–44, and pp. 81–82, 111, 225, 246, 258, 263, 266).

In the beginning of this paper I wanted to inquire into the deep, under-lying reasons that may have drawn Husserl and Hilbert to want to establish a criterion of completeness for formal reasoning. If any definitive answer to my question is possible, finding it would require a vastly more thoroughgoing investigation than is possible here. In particular, a great deal more would have to be known about Husserl's encounter with what Hilbert called "Can-tor's majestic world of ideas."[41] I hope, however, that I have at least suc-ceeded in showing that Husserl's and Hilbert's reasons were tied in with their conviction that the real number system could only be grounded by a logic that countenanced reference to ideal entities. A logic that could not cope with expressions whose, in Husserl's words, "absurdity is mediate, i.e., the countless expressions shown by mathematicians in lengthy indirect demonstrations, to be objectless a priori" could never, they thought, provide secure foundations for knowledge.

I would also like to suggest that approaching Husserl's thought in light of his views on completeness, analyticity, meaning, and identity may also help demystify his phenomenology and so shed light and order where confusion and ineffability have seemed to many to reign. And it could provide keys to understanding some things that have seemed inaccessible about Husserl's thought to those schooled in the logical and epistemological views fashioned by philosophers for whom a rival philosophy of logic and mathematics has occupied a central position in their philosophical views. Husserl was far from ignorant of the developments in mathematics and logic that made *Principia Mathematica* and related systems possible, nor did he turn his back on those developments. He worked long and hard to resolve questions raised by them that are still under discussion today and his exhaustive studies merit study now.

Undertaking such a job, however, is not for the fainthearted for they will find themselves investigating the workings of the human mind, and traffick-ing in intensional phenomena like concepts, essences, properties, and attrib-utes, not to mention courting the a priori and the ideal. Husserl devoted his life to investigating these irksome phenomena which many others have hoped dearly to eradicate by rigorously applying techniques inspired by another logic. In his attempts to find clarity with respect to the central traits of reality, Husserl in fact incorporated into his philosophy almost everything extensionally minded philosophers hoped to ban.

NOTES

* This paper is based on a paper read in April 1992 at a conference on the devel-opment of the foundations of mathematics from 1850–1930 organized by Jaakko Hintikka, which was part of the program of the Boston Colloquium for the Phi-losophy and History of Science. I must thank Dagfinn Føllesdal, who also partic-ipated in the session, for his thorough reading of my text and his insightful sug-gestions as to how to improve it.

This paper was, to my knowledge, the first article-length work to have been written on the subject. Researching it persuaded me of the need to investigate the relationship of Husserl's ideas to those of Georg Cantor before coming to any further conclusions about the relationship between Husserl's and Hilbert's ideas. Chapters 6, 7, and 8 of the present book are the fruit of those efforts.

Chapter 9, written in 1998 and 1999, benefited from that subsequent research on Cantor and Husserl and also from the publication of Husserl's *Logik und allgemeine Wissenschaftstheorie* for the first time in 1996, which has an entire chapter on Husserl's theory of manifolds that is incomparably clearer and more explicit than what is to be found in his other available writings.

In addition, I have since had many exchanges with Jairo da Silva of the Mathematics Department of the University of São Paulo. He has recently convinced me of the importance of the distinction Husserl makes between relative completeness and the absolute completeness of Hilbert. Look for his paper "Husserl's Two Notions of Completeness," forthcoming in *Synthese*. Much research remains to be done on the relationship between Husserl's and Hilbert's ideas.

1. D. Hilbert's "Über den Zahlbegriff" was first published in the *Jahresbericht der Deutschen Mathematiker-Vereinigung* 8 (1900): pp. 180–84, and subsequently as an appendix to post-1903 editions of his *Grundlagen der Geometrie*. I have cited the translation of Hilbert's axioms for arithmetic appearing in M. Kline, *Mathematical Thought From Ancient to Modern Times*, vol. 3 (New York: Oxford University Press, 1972), pp. 990–91.
2. See E. Husserl's *Ideas, General Introduction to Pure Phenomenology* (New York: Colliers, 1962 [1913]), §72 and note, his *Formal and Transcendental Logic* (The Hague: M. Nijhoff, 1969 [1929]), §§28–36, and his *The Crisis of European Sciences and Transcendental Phenomenology* (Evanston: Northwestern University Press, 1970), §9f and note. In these texts Husserl refers back to his discussions in the *Logical Investigations* (New York: Humanities Press, 1970), *Prolegomena*, §§69 and 70 and to the then unpublished material from his Göttingen period, now published in appendices to his *Philosophie der Arithmetik, mit ergänzenden Texten*, Husserliana, vol. XII (The Hague: M. Nijhoff, 1970). Of particular interest is the chapter on manifolds in the recently published *Logik und allgemeine Wissenschaftstheorie*, Husserliana, vol. XXX (Dordrecht: Kluwer, 1996).
   As usual there are some terminological obstacles that make it hard to see the connection Husserl's ideas have with the logical tradition most familiar to readers of English. First of all, for complete and completeness Husserl uses the German words '*definit*' and '*Definitheit*' in the place of Hilbert's '*vollständig*' and '*Vollständigkeit*'. Since in the passages cited above Husserl maintains that his concept of *Definitheit* is exactly the same as Hilbert's *Vollständigkeit*, I have tried to avoid the terminological confusion by translating Husserl's terms by the more familiar 'complete' and 'completeness,' although Husserl translators have understandably chosen 'definite' and 'definiteness.' Second, in the above texts Husserl refers to his theory of complete *Mannigfaltigkeiten*, a term which has been translated by 'multiplicity' or 'manifold'. For Husserl complete *Mannigfaltigkeiten* are the objective correlates of complete axiom systems.
3. S. Bachelard, *A Study of Husserl's Formal and Transcendental Logic* (Evanston, IL: Northwestern University Press, 1968), pp. 59–61; J. Cavaillès, *Sur la logique et la théorie de la science* (Paris: Presses Universitaires de France, 1947),

pp. 70, 73; R. Schmit, *Husserls Philosophie der Mathematik* (Bonn: Bouvier, 1981), pp. 67–86. H. Lohmar, "Husserls Phänomenologie als Philosophie der Mathematik," Ph.D. diss., Cologne, 1987, pp. 151–62; Guillermo E. Rosado Haddock, "Edmund Husserls Philosophie der Logik und Mathematik im Lichte der gegenwärtigen Logik und Grundlagenforschung," Ph.D. diss., Rheinischen Friedrich-Wilhelms Universität, Bonn, 1973; B. Picker, "*Die Bedeutung der Mathematik für die Philosophie Edmund Husserls,*" *Philosophia Naturalis* 7 (1962) 266–355, his Ph.D. diss., Münster, 1955.

4. See for example the note to Husserl's *Ideas* §72.

5. K. Schuhmann, *Husserl-Chronik* (The Hague: M. Nijhoff, 1977), pp. 6–11. I also discuss Husserl's background throughout my *Word and Object in Husserl, Frege and Russell: Roots of Twentieth Century Philosophy* (Athens, OH: Ohio University Press, 1991) and in "Husserl and Frege on Substitutivity," chapter 1 of the present book.

6. L. McAlister, *The Philosophy of Franz Brentano* (London: Duckworth, 1976), pp. 45, 49, 53; A. Osborn, *The Philosophy of E. Husserl in its Development to his First Conception of Phenomenology in the Logical Investigations* (New York: International Press, 1934), pp. 12, 17, 18, 21; M. Dummett, *The Interpretation of Frege's Philosophy* (Cambridge, MA: Harvard University Press, 1981), pp. 72–73, and his *Frege, Philosophy of Language*, 2nd ed. rev. (London: Duckworth, 1981), p. 683; Hill, *Word and Object in Husserl Frege and Russell*, pp. 59–67 and chapter 7.

7. A. Fraenkel, "Georg Cantor," *Jahresbericht der Deutschen Mathematiker Vereinigung* 39 (1930): 221, 253 n., 257; E. Husserl, *Introduction to the Logical Investigations* (The Hague: M. Nijhoff, 1975), p. 37 and notes; J. Cavaillès, *Philosophie Mathématique* (Paris: Hermann, 1962), p. 182; Schmit, pp. 40–48, 58–62; L. Eley, "Einleitung des Herausgebers" to Husserl's *Philosophie der Arithmetik, mit ergänzenden Texten*, pp. XXIII-XXV; *Georg Cantor Briefe*, ed. H. Meschkowski and W. Nilson (New York: Springer, 1991), pp. 321, 373–74, 379–80, 423–24. Two Cantor letters dating from 1895 are published in W. Purkert and H. Ilgauds, *Georg Cantor 1845–1918* (Basel: Birkhäuser, 1991), pp. 206–7.

8. Kline, *Mathematical Thought From Ancient to Modern Times*, vol. 3, pp. 950–56, 960–66; McAlister, *The Philosophy of Franz Brentano*, p. 49; Osborn, *The Philosophy of E. Husserl in its Development to his First Conception of Phenomenology in the Logical Investigations*, p. 18; Husserl, *Introduction to the Logical Investigations*, p. 37.

9. H. Wang, *From Mathematics to Philosophy* (London: Routledge and Kegan Paul, 1974), pp. 145–52 in reference to B. Bolzano's 1837 *Wissenschaftslehre* §§148 and 155. Bolzano's book has been partially translated as *Theory of Science* by R. George (Oxford: Blackwell, 1972) and B. Terrell (Dordrecht: Reidel, 1973).

10. See the Husserl texts cited in note 1.

11. A. Whitehead, *An Introduction to Mathematics* (Oxford: Oxford University Press, 1958 [1911]), pp. 62–64.

12. Appendix VI to Husserl's *Philosophie der Arithmetik, mit ergänzenden Texten*, p. 433 and his *Formal and Transcendental Logic*, §31.

13. Husserl, *Introduction to the Logical Investigations*, pp. 33–36.

14. Hill, *Word and Object in Husserl, Frege and Russell*, pp. 80–95.

15. Good accounts of Husserl's work during the 1890s are given in the editors' introductions to the Husserliana editions of Husserl's *Philosophie der Arithmetik, mit ergänzenden Texten* (vol. XII), *Logische Untersuchungen* (vol.

XVIII) and *Studien zur Arithmetik und Geometrie* (vol. XXI).

16. E. Husserl, *Philosophie der Arithmetik* (Halle: Pfeffer, 1891), p. viii (note this is not the Husserliana edition cited above for the posthumously published material, but Husserl's 1891 book). Hill, *Word and Object in Husserl, Frege and Russell*, pp. 84–86.

17. Husserl, *Philosophie der Arithmetik, mit ergänzenden Texten*, pp. 340–429.

18. Cited in Hill, *Word and Object in Husserl, Frege and Russell*, p. 85. See also D. Willard, *Logic and the Objectivity of Knowledge* (Athens, OH: Ohio University Press, 1984), pp. 115–16.

19. Husserl, *Philosophie der Arithmetik* (1891), pp. 104–5, 132–34. I discuss his arguments in depth in "Husserl and Frege on Substitutivity," chapter 1 of the present book.

20. Husserl, *Logical Investigations, Prolegomena*, p. 179 n. Husserl actually retracted pp. 129–32, not pp. 124–32 as a typographical error in the English edition indicates.

21. Husserl, *Philosophie der Arithmetik* (1891), pp. 130–31.

22. G. Frege, *Philosophical and Mathematical Correspondence* (Oxford: Blackwell, 1980), p. 65, in reference to Frege's article "On Formal Theories of Arithmetic," *Collected Papers on Mathematics, Logic and Philosophy*, ed. B. McGuinness (Oxford: Blackwell, 1984), pp. 112–21.

23. Frege's *Collected Papers*, pp. 118–19. Husserl's own copy of Frege's article is now in the Husserl library at the Husserl Archives in Leuven, Belgium.

24. G. Frege, *Posthumous Writings* (Oxford: Blackwell, 1979), p. 122. See also G. Frege, *Translations from the Philosophical Writings*, 3rd ed. (Oxford: Blackwell, 1980), pp. 22–23, 32–33, 162–213.

25. Frege, *Translations from the Philosophical Writings*, pp. 69–70.

26. Frege, *Posthumous Writings*, p. 123.

27. D. Hilbert, "On the Infinite," *From Frege to Gödel*, ed. J. van Heijenoort (Cambridge, MA: Harvard University Press, 1967), p. 379.

28. Ibid., p. 383.

29. See the introduction to Husserl's *Logical Investigations*, Husserliana, vol. XVIII.

30. See the introduction to *Studien zur Arithmetik und Geometrie*, Husserliana, vol. XXI (The Hague: M. Nijhoff, 1984), p. XII, where a 1901 letter from Husserl's wife is cited.

31. C. Reid's *Hilbert* (New York: Springer, 1970), pp. 67–68 and *Hilbert and Courant in Göttingen and New York* (New York: Springer, 1976) provide anecdotal material about Husserl's time in Göttingen; Schuhmann, *Husserl-Chronik*, p. 10, Husserl's thesis entitled "Beitrag zur Theorie der Variationsrechnung."

32. As Hilbert makes evident in "On the Infinite," *Frege to Gödel*, ed. van Heijenoort pp. 369–92.

33. Frege, *Philosophical and Mathematical Correspondence*, pp. 34–51, and *Gottlob Freges Briefwechsel mit D. Hilbert, E. Husserl, B. Russell* (Hamburg: Meiner, 1980), pp. 3, 47. Also my "Frege's Letters," *From Dedekind to Gödel*, ed. J. Hintikka (Dordrecht: Kluwer, 1995), pp. 97–118.

34. The notes for Husserl's lecture are published as an appendix to *Philosophie der Arithmetik, mit ergänzenden Texten*, pp. 430–506. They are cited in the text as (Husserl 1901). Concerning the invitation see Husserl's wife's letter cited in note 30.

35. E. Husserl, "Review of Melchior Palágyi's *Der Streit der Psychologisten und Formalisten in der modernen Logik*," *Early Writings in the Philosophy of Logic*

and Mathematics (Dordrecht: Kluwer, 1994), p. 201.

36. Husserl, *Introduction to the Logical Investigations*, pp. 36–38, 48. *Formal and Transcendental Logic*, pp. 184–85, 225.
37. van Heijenoort, ed., p. 392.
38. Ibid., pp. 464–65, and 376.
39. Frege, *Translations From the Philosophical Writings*, pp. 22–23, 120–21, 141n., 146n., 159–61, for example.
40. I discuss this at length in my *Word and Object in Husserl Frege and Russell*, especially chapter 4, and in "Husserl and Frege on Substitutivity," chapter 1 of the present book.
41. *From Frege to Gödel*, van Heijenoort, ed., p. 437. Since I wrote this I myself have investigated the relationship between Husserl's and Cantor's ideas. See chapters 6, 7, 8 and 9 of the present book.
42. Husserl, *Logical Investigations*, First Investigation, pp. 293–94.

11

Guillermo E. Rosado Haddock

TO BE A FREGEAN OR TO BE A HUSSERLIAN: THAT IS THE QUESTION FOR PLATONISTS

1. Introduction

There is a historiographical myth or tale in analytic circles according to which in his youth Husserl was a very naive philosopher who in his *Philosophie der Arithmetik*[1] of 1891 not only propounded an extreme form of psychologism but also dared to criticize the almighty Frege's views as presented in *Die Grundlagen der Arithmetik*[2] of 1884. According to the same tale, it was Frege's 'devastating' critique of Husserl's book in 1894 and the study by Husserl of other of Frege's writings which were responsible for Husserl's abandonment of psychologism in the first volume of his *Logische Untersuchungen*[3] of 1900/1901 and his embracing of Frege's views on logic, mathematics and their relationship, and of Frege's distinction between the sense and reference of expressions in the First Logical Investigation. Husserl, however, so says the tale, fell once more out of grace into psychologism in the second volume of *Logische Untersuchungen* and never again freed himself from such a pernicious addiction. To this historiographical myth have adhered many influential scholars in the analytic tradition, e.g., Evert W. Beth in *The Foundations of Mathematics*,[4] Michael Dummett in *Frege: Philosophy of Language*,[5] Dagfinn Føllesdal in *Husserl und Frege*,[6] and, of course, almost every Fregean scholar that has ever mentioned this issue, e.g., Hans Sluga[7] and Christian Thiel,[8] to name just two of the most distinguished. It is then no mystery that Husserl's views on logic and mathematics have been completely ignored in the analytic tradition.

The historiographical myth has been challenged in my dissertation of 1973[9] and especially in my paper "Remarks on Sense and Reference in Frege and Husserl,"[10] and also by J. N. Mohanty in various writings,[11] and more recently and forcefully by Claire Ortiz Hill in her *Word and Object in Husserl, Frege and Russell*[12] and in other writings. The result of such investigations is essentially the following: (1) *Philosophie der Arithmetik*, although published in 1891, represents Husserl's views at most up to 1890; (2) Husserl made the distinction between the sense and reference of expressions around 1890, and it is present in his review of the first volume of Ernst Schröder's *Vorlesungen über die Algebra der Logik* also published in 1891, as Frege himself acknowledged in a letter to Husserl of May of that same year;[13] (3) Husserl's views on logic and mathematics as presented in *Logische Untersuchungen* and other later writings were developed from 1890 to 1895 with total independence of Frege, but under the influence of Bolzano, Lotze, and others, and of the mathematical work of Riemann, Cantor, and others, and are clearly distinct from Frege's; (4) there was no conversion to psychologism in the second volume of *Logische Untersuchungen* and later writings. By the way, as Claire Ortiz Hill has shown,[14] Husserl was not the propounder of a naive extreme psychologism in *Philosophie der Arithmetik* as Frege and his uncritical followers would like us to believe. But even if that were the case, it is a very unusual piece of scholarship to consider only a philosopher's early views on a subject while completely ignoring his mature views. If Kantian scholars from the very beginning had examined only Kant's pre-critical writings, we would very probably never had learnt about his duly famous views on space and time in his critical philosophy.

In this essay I will discuss some Husserlian views on logic, mathematics, and related issues which are clear alternatives to well known Fregean views. Thus, it seems adequate to say a few words about these Fregean views which could serve as a sort of motivation for the discussion of their alternatives. In *Die Grundlagen der Arithmetik*, the first volume of *Grundgesetze der Arithmetik*,[15] and elsewhere Frege argued on behalf of Platonism in mathematics and of the logicist thesis according to which mathematics (geometry excluded) is but a well-developed branch of logic. Thus, mathematical concepts should be definable in terms of logical ones, and mathematical theorems should be obtained, with the help of definitions, from logical axioms. In particular, the concept of finite cardinal number should be defined in terms of logic. On this issue we will simply mention—without any discussion—Frege's two most serious attempts to define the concept of number in *Die Grundlagen der Arithmetik*, namely:

(1) The number that corresponds to the concept F is the same as the number that corresponds to the concept G if and only if the concept F is equinumerous to the concept G.

(2) The number that corresponds to the concept F is the extension of the concept "equinumerous to the concept F."

(As is well known, Frege rejected the contextual definition [1] and adopted definition [2].) In *Grundgesetze der Arithmetik* Frege once more conceived numbers as extensions, which in this work are considered as presumably special cases of what he now conceived as the foremost logical objects, namely, courses-of-values. He introduced courses-of-values in the notorious Principle V, or Basic Law V, namely (in a non-Fregean notation)[16]:

$$x\!f(x) = y\!g(y) :=: (\forall z)\,(f(z) = g(z))$$

This principle, which has a great resemblance to the two rival definitions of (cardinal) number mentioned above, expresses a sort of identity between an identity statement between courses-of-values and the generalization for all arguments of the identity of the values of the corresponding functions. There are plenty of issues related both to Principle V and to the two above mentioned attempts to define the concept of number in *Die Grundlagen der Arithmetik*, as attested to by the bulk of the secondary literature. But we cannot dwell on those issues here.

On the other hand, although Frege propounded mathematical ontological Platonism, he never seriously tried to explain how it is that we have knowledge about mathematical entities and their properties. There was no epistemological Platonism to complement his ontological Platonism. This was so not because he advocated a sort of Kantian epistemology, as some Fregean scholars would like us to believe,[17] but because he was afraid of falling into the abyss of psychologism. It should be mentioned here, however, that in *"Der Gedanke"*[18] of 1918, when he was on the brink of seriously questioning his views, he made some timid remarks concerning our knowledge of abstract objects like thoughts. Of course, later, in 1924/25, he made more explicit and valiant remarks about epistemological matters. But by then he had abandoned his logicism and Platonism, and was trying to make a new beginning as a recently converted Neo-Kantian.

Finally, with respect to the distinction between sense and reference presented by Frege in *"Funktion und Begriff"*[19] and expounded with much more detail a year later in *"Über Sinn und Bedeutung,"*[20] in the first volume of *Grundgesetze der Arithmetik*[21] and in many other writings, it should be mentioned that for him the sense of a statement (i.e., a declarative sentence) is a thought and its reference a truth value, namely, the true or the false. Thus, there is no other entity between the thought and the truth value of a statement. As is well known, in *"Über Sinn und Bedeutung"*[22] Frege argued both that the thought cannot be the referent of statements—and is, therefore, its sense—since it is not invariant under substitution of expressions for expressions with different sense but the same referent, and that only the truth value remains invariant under such substitutions.

From the 1890s onwards, Husserl was a mathematical Platonist of a different sort than Frege, and not only rejected logicism but also other sorts of reductionism like set-theoreticism. Moreover, he was not only an ontological Platonist but an epistemological Platonist, who developed a detailed account of our acquaintance with abstract entities. This account has been totally ignored in the literature, although there are some recent caricatures attributed to Husserl on this issue.[23] Finally, Husserl not only obtained the distinction between sense and reference with complete independence of Frege—as Frege himself acknowledged[24]—but his distinctions in the case of statements are (i) more detailed than Frege's, (ii) more fruitful than Frege's, and (iii) serve to explain some confusions incurred by Frege. In the rest of this essay I will expound Husserl's views on logic and mathematics, his epistemology of mathematics, and his distinction between sense and reference with respect to statements.

2. Husserl's Views on Logic and Mathematics

As Frege, Husserl conceived logic and mathematics as intimately related. The relation envisaged by Husserl, however, was quite different from that envisaged by Frege. Instead of being a sort of extension of logic, mathematics (physical geometry excluded, as in Frege's case) was for Husserl a sort of ontological correlate of logic or, to use some colorful expressions, it was not logic's daughter, as for Frege, but logic's ontologically fat sister discipline. Both logic and mathematics were analytic in Husserl's usage of the word 'analytic', which is clearly different from Frege's. (For Husserl a statement is analytic if it is not only true, but would remain true after complete formalization, whereas for Frege a statement is analytic if it is derivable from the logical axioms.) Physical geometry, on the other hand, was for Husserl as synthetic a priori as for Frege. Husserl's views on logic and mathematics did not suffer any major change from the mid 1890s onwards, not even after he had embraced transcendental phenomenology. If you compare his presentation of such views in *Logische Untersuchungen* and in his *Einleitung in die Logik und Erkenntnistheorie*,[25] his somewhat brief comments in *Ideen I*[26] of 1913, and his detailed treatment in *Formale und transzendentale Logik*[27] of 1929, the only significant changes in this later work compared to the others are a clearer separation between syntax and semantics and a premature optimism with respect to the prospects of establishing a sort of completeness of all logico-mathematical systems.[28] (As many of his contemporaries, before Gödel's and Tarski's revolutionary writings,[29] Husserl did not distinguish clearly between deductive [or syntactic] completeness and semantic completeness.) I am not going to dwell here on such particular issues, but rather concentrate on the invariant bulk of his views.

According to Husserl,[30] the first and most basic stratum of logic is a sort of logico-grammatical stratum, a sort of morphology of propositions, as he calls it in *Formale und transzendentale Logik*. This stratum, which is a sort of propaedeutic for the remaining strata, is the study of the constitutive concepts, the building blocks of every theory. They are, firstly, the fundamental meaning categories, like those of proposition and concept, the forms of combination of pre-propositional elements to form what we now call 'atomic propositions', the combination of those atomic propositions by means of logical connectives, like conjunction, disjunction, implication, etc., to form compound propositions, and the laws that govern such compositions, which can be iterated indefinitely to produce compound propositions of any finite degree of complexity. Although Husserl's terminology sounds semantic, his intentions are clearly syntactic, as can be learnt from his more detailed presentation in *Formale und transzendentale Logik*. It should be clear by now that this stratum is what Carnap much later called 'syntax', and the laws that govern the composition of propositions are essentially what Carnap called 'formation rules'[31] and is nowadays, but not in 1900, part of the folklore of logic texts. Husserl also called such laws 'laws that protect against nonsense' and contrasted them with the logical laws in the strict sense, which are the laws that protect against (formal) countersense or contradiction.

The second stratum of logic is precisely what in a strict sense is called 'logic', namely, the study of the laws and rules that govern deduction, and which includes not only the traditional syllogistic, but also the more fundamental propositional logic and other less elementary parts of logic, which Husserl does not specify.[32] What is clear, however, is that for Husserl traditional syllogistic was a very small part of logic. Husserl conceived this second stratum of logic as syntactic, not as semantic, since its laws are essentially laws of deduction and, as Husserl used to say[33] 'laws that protect against (formal) countersense'. Thus, this stratum of logic is, to use Carnap's terminology, the stratum of 'logical syntax' and its laws of deduction what Carnap called 'transformation rules'.[34] As we mentioned earlier, *in Formale und transzendentale Logik* Husserl added to this logical syntax—which he called 'apophansis' in this late work[35]—a sort of semantical extension, which he called 'logic of truth' and which would consist of the addition to the apophansis of the notion of truth and related (semantical) concepts. Husserl says very little about this logic of truth, but the clear distinction between syntax and semantics and his conception of semantics as an extension of the apophansis seems once more to anticipate Carnap and even Tarski.[36]

Husserl thought that parallel to the meaning categories there were other categories, the formal-ontological categories, which are variants of the fundamental ontological notion of the 'something whatsoever' or 'something no matter what' (*Etwas-überhaupt*). Among the examples of formal-ontological categories given by Husserl we find the notions of cardinal number, ordinal number, set—which he sometimes called 'collection'—relation and whole and part. The examples, however, are not as important as the following two points: (1) the formal-ontological categories are for Husserl the building blocks of mathematics; (2) Husserl conceives mathematics as based on a plurality of formal-ontological categories of equal fundamentality, thus, rejecting any sort of reductionism to any sole formal-ontological category. (The rejection of any sort of reductionism to sets or to a fundamental sort of number appears in Husserl as early as 1890 in a letter to his friend and mentor Carl Stumpf and published just recently as an appendix to the first part of his *Studien zur Arithmetik und Geometrie.*)[37]

On the building blocks of the formal-ontological categories stands the whole of mathematics. The formal-ontological categories originate the most fundamental mathematical theories as, e.g., the theory of sets, or the theory of cardinal numbers or the theory of ordinal numbers, which have the corresponding formal-ontological category as its basic notion, whereas the remaining mathematical theories are obtained either as specializations of the more fundamental and general theories or as combinations or variations of them. Husserl's views on mathematics are clearly very similar to those of the Bourbaki school. The formal-ontological categories originate—to use Bourbaki's terminology[38]—mother structures (or mother theories) and the remaining mathematical structures are obtained by specialization, combination, or variation of them. Husserl's views on the relation between logic and mathematics, however, differ essentially from those of the Bourbaki school. For Husserl, there is a sort of parallelism between logic and mathematics. The mathematical theories are in some sense an ontological correlate of the logical ones. Both are analytic, and are based ultimately on two correlative notions, namely, the notion of meaning and the notion of object in its utmost generality, the '*Etwas-überhaupt*', the something whatsoever. Moreover, the mathematical and logical theories based immediately on the formal-ontological and meaning categories constitute the all embracing theoretical ground for each and every theory, including the non-logico-mathematical ones. This should not be understood as that every theory, or even every logical or mathematical theory, presupposes each and every fundamental law—as would happen with the mathematical disciplines if reductionists were right—but that each theory and, especially, each logical or mathematical theory presupposes some of the fundamental laws originating in the two sorts of fundamental categories. In this way, we obtain, according to Husserl, a 'science of the conditions of possibility of any theory'.[39]

Husserl goes still one step further in his conception of the logico-mathematical theories. The former stratum consisted of the fusion of the logical and mathematical theories as built on the basis of a common all-embracing ground of fundamental meaning and formal-ontological categories and the fundamental theories immediately based on them. The uppermost stratum considers a priori all the possible forms (or sorts) of theories and the laws that govern their relations. It does not consider merely the possibility of a (presumed) theory but the forms of all possible theories. These different forms of possible theories are not unrelated, and it is an important task of this uppermost stratum to investigate the laws that govern their lawful connections with one another and the transformations of some forms into others by producing some variations in their fundamental aspects. This stratum investigates the general propositions that lawfully regulate the development of new forms from others, their connection to build other forms and the transformation of one form into another. It should be clear, as Husserl underscores,[40] that these laws are of a different nature than the logical and mathematical laws of the previous stratum. They are propositions of a meta-logical or metamathematical nature that, e.g., establish how to transform a mathematical structure of a certain form into another of a different form, or how to combine two or more structures to form a more complex one. In fact, Husserl considers[41] that the objectual correlate of the concept of the form of a possible theory is the concept of the form of a possible region of objects, which he calls, following the mathematical usage in those days, a multiplicity or manifold (*Mannigfaltigkeit*). It cannot be a surprise that he sees the investigations of Riemann and Helmholtz on n-dimensional Euclidean and non-Euclidean multiplicities, Lie's investigations of transformation groups and Cantor's investigations on ordinal and cardinal numbers as steps in the direction of this 'supreme objective' of the logico-mathematical theories.[42] Husserl calls this uppermost stratum 'the theory of forms of possible theories' and also 'the doctrine of multiplicities',[43] since it is to be the science that studies the sorts of possible theories and the relations that connect them according to laws. Thus, it should be clear that all actual and possible theories would be considered by Husserl as specializations of the corresponding forms of theories and, correlatively, any actual or possible region of objects a specialization of the corresponding form of a multiplicity.

There should be no doubt by now not only that Husserl's views on logic and mathematics were very different from Frege's, but also that they correspond much better to what has been the development of mathematics in this century—even though the full uppermost level may remain forever as a Kantian regulative idea. General Topology, Universal Algebra, Category Theory, and other mathematical disciplines can be seen as important partial realizations of Husserl's views on mathematics as ultimately the theory of all forms of possible multiplicities or, to use a more frequent term nowadays, forms of possible structures. Finally, with respect to the relation between logic and mathematics, time has shown that Frege was definitely wrong and that Husserl was probably right. Reductionism of mathematics to logic, or even to set theory, is nowadays, in the era of categories and toposes, not more than an anachronism.

3. Husserl's Views on Mathematical Knowledge

Let us consider now Husserl's epistemology of mathematics or, to use a more Husserlian terminology, let us describe how it is that mathematical objects are 'constituted'. Husserl treats this problem in a very detailed fashion in the second part of the Sixth Logical Investigation titled—as a clear reminder of Kant's former attempt—'Sensibility and Understanding'.[44] As Kant, Husserl tries to explain how it is that mathematical knowledge is not founded on experience even though all our knowledge seems to originate with experience. Husserl's almost totally ignored answer—ignored even by those who are presumably expounding Husserl's views—is, however, clearly different from Kant's.[45]

Consider very simple atomic empirical statements like 'The green pencil is on the desk' or 'John is taller than Charles'. It will probably be clear to anybody that to know the truth value of either sentence we usually have some perceptions, e.g., of the desk and of the objects on the desk, respectively, of John and Charles side by side. We have some sense perceptions of the desk, the green pencil, John and Charles, and conclude—on the basis of such perceptions—either that the sentences are true or that they are false, e.g., that the first one is false but the second is true. The corresponding perceptions fulfill (or satisfy) the statements or they do not fulfill them. However, nothing in our sense perceptions corresponds to the formal elements of the statements or of any other statement, to the 'on' and to the 'is taller than', to the 'below' or 'above' or 'at the side of', to the 'not', to the 'or' or to the 'and', to the 'all' or to the 'some'. Nonetheless, we say that the corresponding perceptions fulfill (or satisfy) or do not fulfill the corresponding statements, and we can clearly distinguish between states of affairs that correspond to what is expressed by sentences like 'Peter or Joe is at the door' and 'Peter and Joe are at the door'. E.g., in the first case, the person who asserts the sentence has seen Peter or Joe at the door, but does not know which of the two is there, whereas, in the second case, she has seen both Peter and Joe at the door. The truth conditions of the two sentences are clearly different. Similarly, the statements 'The cat is under the table' and 'The cat is on the table' have different truth conditions and the perceptions that can fulfill them are also different, although there is no sense perception of precisely those components of states of affairs that correspond to the formal elements in the statements which account for the difference (even in truth value). Thus, even in the most trivial fulfilling perceptions of very simple empirical statements, there are nonsensible—Husserl calls them 'categorial'—elements present which play a decisive role in the fulfillment of those statements. Our perceptions are not of raw materials or of isolated objects and properties, but of objects and properties structured in states of affairs, of which nonsensible—categorial—components are essential constituents.

Precisely, states of affairs, like that described by the relational statement 'John is taller than Charles', are one of Husserl's simplest examples of categorial objectualities. They constitute themselves on the basis of the sensible objects but do not reduce to them. The state of affairs described by the statement 'Charles is shorter than John' is another state of affairs, another categorial objectuality. Both states of affairs are 'categorizations' based on the proto-relation or, as Husserl later usually called it,[46] the 'situation of affairs' that John has more height than Charles. Situations of affairs, however, are not categorial objectualities, but passively constituted complex objectualities. However, what is important for our purposes now is simply that states of affairs are categorial objectualities formed on the basis of noncategorial, sensible objectualities. Another simple example of a categorial objectuality considered by Husserl are the finite collections or sets. When we sensibly perceive—or even sensibly imagine, since in this context their difference is not especially relevant—a plurality of objects, let us say, the objects a, $b, \ldots, z$, the collection (or set) of the objects is constituted in a categorial perception (or categorial imagination, if the objects were simply imagined). (From now on it is better to use Husserl's generic term 'intuition' and speak of 'sense intuition' and 'categorial intuition'.) The new objectuality built on the basis of the objects $a, b, \ldots, z$ is not a sensible objectuality, is not physical, although it is formed on the basis of those physical objects. It is not a gluing of those objects nor any sort of physical manipulation of those objects, but a new kind of objectuality that—like states of affairs—is constituted on the basis of sensible objectualities but does not reduce to it.

Once categorial objectualities are constituted on the basis of noncategorial objectualities, there is no reason to stop there. Categorial objectualities can serve as basis for the constitution of new categorial objectualities, categorial objectualities of the second level, e.g., sets of sets of sensible objects, or relations between sets of sensible objects, or sets of relations between sensible objects. Moreover, the constitution of new categorial objectualities on the basis of objectualities of the preceding level can be iterated indefinitely. In this way, we obtain a whole hierarchy of categorial objectualities immediately founded on objectualities of the preceding level and ultimately on the objectualities of the zeroth level, the sensible objectualities. (It should be clear that parallel to this hierarchy of categorial objectualities there is, on the linguistic side, a hierarchy of statements of each corresponding level. These hierarchies evidently have similarities both with the iterative conception of sets and with simple type theory.)[47]

From what has been said up to this point it does not seem clear at all how it is that categorial objectualities eventually free themselves from any empirical foundation. For this purpose, categorial intuition is complemented by categorial (or formal) abstraction, which is essentially an act of formalization and which can take place—and probably does—even at the first level. (This categorial abstraction should not be confused with the more familiar Husserlian eidetic variation.) If a state of affairs like that described by the statement 'John is taller than Charles' is constituted, one can very well replace the 'material terms' by indeterminates, i.e., variables, and even the relation by any relation whatsoever of similar properties. If we have a finite collection of sensible objects, a finite set of cards, or the objects on the desk, or of any not necessarily related objects, we can replace once more the 'material terms' by variables. This process of categorial (or formal) abstraction—which, after all, is part of the folklore of mathematics—discharges any foundation of mathematical objectualities on experience. To say it briefly: mathematical intuition is categorial intuition plus categorial abstraction. Thus, we can now finally explain, in non-Kantian terms, how it is that mathematical knowledge originates in experience but is not empirical knowledge, not even synthetic a priori knowledge. Categorial objectualities, constituted in categorial intuition, structure themselves ultimately on the basis of sensible objectualities constituted on sense intuition. But those categorial objectualities that are mathematical objectualities or, as Husserl calls them in the Sixth Logical Investigation,[48] 'pure categorial objectualities', are purified by the process of formalization called by Husserl 'categorial abstraction' and do not retain any trace of an empirical foundation.

4. Husserl on States of Affairs and Situations of Affairs

Husserl made the distinction between sense and reference—to use Fregean terminology—probably in 1890. As I mentioned above,[49] the distinction appears already in 1891 in his review of the first volume of Schröder's *Vorlesungen über die Algebra der Logik*, but the completion of Husserl's scheme does not arrive until the Sixth Logical Investigation. In 1891 he had distinguished between the sense and the objectuality referred to by an expression and coincided with Frege fully with regard to proper names but disagreed with him with regard to what he called 'universal names', whose sense was for him a concept and whose referent was the extension of the concept. (As is well known, for Frege the referent of a conceptual word, as he used to say, was a concept, and he remained in complete silence regarding the sense of a conceptual word.) In the First Logical Investigation and especially in §11 of the Fourth Logical Investigation Husserl makes it clear that the referent of (in Husserlian terminology: the objectuality referred to by) a statement was for him, not a truth value as for Frege, but a state of affairs. However, he confuses the notion of a state of affairs with the notion of a situation of affairs—which he is going to distinguish from the former for the first time in the Sixth Logical Investigation—as he recognizes in his *Vorlesungen über Bedeutungslehre*.[50] States of affairs are for Husserl the referents of statements, whereas situations of affairs are a sort of referential basis of states of affairs.

Consider Frege's examples 'The morning star is a planet' and 'The evening star is a planet'. It is clear, as Frege argued[51] that they express different thoughts, since someone could consider one of the statements true, e.g., 'The morning star is a planet', and the other false. Husserl agrees with Frege on this point. Frege, however, also argues[52] that when you substitute in a statement a proper name for another with different sense but the same referent, only the truth value remains invariant and, hence, truth values are the referents of statements. On the other hand, Husserl considers that the referent of both statements 'The morning star is a planet' and 'The evening star is a planet' is the state of affairs that Venus is a planet, which, contrary to Frege's unwarranted conclusion, also remains invariant when one goes from one of the statements to the other. This is no accident: states of affairs remain invariant under any transformation like those considered by Frege in "*Über Sinn und Bedeutung*" namely, such that an expression is replaced by another expression with different sense but the same referent.

Let us consider now an arithmetical example, namely, $5 + 3 > 6 + 1$. The state of affairs referred to by such an inequality is simply that the number eight is greater than the number seven. This state of affairs remains invariant if we substitute the expression '9 - 1' for the expression '5 + 3' or the expression '3 + 4' for the expression '6 + 1'. Consider now the inequality $6 + 1 < 5 + 3$. This inequality is not obtained from $5 + 3 > 6 + 1$ by a simple substitution of an expression for another expression with different sense but the same referent. In particular, the expressions '<' and '>' do not refer to the same objectuality. They both refer to relations, but to inverse relations. We need another sort of stronger transformation to obtain $6 + 1 < 5 + 3$ from $5 + 3 > 6 + 1$ than a mere substitution of an expression for another expression with different sense but the same referent. Those statements refer to different states of affairs. However, apart from the truth value, which they share, e.g., with 'Paris is the capital of France in May 1996', they have something in common that they do not share with such a statement nor with infinitely many other true statements. They have in common a sort of proto-relation of which the two inequalities are categorizations; they have in common the situation of affairs. Situations of affairs also remain invariant under transformations of statements in which expressions are substituted for expressions with different sense but the same reference. They even remain invariant under transformations of statements that could change the state of affairs referred to by the statement. Situations of affairs are, however, 'abstract' in the sense of being precategorial. We usually deal with categorial relations, like those of being taller than between John and Charles or that of being shorter than between Charles and John, not with the proto-relation that John has more height than Charles.

In other papers[53] I have tried to develop this Husserlian distinction between states of affairs and situations of affairs, applying it to mathematical contexts to explain not only what dual statements have in common (apart from their truth value) but also what seemingly unrelated interderivable statements, like the Axiom of Choice and Tychonoff's Theorem have in common (apart from their truth value). We have used the expression 'abstract situation of affairs' not only to underscore the abstractness of the notion but also to make it clear that we were applying the Husserlian distinction to problems not envisaged by Husserl. However, in his *Vorlesungen über Bedeutungslehre*[54] Husserl envisaged a similar extension of the range of applications of his notion, although not to mathematical contexts but to physical ones. Thus, when we say that we have the same physical law, although it is expressed in two different but equivalent ways, what we mean, according to Husserl,[55] is that the situation of affairs is the same, even though the states of affairs referred to by the two expressions are different. As hinted by this application of Husserl's distinction between states of affairs and situations of affairs to physical contexts and our application to mathematical contexts, this distinction, although more difficult to apprehend than the one between sense and reference, promises to be even more fruitful than the latter for the semantics of mathematical and physical statements.

5. Two Applications of the Distinction to Fregean Problems

Finally, I shall illustrate the fruitfulness of Husserl's distinction by applying it to two different, although not totally unrelated Fregean problems. Firstly, I will show how Church's argumentation in *Introduction to Mathematical Logic*[56] on behalf of Frege's thesis that the referents of statements are truth values is not only fallacious, but its fallacious nature can be easily clarified in terms of the Husserlian distinction. Then I will use Husserl's notion of a situation of affairs to adequately assess Frege's notorious Principle V, or Basic Law V, a problem that has created a lot of confusion among Fregean scholars and even in Frege himself. Both problems have the same root, namely, the insufficiency of Fregean official semantics to deal with important semantic issues.

First of all, it should be mentioned that each of the Fregean and Husserlian candidates for the referent of statements remains invariant under different groups of transformations of statements into statements. These groups of transformations are such that the Fregean group, namely, the group of transformations determined by the invariance of the truth value, contains the two Husserlian groups as proper subgroups, whereas the group of transformations determined by the invariance of the situation of affairs is not only a subgroup of the Fregean group, but also contains as a proper subgroup the group of transformations determined by the invariance of the state of affairs. Finally, all three groups of transformations contain as a proper subgroup the group of transformations determined by the invariance of the thought expressed by the statement and which consists only of the identity transformation and—if there are completely synonymous expressions—of transformations between synonymous statements, i.e., statements that differ at most by the substitution of synonymous expressions.

Let us now consider Church's argument:[57]

(1) Sir Walter Scott is the author of *Waverley*.

(2) Sir Walter Scott is the man who wrote twenty-nine Waverley novels altogether.

(3) The number such that Sir Walter Scott is the man who wrote so many novels altogether is twenty-nine.

(4) The number of counties in Utah is twenty-nine.

When we go from (1) to (2), the thought changes but the state of affairs (and, thus, the situation of affairs and the truth value) remains the same. When we go from (2) to (3) not only the thought but also the state of affairs changes (as when we go from '5 + 3 > 6 + 1' to '6 + 1 < 5 + 3'), whereas both the situation of affairs and the truth value remain invariant. When we go from (3) to (4), however, only the truth value remains invariant, whereas the thought, the state of affairs and the situation of affairs change. Hence, only the Fregean group of transformations contains the three transformations. Thus, it should come as no surprise that the product of the three transformations only preserves the truth value. It could not be otherwise. Church has not proven—and it is moreover unprovable—that by transforming statements into statements by way of substituting expressions for expressions with different sense but the same reference only the truth value remains invariant.

Consider now the notorious Principle V of *Grundgesetze der Arithmetik*, according to which the course of values of a function f is identical with the course of values of a function g is the same as the identity of the values of the functions f and g for every argument, in symbols:

$$x f(x) = y g(y) :=: (\forall z)\, (f(z) = g(z)).$$

As in the case of the conditions on definitions with respect to the relation between definiens and definiendum, in some writings Frege says that the left and right hand sides of Principle V have the same sense and in others that they have the same reference. Thus, e.g., in "*Funktion und Begriff*"[58] Frege says that

$$xf(x) = yg(y) \text{ and } (\forall z)(f(z) = g(z))$$

express the same sense, whereas in *Grundgesetze der Arithmetik*[59] he says that they merely have the same referent. This has been the source of a debate among Fregean scholars. Sluga,[60] e.g., renders Frege's Principle V as expressing an identity of sense, whereas Dummett[61] and later scholars like Schirn[62] render Principle V as expressing an identity of reference. In his recent *Frege, Philosophy of Mathematics*,[63] however, Dummett has changed his mind and argues that the identity is one of content. But the notion of content was never a precise technical term in Frege. In his writings before 1890 Frege used the term 'content' as a sort of generic and somewhat vague expression borrowed from the philosophical circles of his time, and used it more or less both for what he later called 'sense' and for what he later called 'reference'. After 1890, the term 'content' is rarely used by Frege, and when it is used it expresses the result of adding the coloring to the sense. Dummett's obscure rendering is not worthy of much discussion, although it should be clear that the arguments against Sluga's rendering would apply with at least equal force to Frege's latter usage of the term 'content'. (It is evident, however, that this is not the usage that Dummett has in mind.) On the other hand, Dummett's earlier rendering, as that of Sluga, although precise, are—as I have shown elsewhere[64] and will repeat here briefly—indefensible.

Firstly, let us consider Sluga's rendering of Principle V, according to which this principle expresses an identity of sense between the expressions— in this case: statements—at either side of the principal identity sign. This interpretation is clearly inconsistent with Frege's views in "*Über Sinn und Bedeutung*" (and also with his older views of *Begriffsschrift*).[65] Specifically, it is inconsistent with the views on sense and reference expressed in his—on this matter—more authoritative writings, namely, "*Über Sinn und Bedeutung*"[66] and *Grundgesetze der Arithmetik*.[67] If Principle V were to express an identity of sense, then, on the basis of Frege's notion of sense in these two writings, it would be trivial, since it would be either of the form 'London is (identical with) London' or at best of the form 'London is (identical with) *Londres*'. Let us recall that, on the basis of those writings, '2 × 2' and '2 + 2' express different senses and, thus, '2 × 2 < 5' and '2 + 2 < 5' also express different senses, as do '5 + 3' and '7 + 1'. Nonetheless, '2 × 2 = 2 + 2' and '5 + 3 = 7 + 1' are not only true but analytically so for Frege— in spite of the fact that the expressions at either side of the identity sign in both identity statements express different senses. Precisely, that is what happens with every true identity statement of the form '$a = b$' (no matter if it is analytic or synthetic) each time that the difference between the names 'a' and 'b' is not a mere (arbitrary) difference between the names—as with 'London' and '*Londres*'. However, on the basis of Sluga's rendering of

Principle V, such identity statements of the form '*a* = *b*' would not only be false but trivially so.

But Dummett's first rendering of Principle V, propounded nowadays by Schirn, is still more incredible. Let us suppose that Principle V expresses an identity of the referent of the expressions at either side of its principal identity sign. The expressions at either side of the principal identity sign in Principle V are statements. Principle V is an identity statement between statements. But since for Frege the referent of a statement is a truth value, under Schirn's rendering, Principle V would merely be expressing that the statements at either side of its principal identity sign have the same truth value, i.e., that both have the true or both have the false as referent. However, if that were so, Principle V would remain totally undetermined and the (logically simple) notion of a course of values would remain completely vague, since no matter if

$$x\!f(x) = y\!g(y)$$

has the true or the false as referent, there are denumerably many sentences (more exactly, thoughts) expressible in Frege's conceptual notation that have the same truth value as

$$x\!f(x) = y\!g(y)$$

If we were to express the sentence 'Paris is the capital of France in May 1996' in conceptual notation, then either that sentence or its negation would have the same truth value as

$$x\!f(x) = y\!g(y)$$

and, on the basis of the Dummett-Schirn rendering, we could substitute that sentence or its negation for

$$(\forall z)\,(f(z) = g(z))$$

in Principle V without any significant loss for Frege's system.

Hence, both Sluga's and Schirn's renderings of Principle V have absurd consequences. This partially explains why Frege seems to have hesitated between the two, although, as we will mention shortly, there is a deeper reason. However, if we use Husserl's semantical tools, the rendering of

Principle V comes out smoothly. Firstly, it should be clear to everybody—except Sluga and his followers—that the statements at either side of the principal identity sign of Principle V have different senses. Moreover, none is obtained from the other by substitution of an expression for another expression having different sense but the same referent. Thus, they do not even refer to the same state of affairs. But if Principle V were true, its two sides would have in common not only the truth value—which they would have in common with denumerably many statements expressible in conceptual notation—but also the situation of affairs. Exactly this rendering applies also to Frege's two last attempts mentioned above at defining the concept of number in *Die Grundlagen der Arithmetik*. In both cases, the definition expresses that the definiens and the definiendum—which are sentences—have in common the same situation of affairs.

It is natural to ask why did Frege not realize that his semantical tools were insufficient to adequately assess Principle V. The answer, which I have given in some detail elsewhere,[68] is that Frege really had two notions of sense, namely, the official one mentioned above and which coincides with Husserl's notion of sense, and an unofficial one which is present in a letter to Husserl of 1906,[69] where he takes 'equipollency' or 'equiderivability' as a criterion of sense identity between sentences, i.e., identity of thought, and in a writing of 1897 published posthumously, and which probably accounts for his assessment of Principle V in "*Funktion und Begriff*" and, on the other hand, renders somewhat puzzling his assertion in "*Der Gedanke*"[70] that a conceptual notation does not need to differentiate between sentences with the same sense. Frege's second notion of sense is essentially his notion of conceptual content of *Begriffsschrift*, whose defining properties also involve a similar tension, and which can be seen as a somewhat unclear forerunner of Husserl's notion of a situation of affairs. Frege characterized the conceptual content as follows:[71]

(i) A sentence in the active mood and its passive have the same conceptual content.

(ii) Two sentences with the same (syntactical) consequences have the same conceptual content.

(iii) A conceptual notation does not need to differentiate between sentences with the same conceptual content. Only the conceptual content is relevant for the conceptual notation.

As is easily seen, (ii) corresponds to Frege's assertion in the letter to Husserl mentioned above and (iii) to his assertion in "Der Gedanke." Moreover, (i) corresponds to another assertion in "*Der Gedanke*."[72] Between (ii) and (iii) there seems to be some tension, but that does not need to concern us here. What really matters is that Frege had two notions of sense, namely, his official one and one obtained from his old notion of conceptual content. Frege, however, confused his two notions of sense and used them interchangeably. The reason for such a confusion probably lies in the fact that—contrary to Husserl—he lacked the notion of a state of affairs, which as we have seen, lies between the thought and the situation of affairs and prevents them from collapsing into each other. Thus, after all, also as a semanticist Frege fell short of being a Husserlian.

NOTES

1. E. Husserl, *Philosophie der Arithmetik, mit ergänzenden Texten*, Husserliana, vol. XII (The Hague: M. Nijhoff, 1970 [1891]).
2. G. Frege, *Die Grundlagen der Arithmetik* (Hamburg: Centenarausgabe, Meiner, 1986 [1884]), introduction by C. Thiel.
3. E. Husserl, *Logische Untersuchungen*, Husserliana, vols. XVIII and XIX (The Hague: M. Nijhoff, 1975 and 1984 [1900/01, 2nd ed. rev.,1913]).
4. E. W. Beth, *The Foundations of Mathematics*, 2nd ed. rev. (Amsterdam: North-Holland, 1965 [1959]), p. 353.
5. M. Dummett, *Frege, Philosophy of Language* (London: Duckworth, 1973), pp. XLII-XLIII and p. 158.
6. D. Føllesdal, "Husserl and Frege," *Mind, Meaning and Mathematics*, ed. L. Haaparanta (Dordrecht: Kluwer, 1994), pp. 3–47, translation of his 1958 Norwegian Masters thesis.
7. E.g., in H. Sluga, *Gottlob Frege* (London: Routledge and Kegan Paul, 1980), p. 2, and especially pp. 39–40 and his "Semantic Content and Cognitive Sense," *Frege Synthesized*, ed. L. Haaparanta and J. Hintikka (Dordrecht: Reidel, 1986), pp. 3–47.
8. E.g., in C. Thiel's Editor's Introduction to the Centenarausgabe edition of Frege's *Die Grundlagen der Arithmetik*, p. LI.
9. "Edmund Husserls Philosophie der Logik und Mathematik im Lichte der gegenwärtigen Logik und Grundlagenforschung," Ph.D. diss., Rheinische Friedrich-Wilhelms-Universität, Bonn, 1973.
10. "Remarks on Sense and Reference in Frege and Husserl," *Kant-Studien* 73, no. 4 (1982): 425–39, chapter 2 of the present book. Although published in 1982, this paper was accepted for publication in 1979.
11. E.g., in J. N. Mohanty, *Husserl and Frege* (Bloomington, IN: Indiana University Press, 1982).
12. C. O. Hill, *Word and Object in Husserl, Frege and Russell* (Athens, OH: University of Ohio Press, 1991). See also her "Frege's Attack on Husserl and Cantor" (chapter 6 of the present book), "Husserl and Frege on Substitutivity" (chapter 1 of the present book), and "Husserl and Hilbert on Completeness" (chapter 10 of the present book).
13. See Frege's *Wissenschaftlicher Briefwechsel*, ed. G. Gabriel et. al. (Hamburg: Meiner, 1976), pp. 94–98.

14. See Hill's "Frege's Attack on Husserl and Cantor," chapter 6 of the present book.
15. G. Frege, *Grundgesetze der Arithmetik*, vols. I (1893) and II (1903), reprinted in one volume (Hildesheim: Olms, 1966).
16. Ibid. See, e.g., p. 36 and also the previous discussion in §10.
17. See on this issue, e.g., Sluga's book cited in note 7 and also the papers published in *Frege Synthesized*, ed. L. Haaparanta and J. Hintikka (Dordrecht: Reidel, 1986), by: J. Weiner, "Putting Frege in Perspective," pp. 9–27; Sluga, "Semantic Content and Cognitive Sense," pp. 47–64; T. Burge, "Frege on Truth," pp. 97–154 and P. Kitcher, "Frege, Dedekind and the Philosophy of Mathematics," pp. 299–343.
18. G. Frege, "Der Gedanke," *Kleine Schriften* (Hildesheim: Olms, 2nd ed., 1991), pp. 342–61.
19. G. Frege, "Funktion und Begriff," *Kleine Schriften*, pp. 125–42.
20. G. Frege, "Über Sinn und Bedeutung," *Kleine Schriften*, pp. 143–62.
21. Frege, *Grundgesetze der Arithmetik*, vol. I, pp. 6–7.
22. Frege, "Über Sinn und Bedeutung," pp. 146–51.
23. See, e.g., R. Tieszen, *Mathematical Intuition* (Dordrecht: Kluwer, 1989).
24. Frege, *Wissenschaftlicher Briefwechsel*, pp. 94–98.
25. E. Husserl, *Einleitung in die Logik und Erkenntnistheorie*, Husserliana, vol. XXIV (The Hague: M. Nijhoff, 1984). This text is based on lectures given by Husserl in 1906 and 1907.
26. E. Husserl, *Ideen zu einer reinen Phänomenologie und einer phänomenologischen Philosophie, Erstes Buch, Husserliana*, vol. III (The Hague: M. Nijhoff, rev. ed. 1976 [1913]).
27. E. Husserl, *Formale und transzendentale Logik*, Husserliana, vol. XVII (The Hague: M. Nijhoff, 1974 [1929]).
28. None of these changes represents the abandonment of any of his views. The first is an addition that results in a clarification, whereas the second is simply a matter of emphasis.
29. See K. Gödel, "Über formal unentscheidbare Sätze in *Principia Mathematica* und verwandter Systeme," reprinted with accompanying translation in his *Collected Works*, vol. 1 (New York: Oxford University Press, 1986), pp. 144–95. See also A. Tarski, "The Concept of Truth in Formalized Languages," *Logic, Semantics and Metamathematics* (Indianapolis: Hackett, 2nd ed., 1983 [1956]), pp. 152–278.
30. Our exposition will for the most part follow that of chapter 11 of the first volume of *Logische Untersuchungen* with occasional references to the other three works cited in footnotes 25–27.
31. R. Carnap, *The Logical Syntax of Language*, expanded English ed. (London: Routledge and Kegan Paul, 1937), pp. 2, 4, 11f. , translation of *Logische Syntax der Sprache*.
32. On this issue, see, e.g., Husserl's *Einleitung in die Logik und Erkenntnistheorie*, pp. 435–36.
33. See, e.g., Husserl, *Logische Untersuchungen*, Fourth Investigation, §14.
34. Carnap, *Logical Syntax of Language*, pp. 2, 4, 27f.
35. See Husserl's *Formale und transzendentale Logik*, chapters 1 and 2.
36. Since Husserl is very schematic, we cannot force the issue of his anticipation of the convenience of extending the syntax to a semantics too far. However, the influence of Husserl on Carnap seems not to be casual and does not limit itself to *Die logische Syntax der Sprache*. It is far more reaching in *Der*

logische Aufbau der Welt (Berlin: Weltkreis, 1928) where Husserl's transcendental phenomenology and, especially, his treatment of the problem of intersubjectivity seem to have had a more decisive influence than Carnap would like us to believe.

37. E. Husserl, *Studien zur Arithmetik und Geometrie*, Husserliana, vol. XXI (The Hague: M. Nijhoff, 1983), pp. 244–51. The writings in this volume are for the most part contemporary to the writing and publication of the 1891 *Philosophie der Arithmetik*.

38. See N. Bourbaki, "The Architecture of Mathematics," *American Mathematical Monthly* 57 (1950): 221–32.

39. Husserl, *Logische Untersuchungen*, vol. 1, chapter 11, §§68–69.

40. Ibid., §69.

41. Ibid., §70.

42. Ibid.

43. Ibid.

44. Husserl, *Logische Untersuchungen*, Sixth Investigation.

45. See note 22 above.

46. See, e.g., Husserl's *Vorlesungen über Bedeutungslehre*, Husserliana, vol. XXVI (Dordrecht: Kluwer, 1987). This book is based on lectures of the summer of 1907. See also his *Erfahrung und Urteil* (Hamburg: Meiner, 5th ed., 1976 [1939]).

47. See my "Husserl's Epistemology of Mathematics and the Foundation of Platonism in Mathematics," *Husserl Studies* 4, no. 2 (1987): 81–102, chapter 12 of the present book.

48. Husserl, *Logische Untersuchungen*, Sixth Investigation §60.

49. See § 1 of the present paper.

50. Husserl, *Vorlesungen über Bedeutungslehre*, pp. 29–30. In his "Meaning and Language" (in *The Cambridge Companion to Husserl*, ed. B. Smith and D. W. Smith [Cambridge, UK: Cambridge University Press, 1995], pp. 106–37), Peter Simons is guilty of the same confusion. See p. 112.

51. Frege, "Über Sinn und Bedeutung," pp. 143–44 and especially p. 148.

52. Ibid., pp. 149–51.

53. See chapters 4, 13, and 14 of the present book and my recent paper: "On the Semantics of Mathematical Statements," *Manuscrito* 19, no. 1 (1996): 149–75.

54. Husserl, *Vorlesungen über Bedeutungslehre*, pp. 101–02.

55. Ibid.

56. A. Church, *Introduction to Mathematical Logic* (Princeton: Princeton University Press, 1956).

57. Ibid., p. 25.

58. Frege, "Funktion und Begriff," p. 130.

59. Frege, *Grundgesetze der Arithmetik*, vol. I, p. 16.

60. Sluga, *Gottlob Frege*, pp. 156–57.

61. M. Dummett, *The Interpretation of Frege's Philosophy* (London: Duckworth, 1981), pp. 335–36 and 531–32.

62. See, e.g., M. Schirn, "Axiom V and Hume's Principle in Frege's Foundational Project," *Diálogos* 66 (1995): 7–20.

63. M. Dummett, *Frege, Philosophy of Mathematics* (London: Duckworth, 1991), pp. 170–76.

64. In "On Frege's Two Notions of Sense," chapter 4 of the present book. See note 52.

65. G. Frege, *Begriffsschrift* (Darmstadt: Wissenschaftliche Buchgesellschaft, 1964 [1879]), pp. 13–15.
66. Frege, "Über Sinn und Bedeutung," pp. 143–51 and 162.
67. Frege, *Grundgesetze der Arithmetik,* vol. I, p. 7.
68. In "On Frege's Two Notions of Sense," chapter 4 of the present book.
69. See Frege's *Wissenschaftlicher Briefwechsel,* p. 102. See also his "Logik 1897" in his *Nachgelassene Schriften,* ed. H. Hermes et al. (Hamburg: Meiner, 2nd ed. 1983 [1969]), pp. 137–63, especially pp. 152–55.
70. Frege, "Der Gedanke," pp. 347–48.
71. Frege, *Begriffsschrift,* p. 3.
72. Frege, "Der Gedanke," p. 348.

REFERENCES

Beth, E. W. *The Foundations of Mathematics.* Rev. ed. Amsterdam: North Holland, 1965.

Bourbaki, N. "The Architecture of Mathematics." *American Mathematical Monthly* 57 (1950): 221–32.

Burge, T. "Frege on Truth." *Frege Synthesized.* Ed. L. Haaparanta and J. Hintikka. Dordrecht: Reidel, 1986, 97–154.

Carnap, R. *Der logische Aufbau der Welt.* 4th ed. Hamburg: Meiner, 1974 (Berlin: Weltkreis, 1928).

———. *The Logical Syntax of Language.* English expanded version. Translation of *Die logische Syntax der Sprache.* London: Routledge and Kegan Paul, 1937.

Church, A. *Introduction to Mathematical Logic.* Princeton: Princeton University Press, 1956.

da Silva, J. "Husserl's Philosophy of Mathematics." Manuscrito XVI, no. 2 (1993): 121–148.

Dummett, M. *Frege: Philosophy of Language.* London and Cambridge, MA: Duckworth and Harvard, 1974.

———. *The Interpretation of Frege's Philosophy.* London and Cambridge, MA: Duckworth and Harvard, 1981.

———. *Frege: Philosophy of Mathematics.* Duckworth: London and Cambridge, MA: Duckworth and Harvard, 1991.

Føllesdal, D. *Husserl und Frege.* 1958. Translated in *Mind, Meaning and Mathematics.* Ed. L. Haaparanta. Dordrecht: Kluwer, 1994, pp. 3–47.

Frege, G. *Begriffsschrift,* 1879. Reprint, Darmstadt: Wissenschaftliche Buchgesellschaft, 1964.

———. *Die Grundlagen der Arithmetik,* 1884. Hamburg: Centenarausgabe, Meiner, 1986.

———. *Grundgesetze der Arithmetik I,* 1893 and II, 1903. Reprinted in one volume, Hildesheim: Olms, 1962.

———. *Kleine Schriften.* 2nd ed. Hildesheim: Olms, 1990.

———. *Nachgelassene Schriften.* Ed. H. Hermes, et al. Hamburg: Meiner, 1969. 2nd ed. 1983.

———. *Wissenschaftlicher Briefwechsel.* Hamburg: Meiner, 1974.

Gödel, K. "Über formal unentscheidbare Sätze der *Principia Mathematica* und verwandter Systeme," 1931. Reprinted in *Collected Works* I, New York: Oxford University Press, 1986, pp. 144–95.

Hill, C.O. *Word and Object in Husserl, Frege and Russell.* Athens, OH: University of Ohio Press, 1991.

———. "Frege's Attack on Husserl and Cantor." *Monist* 77, no. 3 (1994): 345–57. Chapter 6 of the present book.

———. "Husserl and Frege on Substitutivity." In *Mind, Meaning and Mathematics,* ed. L. Haaparanta. Dordrecht: Kluwer, 1994, pp. 113–40. Chapter 1 of the present book.

———. "Husserl and Hilbert on Completeness." In *From Dedekind to Gödel,* ed. J. Hintikka. Dordrecht: Kluwer, 1995, pp. 143–63. Chapter 10 of the present book.

Husserl, E. *Philosophie der Arithmetik.* 1891. Husserliana edition, vol. XII. The Hague: M. Nijhoff, 1970.

———. *Logische Untersuchungen,* 1900–1901. Husserliana edition, vols. XVIII and XIX. The Hague: M. Nijhoff, 1975 and 1984.

———. *Ideen zu einer reinen Phänomenologie und einer phänomenologischen Philosophie I,* 1913. Husserliana revised edition, vol. III. The Hague: M. Nijhoff, 1976.

———. *Formale und transzendentale Logik,* 1929. Husserliana edition, vol. XVII. The Hague: M. Nijhoff, 1974.

———. *Erfahrung und Urteil,* 1939. 5th ed. Hamburg: Meiner, 1976.

———. *Aufsätze und Rezensionen 1890–1910.* Husserliana, vol. XXII. The Hague: M. Nijhoff, 1979.

———. *Studien zur Arithmetik und Geometrie.* Husserliana, vol. XXI. The Hague: M. Nijhoff, 1983.

———. *Einleitung in die Logik und Erkenntnistheorie.* Husserliana, vol. XXIV. Dordrecht: Kluwer, 1984.

———. *Vorlesungen über Bedeutungslehre.* Husserliana, vol. XXVI. Dordrecht: 1987.

Kitcher, P. "Frege, Dedekind and the Philosophy of Mathematics." *In Frege Synthesized,* ed. L. Haaparanta and J. Hintikka. Dordrecht: Reidel, 1986, pp. 299–343.

Mohanty, J. N. *Husserl and Frege.* Bloomington, IN: Indiana University Press, 1982.

Rosado Haddock, G.E. "Edmund Husserls Philosophie der Logik und Mathematik im Lichte der gegenwärtigen Logik und Grundlagenforschung." Ph.D. dissertation. Bonn: Rheinische Friedrich-Wilhelms-Universität, 1973.

———. "Remarks on Sense and Reference in Frege and Husserl." *Kant-Studien* 73, no. 4 (1982): 425–39. Chapter 2 of the present book.

———. "Identity Statements in the Semantics of Sense and Reference." *Logique et Analyse* 25, no. 100 (1982): 399–411. Chapter 3 of the present book.

———. "On Frege's Two Notions of Sense." *History and Philosophy of Logic* 7, no. 1 (1986): 31–41. Chapter 4 of the present book.

———. "Husserl's Epistemology of Mathematics and the Foundation of Platonism in Mathematics." *Husserl Studies* 4, no. 2 (1987): 81–102. Chapter 12 of the present book.

———. "On Husserl's Distinction between State of Affairs (*Sachverhalt*) and Situation of Affairs (*Sachlage*)." In *Phenomenology and the Formal Sciences,* ed. D. Føllesdal, J. N. Mohanty, and T. M. Seebohm. Dordrecht: 1991, pp. 35–48. Chapter 14 of the present book.

———. "Interderivability of Seemingly Unrelated Mathematical Statements and the Philosophy of Mathematics." *Diálogos* 59 (1992): 121–34. Chapter 13 of the present book.

———. "On Antiplatonism and its Dogmas." *Diálogos* 67: (1996): 7–38. Chapter 15 of the present book.

———. "On the Semantics of Mathematical Statements." *Manuscrito* XIX (1996): 149–75.

Schirn, M. "Axiom V and Hume's Principle in Frege's Foundational Project." *Diálogos* 66 (1995): pp. 7–20.

Simons, P. "Meaning and Language." In *The Cambridge Companion to Husserl,* ed. B. Smith and D. W. Smith. Cambridge: Cambridge University Press, 1995, pp. 106–37.

Sluga, H. *Gottlob Frege.* London: Routledge and Kegan Paul, 1980.

——. "Semantic Content and Cognitive Sense." In *Frege Synthesized,* ed. L. Haaparanta & J. Hintikka. Dordrecht: Reidel, 1986, pp. 47–64.

Suszko, R. "The Reification of Situations." In *Philosophical Logic in Poland,* ed. J. Wolenski. Dordrecht: Kluwer, 1994, pp. 247–70.

Tarski, A. "The Concept of Truth in Formalized Languages," 1935 (in German). Translation in A. Tarski, *Logic, Semantics and Metamathematics* 1956, 2nd ed. Indianapolis: Hackett, 1983, pp. 152–278.

Thiel, C. " '*Einleitung*' to the Centenarausgabe edition of Frege's *Die Grundlagen der Arithmetik.*" Hamburg: Meiner, 1986.

Tieszen, R. *Mathematical Intuition.* Dordrecht: Kluwer, 1989.

Weiner, J. "Putting Frege in Perspective." In *Frege Synthesized,* ed. L. Haaparanta and J. Hintikka. Dordrecht: Reidel, 1986, pp. 9–27.

12

Guillermo E. Rosado Haddock

HUSSERL'S EPISTEMOLOGY OF MATHEMATICS AND THE FOUNDATION OF PLATONISM IN MATHEMATICS

1. Introduction[1]

In recent years the so-called Platonistic conception of the nature of mathematical entities, according to which mathematical statements are about entities and relations in a similar way as statements that are concerned with real physical objects—with the only difference that the entities with which mathematical statements are concerned are neither physical, nor possess spatio-temporal properties in an essential way and, thus, are not sensuously perceived—has been the object of severe criticisms, both by propounders of some sort of constructivism[2] and by defenders of the old Anglo-American empiricist tradition recently rebaptized as 'causal'.[3] Certainly, the doctrines of the best known defenders of Platonism, like Cantor, Frege, and Gödel, have the common defect of not having developed an epistemology of mathematics that could explain in a satisfactory way how it is that we have access to the so-called mathematical entities. It is true that in the later writings of Frege (beginning with *Der Gedanke*, but especially in those written after 1920)[4] and in some of Gödel's writings[5] there are sketches of an explanation of our knowledge of mathematical truths, but its insufficient elaboration does not allow its use as an answer to the antiplatonistic criticisms.

Although not so well-known as a philosopher of mathematics, Edmund Husserl defended an essentially Platonistic conception of mathematics; however, contrary to the authors just mentioned, Husserl was especially interested in epistemological problems, and his philosophy of mathematics, although it can be considered separately,[6] has its epistemological foundation in what Husserl called categorial intuition.

Now, the term 'categorial intuition' is in Husserl—especially in the *Logische Untersuchungen* (Husserl 1900/01)—a sort of generic term that requires some clarification if we want to penetrate to the depths of Husserl's epistemology of mathematics. Moreover, the contrast between categorial and sensible intuition in Husserl will not only serve the purpose of somehow demarcating the notion of categorial intuition, but will also serve to show that categorial intuition is nothing mysterious or metaphysical (in the negative sense of this last term), but, on the contrary, has its foundation in sensible intuition. Therefore, although mathematical entities are for Husserl deprived of any 'sensuousness', they are not completely disconnected from sensible intuition.

In this paper I will make a reconstruction and a systematization of Husserl's epistemology of mathematics as based on the notion of categorial intuition. I will follow Husserl's discussion in the Sixth Logical Investigation and in *Erfahrung und Urteil* (Husserl 1939).

1. The Problem of Fulfillment of Formal Constituents of Statements

In the First Logical Investigation Husserl distinguished between acts in which meanings are constituted and acts in which those meanings are fulfilled or realized, and, correspondingly, between the meanings of expressions and the objectualities referred to by those expressions by means of their meanings. Although in the First Investigation Husserl was not completely explicit in this respect, it is clear, e.g., from §11 of the Fourth Investigation and from §60 of *Erfahrung und Urteil*, that statements whose meaning is a proposition (or thought) have as reference a state of affairs.

Now, as Husserl's interest in the First and Fourth Investigations is not epistemological, he postpones the discussion of the fulfillment (or realization) of the meanings of statements, together with other related problems, until the Sixth Investigation, which is a sort of culmination of his efforts in the rest of that philosophical masterpiece. In §§40, 42, 43, and 51 of the Sixth Investigation Husserl tells us that the formal constituents of statements, namely, particles like 'is', 'and', 'or', 'not', 'if', 'then', 'all', 'some', 'none', 'many', 'few', etc. and also numerical determinations and many other expressions (e.g., the relational expressions 'greater than' and 'at the side of') do not have any direct counterpart in sensible perception—although we speak of the fulfillment of the meanings of statements in which those constituents occur. No matter how closely a statement is linked to perception, only the material elements that are found in its terms can have their fulfillment in sensible perception or imagination, i.e. in sensible intuition.[7] On the other hand, formal constituents of statements do not obtain their fulfillment from sensible intuition. To the 'is", to the 'and' and to the 'or', to the 'some', to the 'not' and to the 'is greater than' there does not correspond anything in sensible perception, nor can we represent sensibly in the imagination the intuitive counterparts of those particles. We cannot apprehend with any of our senses, nor paint nor photograph objectual counterparts of such expressions (Husserliana, vol. XIX/2, 688).

However, that does not mean that the meanings of such formal constituents of statements cannot be fulfilled by a corresponding objectuality. Actually, if there were no possible fulfillment of the meanings of such formal constituents of statements, we could not clearly differentiate between the fulfillment of the meanings of the statements 'John and Peter are in the park' and 'John or Peter is in the park', although those two statements have different truth conditions.[8]

2. Sensible and Categorial Intuition

Hence, although we do not sensibly intuit anything that could correspond to the formal constituents of statements, there must be some act of intuition, similar to but different from that of sensible intuition, in which such 'formal' expressions are fulfilled. Husserl calls this sort of intuition, in which the meanings of the formal constituents of statements are fulfilled and in which new categorially formed objectualities are constituted, categorial intuition. In this sort of intuition the categorially formed objectuality is not merely symbolically mentioned—as in the corresponding act in which the meaning of the expression used to refer to it is constituted—but actually intuited. It is in this sort of act that collections, indeterminate pluralities, totalities, numbers, states of affairs, etc. become objects (Husserliana, vol. XIX/2, 672). Such objectualities are based on the sensibly given, but are not to be reduced to the sensibly given. Similarly, the categorial act in which such objectualities of the understanding are constituted is based on acts of sensible intuition, but is not to be reduced to them. In an act of sensible intuition the constituted object is apprehended in a simple way, whereas in an act of categorial intuition the constituted object is constituted in founded acts that connect what is given in the founding (sensible) acts (Husserliana, vol. XIX/2, 674). Moreover, in an act of sensible intuition the object is receptively apprehended, whereas in an act of categorial intuition the object is never apprehended in a purely receptive way, but requires the intervention of the spontaneity of the understanding. Hence, Husserl calls the objects given in categorial intuition not only categorial or syntactical objectualities, but also objectualities of the understanding (Husserl 1939, §§58*f*).

Thus, Husserl clearly distinguishes in a general way between sensible and categorial intuition, and calls the objects given in sensible intuition objects of lower level and the objects given in categorial intuition objects of higher level (Husserliana, vol. XIX/2, 674). Moreover, every act of sensible intuition (i.e., perception or imagination) can appear, either alone or together with other acts, as founding new acts based on it, in which new objectualities are constituted that could not have been given in any founding act. But it should be emphasized that the objectuality constituted in the categorial founded acts is 'built' on the objectualities given in the founding acts, and can only be given as such a founded objectuality. 'It is in such founded acts that the categorial in intuition and knowledge lies, and predicative thought finds its fulfillment' (Husserliana, vol. XIX/2, 675).

Now, it is important to underscore—as Husserl does in §61 of the Sixth Investigation—that the objectualities of the understanding are not objects in the primary sense of 'being', since they are not possible objects of sensible intuition. The categorial act in which they are constituted, neither modifies, nor affects, nor transforms the sensibly given, since that would be a sort of distortion; rather, it 'builds' on the sensibly given. Categorial intuition neither glues together nor links sensible objects to produce a new sensible whole. If this were the case, the originally given in sensible intuition would be modified, and categorial intuition would be a falsifying reorganizing of the sensibly given. In such a case, the result would be a new sensible object, although different from those of the founding acts. But what is constituted in a categorial act, although founded in the sensibly given, is not only an objectuality of a higher level than the sensible objects of its founding acts, but an objectuality of a different sort, a nonsensible objectuality.

3. Examples of Categorial Objectualities

The two typical examples of categorial objectualities (or objectualities of the understanding) considered by Husserl both in the Sixth Investigation and in the second part of *Erfahrung und Urteil* (§§59ff.) are states of affairs and collections (or sets). With respect to the first of these, one can distinguish at least two cases, although the analysis is very similar in each case. Thus, a state of affairs can be constituted when we (mentally) detach a part or moment[9] from the whole to which it belongs, e.g., when we predicate of a book its being red or blue. In sensible intuition the whole is passively constituted with its parts and moments. But the relation that is constituted when we (mentally) detach a part or moment to bring out its connection to the whole, as, e.g., in the statements 'The book is blue' or 'Being blue is a property of the book' is of categorial nature. The sensibly given, namely, the book with its parts and moments, includes the passive proto-relation (in *Erfahrung und Urteil* Husserl prefers the expression 'situation of affairs' [*Sachlage*]) of the book with its blue moment, and on this proto-relation two states of affairs are founded, which are the different objectualities that are referred to by the statements 'The book is blue' and 'Being blue is a property of the book'.

Something similar occurs when we connect in a relation the objects of two (or more) sensible intuitions to form new states of affairs. Thus, e.g., we can link two objects A and B to form the states of affairs that A is bigger than B and B is smaller than A. In sensible intuition we have two objects A and B and the situation of affairs (*Sachlage*) that A has a greater size than B, and on the basis of the sensibly given two different states of affairs are constituted, namely, those which are referred to by the equivalent statements 'A is bigger than B' and 'B is smaller than A'. Something similar occurs in the case of two or more objects that lie side by side or one above the other. The sensibly given are the objects and the existing situation of affairs, whereas the corresponding states of affairs that are referred to by the statements 'A lies above B' and 'B lies below A' are objectualities of the understanding constituted in a categorial intuition founded on sensible intuitions.

It is important to underscore here that in some sense situations of affairs are also founded objectualities, since they are complexes of simple sensible objects. But they are not the object of any sensible intuition. Although they are not properly objectualities constituted by the spontaneity of the understanding, i.e., they are not categorial objectualities, in receptive sensible intuition we do not have them thematically as objectualities. Situations of affairs appear, rather, as mere passively constituted foundations of different states of affairs. Now, once the states of affairs are constituted and objectified in predications, they can be objectively apprehended as the situation of affairs underlying two or more states of affairs. For example, once a relation R and its inverse relation R^{-1} are constituted, one can apprehend as object the corresponding situation of affairs as the 'abstract' invariant proto-relation that underlies both of them.

Another typical example of an objectuality of the understanding considered by Husserl—especially in §61 of *Erfahrung und Urteil*—is that of a (finite) set or collection. In sensible intuition not one but many objects can be given at the same time and even as belonging together—although, as Husserl remarks (Husserl 1939, §62, p. 297), the objects brought together in a collection do not need to have any sensible link between them, e.g., that of similarity or spatio-temporal contiguity. In such a case, the plurality of objects is given, like the trees in a park, or the seats in a classroom, or the books on a shelf. But the collection (or set) of trees, or seats, or books is not sensibly given. The collection is not a sensibly perceived objectuality nor can it be represented sensibly in the imagination. It is an objectuality of the understanding constituted in a categorial intuition, built on sensible intuitions, in which the objects that belong to the collection are constituted. Sets—and similarly states of affairs—do not 'dissolve' into the sensible objects on which they are founded, but are objectualities of a different sort, namely, of a categorial sort, and precisely as such can only be given as founded objectualities, as founded on the objects that are their members. Certainly, as in the case of states of affairs and situations of affairs, there is in some sense a plurality given in receptivity, but it is only in a categorial intuition that the set as such is constituted (Husserl 1939, §61, p. 292).

4. Sorts of Categorial Intuition

In all the examples of categorial intuitions and the corresponding categorial objectualities considered above, the objects of the founding sensible acts are in some sense incorporated as constituents in the objectuality that is constituted in categorial intuition. But in the Sixth Investigation (§§41, 47, and 52), Husserl also considers categorial objectualities which are founded on sensible objectualities but do not incorporate these as constituents. This is the case of what Husserl in the Sixth Investigation calls generalization, in which general objects or 'species' are constituted. In this case, the object given in the founding sensible intuition is only an instantiation or example of the species, but not one of its constituents. A similar situation occurs in the case of what Husserl calls (Husserliana, vol. XIX/2, 676) the singular indeterminate conception, for which the object given in sensible intuition, e.g., the concrete triangle sketched on the blackboard, is only a mere illustrative aid, not a constituent of the constituted objectuality. In both cases objectualities of higher level, founded on the sensibly given, are constituted, but this foundation is of a different sort than the foundation of the examples considered before.

In the act of generalization or generalizing abstraction a general object, the species, is constituted, and this species is one, in contrast to the unbounded plurality of singular objects of the 'same species' that can serve as basis of the categorial act and can be constituted in the founding sensible intuitions. By means of the comparative variation of the founding sensible acts and its corresponding referents, we become conscious of the identity of the species as the general object of which the objects of the plurality of possible founding acts are mere instantiations. On the other hand, in the act of singular indeterminate conception a singular but arbitrary object of a determinate species is given and, thus, not—as in the previous case—the species or idea of a triangle, nor any determinate singular triangle belonging to that species, but any triangle whatsoever.

It must be said here, however, that in *Erfahrung und Urteil* Husserl seems to use the expression 'objectuality of the understanding' (and thus 'categorial objectuality' and 'syntactical objectuality') in a somewhat more restricted sense that excludes the so-called general objects or species. In §64 of that work Husserl underscores the idea that the irreality of the objectualities of the understanding has to be distinguished from the generality of the species.

Although—as we have seen—in the Sixth Investigation Husserl stresses some differences between the constitution of species and that of other non-sensible objectualities, in *Erfahrung und Urteil* the contrast between these sorts of objectualities is radicalized. A species is such that it can be instantiated in different objects—e.g., a color in different colored objects—and each of these objects has its individual moment of the species. The species—e.g., a color—can be apprehended only because a variety of different individual moments of the species is given; we compare them, and then we abstract the generic-universal by varying the examples perceived or imagined. But to apprehend a number (or state of affairs or collection) we do not need any comparison of supposed individual moments, nor any generalizing abstraction. The number 5 is identically the same object referred to in an unlimited plurality of acts, and not something obtained by comparing the objects referred to in those acts and then applying a so-called generic abstraction.

This restriction of the expression 'objectuality of the understanding' in *Erfahrung und Urteil*, although a clear modification of Husserl's views in *Logische Untersuchungen*—see, e.g., the Second Investigation—does not affect our discussion, since we are interested here almost exclusively in mathematical objectualities (see §§6 and 7 below) or, as I will also call them, pure categorial objectualities, and these are clearly different from species.

5. Levels of Categoriality

Now, the sorts of categorial acts can diverge in many directions. First of all, I have not so far emphasized the difference between the two basic sorts of sensible intuition, namely sensible perception and sensible imagination, since such a distinction is irrelevant for many purposes. It should be noticed, however, that an act founded on two or more simple acts can be founded either only on sensible perceptions, or only on sensible imaginations, or, in a mixed way, partly on sensible perceptions and partly on sensible imaginations (Husserliana, vol. XIX/2, 675). If the number of sensible acts is greater than two, this offers a diversity of combinations. Although Husserl is not here explicit in this regard, it is clear that for a categorial intuition to be a categorial perception, all its founding acts should be perceptions.

More interesting, however, is the complication that arises when categorial acts serve as founding acts of new categorial acts of higher level (Husserliana, vol. XIX/2, §§46, 59 and 60), as, e.g., when we establish a relation between two states of affairs or two sets. Since these new categorial acts of second level can also serve as founding acts of new categorial acts of third level, and so on indefinitely, we obtain a sequence of levels of foundation of categorial acts (Husserliana, vol. XIX/2, 675), that can be still more complicated if not all the immediately founding acts of a categorial act belong to the same level. In this manner we obtain a whole hierarchy of types of categorial acts, in whose zero level lie the sensible acts and in whose first level lie those categorial acts that we have been considering, all of whose founding acts are sensible. This hierarchy of types of categorial acts is ruled by a priori laws that are 'the intuitive counterpart' of the logico-linguistic laws studied by Husserl in the Fourth Investigation. Actually, in either case the problem of truth does not play any role. The laws that govern the hierarchy of types of categorial intuitions 'do not directly say anything about the ideal conditions of possibility of an adequate fulfillment of meanings' (Husserliana, vol. XIX/2, 711). They are only laws of the pure doctrine of forms of intuitions, that regulate its primitive types, its forms of complication, and the sequence of ever more complicated forms of intuitions obtained by iterating the forms of complication (Husserliana, vol. XIX/2, 711). Restrictions enter the scene, however, as soon as one considers not only the mere syntactic possibility of forms of intuitions, but also the logical possibility of the objects of such intuitions.[10]

The possibility of taking no matter which categorial acts as founding acts of categorial acts of higher level and the corresponding possibility of 'expressing' these acts in corresponding meanings, lead to a relative distinction, both on the side of intuitions and on the side of meanings, between form and matter (Husserliana, vol. XIX/2, 711). In this relative sense, the objectuality constituted in a categorial act, on the basis of other objectualities that served as the material for its constitution, can serve as the material for the constitution of other objectualities of still higher level in new categorial acts. On the side of statements, the terms that are the matter of statements, correspond to the objects of founding acts, and it is to those terms that one has to look for any contribution made by sensibility. But since the objects of the founding acts can themselves be categorial objectualities, we have to inquire about their constitution. Actually, if we are interested in the fulfillment in intuition of the meaning of a statement, we have to inquire about the meanings of its terms, and if these are fulfilled by categorial objectualities, we have to inquire further about the objectualities that correspond to its founding acts. In this manner we have to continue descending the hierarchy of founded acts that serves as basis of the categorial act in which the state of affairs referred to by the statement under discussion is constituted, until we arrive at simple objects, which are the objects constituted in the simple acts of intuition that serve as ultimate foundations of such a categorial act (Husserliana, vol. XIX/2, 712).

6. Pure Categoriality and Mathematical Intuition

According to Husserl (Husserliana, vol. XIX/2, 712), everything categorial is based on sensible intuition, although the link with sensible intuition can vary in many ways. Now, Husserl calls 'sensible' only the simple acts, whereas he calls every founded act 'categorial', no matter if it is immediately or mediately based on sensibility. In the universe of categorial acts, however, Husserl distinguishes (Husserliana, vol. XIX/2, 713) between pure categorial acts, which are acts of the pure understanding, and mixed categorial acts, which are acts of the understanding mixed with sensibility.

Just as the generalizing abstraction, whose object is the species or idea, although necessarily based on individual intuition, does not refer to something individual, so there exists the possibility of general intuitions that refer neither to something individual nor to something sensible. Thus, Husserl distinguishes between sensible (generalizing) abstraction and pure categorial (generalizing) abstraction. The first gives us sensible concepts and sensible concepts mixed with categorial forms (or, briefly, mixed concepts). To the first group belong, e.g., the concepts of house, color, and wish, whereas to the second group belong, e.g., the concepts of coloring, virtue, and parallel axiom. Categorial abstraction, on the other hand, gives us purely categorial concepts like, e.g., the concepts of relation, set, number, and generally, all such concepts called by Husserl (see, e.g., Husserliana, vol. XVIII, 245 and Husserliana, vol. III/I, 27) formal-ontological categories and the derived concepts obtained from them. Sensible concepts—whether purely sensible (like house) or mixed (like parallel axiom)—have their ultimate foundation in sensible intuition. On the other hand, categorial concepts have their foundation in categorial intuitions, and with exclusive reference to the categorial form of the whole object categorially formed.

But this takes us to the origin of formal-ontological categories and to the problem of the intuition of mathematical entities. Given a categorial intuition of a relation, pure categorial abstraction directs itself to the form of the relation, leaving aside everything material in the related objects, considering them as mere indeterminate points of the relation. Thus, given a categorial intuition of the relation of 'being bigger than' between the sensible objects A and B, pure categorial abstraction directs itself to the relation, leaving the objects related as mere indeterminate points of the relation completely void of any individualizing traits. Similarly, given a categorial intuition of a set, pure categorial abstraction directs itself to the form of the collection, leaving the members of the set completely indeterminate. In this manner the formal-ontological categories of relation and set, respectively, are constituted. Thus, we can say that the intuition of mathematical entities, or, briefly, mathematical intuition, is categorial intuition purified by pure categorial abstraction. In this sense both pure logic and pure mathematics are purely categorial, since they do not contain any sensible concept in their whole theoretical foundation. In both of them the 'terms' remain purely indeterminate, and are usually represented by mere indicators (i.e., variables).

Once the formal-ontological categories, i.e., the primitive mathematical concepts, are constituted, the other mathematical entities are constituted in new pure categorial acts of higher level, in which the objectualities that serve as foundations are left completely indeterminate. In this manner the whole spectrum of mathematical entities is constituted, built on the basis of the formal-ontological categories which were constituted in a pure categorial abstraction that left indeterminate the sensible material on which a categorial intuition of first level was based. Thus, without contradiction, we can say that everything categorial is based on sensible intuition, and, on the other hand, that the concepts of pure mathematics are purely categorial, i.e., that they do not have any trace of something sensible in their constitution.

7. The Laws of Categorial Intuitions

In theoretical thought categorial intuitions act as real or possible fulfillments or frustrations of meanings, and confer on the statements, depending on whether the first or second is the case, the truth value 'true' or 'false' (Husserliana, vol. XIX/2, 720). Thus, the pure laws that rule over categorial intuitions are, for Husserl, the pure laws of thought in the strict sense.

To begin with the discussion of these laws, we should first of all notice with Husserl (Husserliana, vol. XIX/2, 716) that there is a great freedom in the application of categorial forms in the constitution of new objectualities of the understanding, since that which acts as the material to which a determinate categorial form is applied does not determine the categorial form.

But this does not mean that the formation of new categorial objectualities is not ruled by laws that in some sense restrict this freedom. First of all, one should notice that categorial objectualities are constituted only in founded acts and never in acts of the lowest level. Moreover, categorial objectualities cannot be constituted on the basis of just any foundation, even though 'we can think—understood as merely signifying—any relation between any points of reference, and, more generally, any form on the basis of any material' (Husserliana, vol. XIX/2, 717). As we shall see below, there is no complete parallelism between the laws of formation of categorial intuitions and its objects, and the laws of formation of meanings, since the former have restrictions that are totally foreign to the latter.

Now, the ideal laws that rule over the possibilities and impossibilities of categorial objectualities belong to them *in specie*, and, thus they belong to the formal-ontological categories and their derived concepts. Such laws regulate the possible variations to which the categorial objectualities can be submitted, while the material that corresponds to their foundations is left fixed. They restrict the variety of reorderings and transformations of categorial objectualities, and since the peculiarity of the pertinent materials is totally irrelevant, such laws have the character of purely analytic laws. Owing to this, on the side of expressions, algebraic symbols are used as bearers of totally indeterminate and arbitrary representations to express such materials. And for gaining insight into such laws, 'any categorial intuition would suffice that puts before our eyes the possibility of the categorial objectuality concerned' (Husserliana, vol. XIX/2, 718).

These purely analytic laws are precisely those that rule over the possibility of mathematical entities, since, as Husserl observes (Husserliana, vol. XIX/2, 718–19), 'the ideal conditions of the possibility of categorial intuitions are correlatively the conditions of the possibility of its objects, i.e., of categorial objectualities generally'. It is essentially the same thing to say that a determinate categorial objectuality is formed in such and such a manner, and to say that a categorial intuition is performable in which such an objectuality is constituted on the basis of such and such corresponding founding intuitions.

Now, such analytic laws do not say anything about the categorial acts performable on the basis of sensible intuitions that serve as ultimate foundations. All sorts of contents can serve as material for any categorial intuition. But such analytic laws can teach us that, when a given material assumes some given form, then there is a fixed set of other forms that such a material may assume. In other words, 'there is an ideally closed set of possible transformations of a given form into other forms' (Husserliana, vol. XIX/2, 720).

8. The Parallelism with Pure Grammar and the Possibility of Paradoxes

As I have said more than once (see §§5 and 7 above), the laws of formation of categorial intuitions, and correlatively the laws of categorial objectualities, run parallel to the laws of formation of meanings. To all categorial acts with their categorially formed objectualities there can correspond mere meaning acts. But, as we know from the First Investigation, a meaning can be constituted in a meaning act without there being an intuition that fulfills it. Moreover, the region of meanings is more inclusive than the region of intuitions in which their possible fulfillments can be constituted, since on the side of meanings there exists an unlimited plurality of complex meanings which are 'impossible', i.e., they are complexes of meanings linked in such a way as to be a complex unitary meaning, but such that the possibility of a corresponding objectuality in (categorial) intuition is excluded. Therefore, there does not exist a complete parallelism between the hierarchy of sorts of meanings and the hierarchy of sorts of objectualities of the understanding. Such a parallelism exists at the lowest level, i.e., that of sensible intuition and their corresponding meanings, but since there is a greater freedom for linking meanings to form complex meanings than for linking objectualities to form complex (categorial) objectualities, it is not the case that to every sort of meaning there corresponds a categorial objectuality. In other words, since one speaks of compatibility or incompatibility only in the region of the complex, to each simple meaning there corresponds an objectuality, but not to each complex meaning (Husserliana, vol. XIX/2, 729).

Thus, here originates the possibility of (analytic) contradictions and, especially, the possibility of the so-called paradoxes. From the point of view of the formation of meanings it is legitimate to form meanings such as 'the greatest ordinal' or 'the greatest cardinal number', or 'the set of all sets that do not contain themselves as elements'. But the objectualities that would correspond to such meanings are impossible, i.e., they cannot be constituted in any categorial intuition (see Appendix 2 below).

The divergence between the hierarchies of meanings and of categorial intuitions with their corresponding categorial objectualities leads Husserl (Husserliana, vol. XIX/2, §§63f.) to a distinction between what he calls proper and improper laws of thought. The laws of proper thought rule over the possibility of forming new categorial acts and their corresponding categorial objectualities. The laws of improper thought rule over the possibility of forming complex unitary meanings. Correspondingly, Husserl calls improper acts of thought all those acts in which meanings are constituted, and proper acts of thought all the corresponding (possible) intuitions. Now, since in the region of formation of meanings we are free to form any complex meaning, even when no object could correspond to it, if we want to avoid not only nonsense, but also formal (and material) countersense, we have to restrict the more extensive region of improper thought to conform to the objective possibility of fulfillment by some categorial intuition. If we restrict our consideration to formal countersense, the laws so obtained, namely, the pure laws of the validity of meanings, are precisely the pure logical laws in the strict sense, which, according to Husserl (Husserliana, vol. XIX/2, 723), have to run parallel to the pure analytic laws that rule over the formation of categorial objectualities, and are no less analytic than these. Such pure logical laws in the strict sense rule over the possibilities, determined in a purely categorial manner, of connection and transformation of meanings, without affecting the possibility of fulfillment of meanings, or, as we could also say, salva veritate. As in the case of the other two sorts of laws considered above, namely, those of proper thought and those of improper thought in its full, i.e., nonrestricted sense, in such laws the material does not play any role, and 'material meanings' are substituted by algebraic signs that mean in an indirect and completely indeterminate way (Husserliana, vol. XIX/2, 724). It is in this precise Husserlian (neither Kantian, nor Fregean, nor Carnapian) sense that all such laws are analytic.[11] Here we are concerned with the pure conditions of the objective possibility of meanings, i.e., the possibility of meanings having a referent, and this leads us to the conditions of possibility of categorial intuitions. Thus, although these logical laws of the validity of meanings are not identically the same as the laws of categorial intuitions (the laws of the possibility of constitution of categorial objectualities), they follow closely in their steps.[12] Actually, the laws of the possibility of constitution of categorial objectualities are precisely the laws of mathematical existence.

Now, usually our thought operates partly intuitively and partly symbolically (i.e., without corresponding intuition). Husserl calls (Husserliana, vol. XIX/2, 725) complex acts that are partly intuitive and partly symbolical improper categorial intuitions. In such a case the objectual correlate of such an act is represented only improperly and its possibility is not secured. But this takes us once more to the problem of the possibility of contradictions and, especially, of paradoxes, since only when the a priori possibility of the constitution of each of the founding objectualities that serve as 'material' foundations of different levels of a complex categorial objectuality is established is the possibility of the constitution of such an objectuality also established. Hence, although our thought operates partly symbolically, the fulfillment of every complex non-contradictory meaning must in principle be possible.

APPENDIX 1

The Assessment of Categorial Objectualities in *Erfahrung und Urteil*

As is well known, from 1905 onwards there is a reorientation of Husserl's philosophy in the direction of transcendental phenomenology, and it is pertinent to determine if this reorientation somehow affects the present discussion.[13] It is reasonable to think that even when in the *Logische Untersuchungen* —some would like to say in its first volume only— Husserl had sustained a Platonist conception of mathematics, in his later philosophy he must have inclined himself to a sort of constructivism that had remained latent since his *Philosophie der Arithmetik*. It seems appropriate to try to illuminate this situation by considering Husserl's discussion of the objectualities of the understanding in §64 of *Erfahrung und Urteil*, a text on which Landgrebe was working under Husserl's supervision at the time of the latter's death and which would thus include any modification of Husserl's conception of mathematical entities produced by any of Husserl's reorientations of his philosophy after 1900.[14] We will see, however, that Husserl's assessment of mathematical (and other categorial) objectualities in *Erfahrung und Urteil* does not lead to any sort of constructivism, but at most to a refinement of his Platonistic conception. Since what is true in general of the objectualities of the understanding—even in the restricted sense of this expression in *Erfahrung und Urteil*[5]— is also true in particular of mathematical entities, in what follows I will often refer only to mathematical objectualities, which are my main interest and for which Husserl's assessment seems easier to defend than for categorial objectualities in general.

For Husserl in *Erfahrung und Urteil* the principal difference between individual objects given in sensible intuition and categorial objectualities lies in their different relation to temporality. Certainly, for Husserl all objects have a determinate relation to the internal time of consciousness, in which all acts of consciousness are constituted. But, whereas individual objectualities are also linked to physical objective time and to its objective temporal points, the objectualities of the understanding are not so linked. They are unreal objectualities, if by 'real' we understand 'real physical'. Real individual objects of sensibility are individualized by their appearance in the objective point (or interval) of physical time, which is represented in internal time. But the time points that individualize objects of the lowest level do not play a similar role in the case of objects of higher level (immediately or mediately) founded on them, and a fortiori do not play any role in the case of mathematical objectualities, whose link with sensible individual objects—as explained in §6 above—is particularly thin. Mathematical objectualities are not bounded to any temporal point or interval of points. In contrast to real objects, they are irreal in the sense of being anywhere and nowhere. They can appear in any spatio-temporal coordinates—even in different spatial coordinates at the same time—and be the same objectuality. One can say that mathematical objectualities existed before they were discovered, and that the laws based on them were valid before being discovered, in the sense that both mathematical objectualities and the laws that rule over them can be generated at any time point by subjects with the capacity to constitute the first and to recognize the validity of the second. In this way the existence of mathematical objectualities is omnitemporal, and one can say generally that mathematical (and other unreal) objectualities existed even before someone had constituted them. Certainly, when they are actualized or constituted, they enter spatio-temporal facticity, i.e., they locate themselves spatio-temporally, but this insertion in spatio-temporal facticity does not individualize them.

The intemporality, the everywhere and nowhere of mathematical objectualities, is a special form of temporality, namely, omnitemporality, and this form of temporality is clearly different from that of real objects. Mathematical objectualities exist in any time, and do not have any link to objective spatio-temporal points or intervals of points that correspond to the internal time in which the categorial acts that constitute them are themselves constituted.

APPENDIX 2

Some Nonobjectualities of the Understanding

In §5 I have shown how categorial objectualities are constituted, and in §8 I have underscored the thesis that the parallelism between the formation of complex meanings and the formation of complex categorial objectualities breaks down, since some restrictions intervene in the latter case that are completely unknown in the first. In this Appendix I will make use of such restrictions to show that some supposed mathematical entities that have originated well-known paradoxes in set theory cannot be constituted in any categorial intuition and, thus, cannot have any mathematical existence on the basis of the epistemology of mathematics that I have extracted from Husserl's texts.

    1. Russell's Paradox: If it were true—as was thought to be true in naive set theory—that every property determines a set, one could consider the set R of all sets x such that $x \notin x$. Thus, $R = \{x/x \notin x\}$. One could then ask if the set R is or is not an element of itself, i.e., one could ask if $R \in R$ or $R \notin R$. Now, if $R \in R$, i.e., if $R \in \{x/x \notin x\}$, then $R \notin R$. On the other hand, if $R \notin R$, then R satisfies the condition required for membership in R, since $R = \{x/x \notin x\}$. Hence, $R \in R$. Thus, if $R \in R$ then $R \notin R$, and if $R \notin R$, then $R \in R$.

Now, it is easy to observe that the Russell set R cannot be constituted at any of the levels in the hierarchy of categorial intuitions, since, although new sets can be constituted at every level of the hierarchy, they have as members only objectualities constituted at lower levels of the hierarchy. Hence, it can never be the case that a set is a member of itself. More generally, for any categorial (and, in particular, for any mathematical) objectuality, it is never the case that it applies to itself, since its 'arguments' have to be constituted at lower levels of the hierarchy.

    2. Cantor's Paradox: As is well known, two sets have the same cardinality if there exists a bijection between them, and a set A has a greater cardinality than a set B if there exists a bijection between B and a (proper) subset of A but no bijection between A and B. Cantor's Theorem establishes that the power set of a set C (i.e., the set of all subsets of C) has a greater cardinality than C. Thus, beginning with the set of natural numbers, the power set construction allows the formation of sets of ever greater infinite cardinality. Now, in naive set theory one assumes that there exists a set V of all sets. Since V is the set of all sets, it must have cardinality equal to or greater than that of any set C, since any other set D is a subset of V. Now, by the power set construction, one can obtain

the power set P(V) of V, and by Cantor's Theorem the cardinality of P(V) is greater than the cardinality of V. But since P(V) is a set, its cardinality must be equal to or less than the cardinality of V. Thus, we obtain a contradiction.

Now, a set like V, i.e., the set of all sets, cannot be constituted at any stage of the hierarchy of categorial objectualities, since new sets can be constituted at each stage of the hierarchy, no matter if the hierarchy continues in the transfinite. The process of constitution of categorial objectualities—and, particularly, that of sets—is iterable without limit.[16] Thus, since the set of all sets cannot be constituted in any categorial intuition, Cantor's Paradox is also blocked.

In the same vein one can show that the supposed entities that originate the rest of the paradoxes of naive set theory cannot be constituted in any categorial intuition.

APPENDIX 3

Some Objectualities of the Understanding

According to Husserl's philosophy of mathematics, mathematics—geometry excluded—is formal ontology, i.e., an ontology based on the completely formal concept of 'something in general' (*Etwas überhaupt*). The fundamental concepts of mathematics, the so-called formal-ontological categories, are formal variations of the concept of 'something in general' that give rise to the fundamental mathematical structures, each based on one of those categories. Among these categories Husserl includes the concepts of set and relation, together with, e.g., those of whole and part, and of cardinal and ordinal number.

As is well known, in current discussions of set-theoretical foundations of mathematics not all such notions are considered equally fundamental.[17] The notion of set is usually considered as the most fundamental notion of mathematics and all other basic notions are defined by means of it. Such fundamentality of the notion of set has been recently questioned by mathematicians working in category theory.[18] However, since our interest here is to show how it is that mathematical entities are constituted in mathematical intuition, we can avoid discussions about the foundations of mathematics, and will simply take as fundamental mathematical objectualities two of Husserl's favorite examples, namely, sets and relations. Thus, I will give some sketchy indications of how some other mathematical entities are constituted in mathematical intuition on the basis of those two fundamental ones.

As was shown in §3 above—see also §6—both collections or sets and relations can be constituted in categorial intuitions of the first level, and since we can abstract from the peculiar nature of the members of the set and of the terms of the relation, the formal-ontological categories of set and relation are constituted in a mathematical intuition of the first level. Thus, given two or more sets, relations between them, e.g., bijective correspondences between sets or between sets and subsets of other sets can be constituted in mathematical intuitions of the second level—mathematical since the peculiarity of the members of the correspondence pairs is here abstracted from. Finite cardinal numbers can then be constituted in mathematical intuitions of the third level as equivalence classes of sets whose members are related by such bijective correspondences.

Once sets have been constituted in mathematical intuitions of the first level, nonempty intersections, unions, relative complements, and symmetric differences of sets can be constituted in mathematical intuitions of the first level. On the basis of the relative complement, the empty set can be constituted as the relative complement of a set to itself in a mathematical intuition of the first level. But once the empty set is constituted, intersections in the more general sense—that does not exclude the possibility of being empty—can also be constituted in mathematical intuitions of the first level.[19] Moreover, families of sets or of subsets of a given set, and, in particular, the power set of a given set, can be constituted in mathematical intuitions of the level immediately higher than the level in which the given set (or sets) is (are) constituted, and such a process of formation of new sets in acts of mathematical intuition of ever higher level can be iterated indefinitely. But once a set, its power set, the empty set, and the intersection and union of sets have been constituted, filters and ultrafilters, and ideals and prime ideals can be constituted in new mathematical intuitions of the same level as the power set of which they are subfamilies.

In this manner all (legitimate) mathematical entities of classical mathematics should in principle be constituted in mathematical intuitions of ever higher level. The legitimacy of a mathematical entity would be established by tracing the genesis of its constitution in the hierarchy of types of mathematical intuitions, since, as we have seen above, paradoxical entities cannot be constituted in any mathematical intuition. This program, even if successful, would not, however, eradicate the possibility of contradictions in mathematics, since it is always possible for a mathematician to attribute contradictory properties to a legitimate mathematical entity, but the danger of contradictions would be considerably diminished.

NOTES

1. References to Husserl's works published in the Husserliana edition are the standard ones. Cited as Husserl 1939 is *Erfahrung und Urteil*, 5th ed. rev. (Hamburg: Meiner, 1976).
2. See, e.g., M. Dummett's "The Philosophical Basis of Intuitionistic Logic," *Philosophy of Mathematics*, 2nd ed. rev., ed. P. Benacerraf and H. Putnam (Cambridge: Cambridge University Press 1983), pp. 97–129.
3. See, e.g., P. Benacerraf's "Mathematical Truth," *Philosophy of Mathematics*, ed. P. Benacerraf and H. Putnam, pp. 403–20.
4. See G. Frege, *Nachgelassene Schriften*, 2nd ed. (Hamburg: Meiner, 1983), pp. 282–302. For "*Der Gedanke*" see G. Frege, *Kleine Schriften* (Darmstadt: Wissenschaftliche Buchgesellschaft, 1967; 2nd ed., Hildesheim: Olms, 1990), pp. 342–62.
5. See K. Gödel's "Russell's Mathematical Logic," *Philosophy of Mathematics*, ed. P. Benacerraf and H. Putnam, pp. 447–69.
6. For an exposition of Husserl's philosophy of mathematics, see my doctoral dissertation "Edmund Husserls Philosophie der Logik und Mathematik im Lichte der gegenwärtigen Logik und Grundlagenforschung," Rheinische Friedrich-Wilhelms-Universität, Bonn, 1973. See also R. Schmit, *Husserls Philosophie der Mathematik* (Bonn: Bouvier, 1981).
7. Perception and imagination are, for Husserl, the two principal sorts of intuition, but their difference is unimportant for most of our discussions.
8. The interested reader can compare our treatment of the issues discussed in this and the next two sections with that of Adolf Reinach in "*Zur Theorie des negativen Urteils,*" (1911), translated in *Parts and Moments*, ed. B. Smith (Munich: Philosophia, 1982), pp. 315–77. See especially Part II, pp. 332–54.
9. For Husserl's distinction between part and moment, see his *Logische Untersuchungen*, Husserliana, vol. XIX/I, Third Investigation and *Parts and Moments*, ed. B. Smith, especially the first of the "Three Essays in Formal Ontology" by P. Simons and the introductory essay "Pieces of a Theory" by B. Smith and K. Mulligan.
10. Such restrictions will be discussed in §§7 and 8. In this context Husserl is considering the possibility of a morphology of intuitions in analogy to the morphology of meanings considered in pure logical grammar (see, e.g., his *Logische Untersuchungen*, Husserliana, vol. XIX/I, Fourth Investigation and his *Formale und transzendentale Logik* Husserliana, vol. XVII, chapter 1, §13), and for which even the formal possibility of fulfillment (and, thus, the avoidance of countersense) is unimportant. However, although the distinction in the realm of meanings between the levels determined by the laws that avoid nonsense and those that avoid countersense (the logical laws) has been particularly fruitful and generally accepted, the analogous distinction in the realm of intuitions seems somewhat artificial.
11. For Husserl, a statement is analytically true if it is true and can be formalized salva veritate. See his *Logische Untersuchungen*, Husserliana, vol. XIX/I, Third Investigation §12.
12. For Husserl, there is (or should be) a sort of parallelism between logic and mathematics. See his *Formale und transzendentale Logik*, Husserliana, vol. XVII, chapters 2–3, and *Logische Untersuchungen*, vol. 1, Husserliana, vol.

XVIII, chapter 11, and my dissertation "Edmund Husserls Philosophie der Logik und Mathematik im Lichte der gegenwärtigen Logik und Grundlagenforschung" cited in note 6.

13. Supposedly in the 1930s there occurred another reorientation of Husserl's thought.

14. Although *Erfahrung und Urteil* was prepared for publication, possibly with stylistic alterations, by Ludwig Landgrebe on the basis of Husserlian manuscripts, I do not have any misgivings concerning the authorship of the ideas expounded in that work. In particular, Husserl's conception of ideal entities in *Erfahrung und Urteil* is, in the present writer's opinion, a 'natural consequence' of a reexamination of his conception in *Logische Untersuchungen* in the context of his later philosophy.

15. See §4 above.

16. The first set of infinite level would be obtained by constituting the set of all categorial objectualities constituted at any finite level.

17. The concepts of part and whole are not even generally acknowledged as mathematical concepts. See, however, *Parts and Moments,* ed. B. Smith, cited in note 9.

18. Both for set-theoretical and for categorial foundations of mathematics, see W. S. Hatcher's excellent book *The Logical Foundations of Mathematics* (Oxford: Pergamon Press, 1982). (This book is a revised version of Hatcher's earlier *Foundations of Mathematics* (London: W. B. Saunders & Co., 1968.)

19. I do not intend to be completely rigorous here, especially when considering intuitions of the same level. Thus, I am not excluding any other more 'natural' ordering in the generic constitution of such objectualities. Alternative orderings—even those for which the notions of set and relation are not fundamental—are allowed, provided that type restrictions in the hierarchy of mathematical objectualities are not violated. Moreover, it should be underscored that in this (and the previous) appendix I use freely the term 'set', although only the constitution of finite sets has been discussed in the main text.

13

Guillermo E. Rosado Haddock

INTERDERIVABILITY OF SEEMINGLY UNRELATED MATHEMATICAL STATEMENTS AND THE PHILOSOPHY OF MATHEMATICS

1. The Problem

Interderivability of mathematical statements on the basis of certain assumptions or, briefly, mathematical equivalence of two statements in a determined area of mathematics is a relatively common phenomenon in the mathematical sciences. That two mathematical statements S and S' in general topology—or in group theory, or in number theory—are interderivable, is a familiar mathematical phenomenon. That a group G has property P if and only if it has property P', or that a topological space T has property Q if and only if it has property Q' seems completely natural both to the mathematician and to the philosopher of mathematics.

However, that mathematical statements of the most diverse areas of mathematics and apparently speaking about very different things, are interderivable, does not seem so natural. That a mathematical statement that speaks about well-ordered sets is interderivable under Zermelo-Fraenkel set theory (from now on ZF),[1] with, e.g., a statement about vector spaces, or with a statement about topological spaces, or with a statement about the cardinality spectrum of models of sets of first-order sentences, seems bizarre and at first sight highly implausible.

But precisely that seemingly implausible situation occurs in classical mathematics, and its philosophical significance has been neglected. Under ZF, the Ultrafilter Theorem, namely, the statement that says that every filter on a set can be extended to an ultrafilter, is interderivable, e.g., with the following three statements:[2]

(1) On each infinite set there is a two-valued additive measure such that each singleton has measure zero.

(2) The product of any family of compact Hausdorff spaces is a compact Hausdorff space.

(3) *Compactness Theorem for first order logic*: A set S of first-order sentences has a model if and only if every finite subset of S has a model.

For anyone acquainted with the concepts occurring in those four mathematically equivalent statements, it is clear that these statements speak about different and seemingly unrelated things.

A still more dramatic, but essentially similar, situation is that of the interderivability, under ZF, of the Axiom of Choice with many other mathematical statements in the most diverse areas of mathematics (and logic). The Axiom of Choice states that, given any family **F** of non-empty pairwise disjoint sets, there exists a function f such that for each set S in **F**, $f(S) \in S$, i.e., the function selects from each set S in **F** a representative of S. As a way of further illustrating the interderivability phenomena between seemingly unrelated mathematical statements, let us consider the following list of a few of the many mathematical equivalents of the Axiom of Choice:

(1) *Well Ordering Theorem*: Every set can be well-ordered.

(2) *Zorn's Lemma*: If **F** is a family of sets such that the union of every chain **C** $\subseteq$ **F** is in **F**, then **F** contains a maximal set under inclusion.

(3) Every lattice with a unit and at least another element has a maximal ideal.

(4) If **V** is a real vector space, then for every subspace S of **V** there is a subspace S^* of **V** such that $S \cap S^* = \{0\}$ and $S \cup S^*$ generates **V**.

(5) *Löwenheim-Skolem-Tarski Theorem*: If a countable set of first order sentences has an infinite model, then it has a model of each infinite cardinality.

(6) *Tychonoff's Compactness Theorem*: The product of any family of compact topological spaces is a compact topological space.

For any one familiar with the concepts occurring in the Axiom of Choice and its six mathematical equivalents cited immediately above, it is clear that they all speak about very different things, e.g., about choice functions, well orderings, lattices, real vector spaces, compact topological spaces, and the cardinality spectrum of models of countable sets of first order sentences.

The problem that I would like to consider in this paper and that seems not to have received its due attention by philosophers of mathematics is whether such interderivability phenomena of seemingly unrelated mathematical statements have any relevance for the philosophy of mathematics or are completely neutral with respect to our choice of a particular philosophy of mathematics. More explictly, I ask whether all philosophies of mathematics can fare well with respect to these phenomena.

2. A Brief Examination of Some Philosophies of Mathematics

It would be a formidable and rather boring task to examine in detail the wide variety of philosophies of mathematics with respect to the problem in which we are interested here, namely, rendering philosophically intelligible the interderivability phenomena of seemingly unrelated mathematical statements. However, some brief remarks seem appropriate in order to motivate defenders of at least some philosophies of mathematics to try to assess such interderivability phenomena from the standpoint of their respective conceptions of the nature of mathematics. The burden of the proof is on their side. Of course, they can ignore that burden simply by rejecting those phenomena as not belonging to mathematics as they conceive it. But in this way they would be depriving mathematics of some of its philosophically most puzzling results.

(*a*) For some philosophies of mathematics, however, there is no other possibility than to explicitly reject or simply ignore such interderivability phenomena. A formalist philosophy of mathematics like the one timidly advocated by Paul Benacerraf at the end of his "What Numbers Could Not Be,"[3] which does not accept the existence of numbers, but only of number words—and, hence is not very different from the formalism criticized by Frege a hundred years ago—would surely not accept the existence of, e.g., topological spaces, vector spaces, lattices, or ultrafilters, but at best that of topological space words, vector space words, lattice words, etc., whatever that may mean. Even the formulation of the statements in the two lists of mathematically equivalent statements of §1 would seem extremely bizarre, and their mathematical equivalence would be completely unintelligible.

(*b*) Contemporary nominalists are probably ontologically more liberal than the (sorts of) formalists that we have been considering. According to Eberle,[4] contemporary nominalists postulate the existence of concrete individuals, but reject the existence of numbers, classes, or other abstract entities. Of course, ultrafilters, lattices, vector spaces, topological spaces, and such, as they are understood in classical mathematics, are clearly abstract entities. Nominalism has to deprive classical mathematics of some of its dearest parts, and at best try to construct a parallel mathematics—as Lesniewski did with his mereology as a possible substitute for set theory. But then the interderivability phenomena that we have been considering remain completely unintelligible for nominalism, since the statements that are proved to be mathematically equivalent speak about entities that are based on the classical notion of set and not on its mereological surrogate or on any other sort of nominalist substitute, which cannot have the same properties as sets. (E.g., in set theory—and in general topology, which is based on it—one distinguishes between a unit set and its unique member, whereas in mereology a mereological unit class is identified with its unique member—or part, since mereology is a part-whole theory).

(*c*) Constructivism in mathematics is a particularly popular philosophical trend. The most basic difficulty with constructivism as a philosophy of mathematics, however, seems to lie in the term 'constructivism' itself. There does not seem to be any generally accepted precise meaning of that term. According to Beeson,[5] by a constructivist philosophy of mathematics one understands a philosophy of mathematics based on the following two basic principles:

(1) to say that "*x* exists" means that there is a way to explicitly find *x*;
(2) the word 'truth' does not have any a priori meaning, and a sentence is called 'true' just in case that a proof of it can be found.

The preceding characterization, however, is not sufficiently informative, since not only the philosophical preconceptions that underlie different constructivist schools can vary, but their notions of a mathematical proof can be very different and, thus, the results that the different constructivist schools consider mathematically sound can diverge essentially. Even in the most important constructivist school of this century, namely, intuitionism, there has been some disagreement concerning what should be considered mathematically sound.[6] Of course, if the philosophical preconceptions and the notion of a mathematical proof are too restricted, some of the entities spoken about in the statements belonging to either of the two lists of mathematically equivalent statements of §1 could not be constructed and many of the theorems related to those entities could not be obtained. E.g., the Axiom of Choice itself would certainly be considered mathematically inadmissible and the talk about different infinite cardinalities in the Löwenheim-Skolem-Tarski Theorem would be regarded as meaningless by many constructivists. Thus, if constructivism is understood in a too restrictive way, there is no hope of philosophically assessing the interderivability results under discussion. For such constructivisms those interderivability phenomena would be almost as unintelligible as for formalists and nominalists.

I am not interested, however, in restricting in any way the notion of constructivism to win my case. Thus, let us suppose that there is a constructivism so liberal that it allows its defenders to acknowledge the existence of the same entities and to obtain the same theorems as classical mathematicians—

with the only difference that some divergent methods are required, since they have to 'construct' the mathematical entities. Even in such a case, the interderivability results under discussion would remain a complete mystery for them. Since the entities spoken about in the two lists of mathematically equivalent statements of §1, namely, lattices, ultrafilters, topological spaces, vector spaces, cardinality spectra of models of sets of first-order sentences, etc. differ considerably and have so diverging properties, even if all of them can be constructed and all corresponding theorems about them proved, the interderivability results would be philosophically as unintelligible for such a liberal constructivism as if, e.g., the statements 'Paris is the capital of France' were interderivable with the statement 'Plato was Aristotle's teacher.' What would be constructed by such a liberal constructivism are lattices, ultrafilters, topological spaces, etc., which are clearly very different mathematical entities. The interderivability of, e.g., the Löwenheim-Skolem-Tarski Theorem, Tychonoff's Theorem, and Zorn's Lemma is in need of a philosophical assessment. But an adequate assessment cannot be made if one assumes that all entities spoken about in such mathematical statements are constructed by (the community of) mathematical subjects. Hence, we have to conclude that no matter how liberal a constructivist philosophy of mathematics might be, it is incapable of an adequate philosophical assessment of the interderivability phenomena under discussion.

(*d*) After Frege's[7] and Husserl's[8] critiques of Mill's empiricism in mathematics and logic, one might have thought that empiricism concerning these disciplines would not reenter the philosophical scenario. Even logical empiricists clearly restricted their empiricist claims to other areas of science. In the last few decades, however, and probably under the influence of some remarks by Quine,[9] empiricism in logic and mathematics has rather shyly reappeared. Thus, it seems appropriate to examine how well empiricism does fare with respect to the problem that we have been discussing, namely, the philosophical assessment of the interderivability of seemingly unrelated mathematical statements.

Contrary to the other currents in the philosophy of mathematics briefly considered above, mathematical empiricism usually does not renounce from the outset large parts of classical mathematics. The basic problem with empiricism of whatever sort is to formulate a convincing theory of how it is that we come to consider so highly abstract mathematical entities as ultra-products, toposes, and many others on a more or less thin empirical basis, and how it is that on such a basis we obtain mathematical knowledge about such entities. The burden of the proof is clearly on the side of mathematical empiricists. Moreover, if the empirical data accepted are so thin as those of (the sort of) behaviorism presupposed by Quine,[10] there is no hope of completing the task. If the empirical basis is such as to allow the so-called indeterminacy of translation[11] and the so-called subdeterminacy of physical theories,[12] there is certainly no possibility of explaining our acquaintance with ultraproducts, topological spaces, algebras, etc. and our mathematical knowledge concerning them. Such entities and their properties are much farther away from any empirical basis than languages or physical theories. Moreover, our mathematical knowledge presupposes language (or some sort of symbolic system of representing concepts), and the collective mathematical knowledge of the mathematical community presupposes the translatability of mathematical texts. Our mathematical knowledge not only cannot be obtained from such a thin empirical basis, but—if Quine's indeterminacy thesis of translation is correct—is hardly compatible with it.

Some mathematical empiricists, like Kitcher,[13] would probably say that the empirical basis does not have to be so thin and that one should allow the mathematical subject more freedom for constructing mathematical entities. Kitcher even speaks about an idealized mathematical subject.[14] In such a case, it seems pertinent to ask if this idealized mathematical subject operates on a purely empirical basis, i.e., without any sort of categorial device in Husserl's sense. Moreover, Kitcher says that mathematics is concerned with structures present in physical reality.[15] One should ask Kitcher to point to a physical structure that has some resemblance with, e.g., an ultraproduct, and one should urge him to construct such a mathematical entity on a purely empirical basis. One could continue arguing in this direction against Kitcher, who, e.g., says that arithmetic owes its truth to the structure of the world.[16] For even if the physical world were completely different from ours, mathematical and, hence, also arithmetical theorems would continue to be true.

However, I am not interested here in such more or less traditional arguments against empiricism. Let us assume, contrary to all available evidence, that mathematical empiricists succeed in constructing all entities of classical mathematics on a purely empirical basis—i.e., without any unacknowledged nonempirical tools—and that they are capable of proving all theorems of classical mathematics. Even in such a very improbable case, the interderivability phenomena of seemingly unrelated mathematical statements would remain completely unintelligible for the mathematical empiricist. For surely the sense data (the physical basis or whatever that may be) that would serve as the empirical foundation in the genesis of lattices, topological spaces, vector spaces, cardinality spectra of models of sets of first-order sentences, etc. would have to be very different. Moreover, the properties or relations attributed to those entities in the mathematically equivalent statements that we have been considering are very different, and since for a genuine mathematical empiricist they too must be empirically founded, their empirical foundations would also have to diverge. Hence, there is no way for (a genuine) mathematical empiricism to explain the interderivability phenomena under discussion. They are for mathematical empiricism as puzzling as for any of the other philosophies of mathematics considered above.

It seems unnecessary to examine every sort of non-Platonist philosophy of mathematics with respect to the interderivability phenomena under discussion. In any other case one can argue essentially in the same way to show the inadequacy of such a philosophy of mathematics to assess the interderivability results of seemingly unrelated mathematical statements. Hence, either such mathematical results are simply incapable of any philosophical assessment, or we have to accept a sort of mathematical Platonism as the correct philosophy of mathematics.

But there can be more than one sort of mathematical Platonism, and even if the second member of the former exclusive disjunction is true, that does not entail that any mathematical Platonism can adequately assess the interderivability phenomena.

3. On Platonisms

For a philosophy of mathematics to be defensible, it has to be complemented by an epistemology of mathematics, i.e., an explanation of how it is that we come to have knowledge about mathematical entities. Constructivist philosophies of mathematics like those of Kant and Brouwer seem almost inseparable from their corresponding epistemologies. The main defect, however, of most Platonist philosophies of mathematics is precisely that they have not developed an accompanying epistemology of mathematics. Thus, even if they were to correctly assess the nature of mathematics and adequately resolve the riddle of mathematical entities, the foundational (not historical) genesis of mathematical knowledge would turn into a new puzzle. Of the defenders of a sort of Platonism in this century, only Husserl seems to have sufficiently developed an epistemology of mathematics.[17]

However, a Platonist philosophy of mathematics not only is in need of an accompanying epistemology of mathematics, but presupposes a semantics appropriate for mathematical statements. Without an adequate underlying semantics, a Platonist philosophy of mathematics does not seem to go much farther in explaining the interderivability phenomena than the philosophies of mathematics considered in §2 above. Of course, a semantics adequate for mathematical statements and a basically correct epistemology of mathematics are no complete guarantee of the correctness of a philosophy of mathematics, although, taken together, they seem to be necessary conditions both of the correctness and of the rational credibility of a Platonist philosophy of mathematics.

Let us consider briefly Frege's philosophy of mathematics. As is well known, Frege defended both a Platonist and a logicist conception of mathematics (with the exclusion of geometry)—which should be clearly separated from each other. Thus, he not only conceived mathematical entities, e.g., numbers, logical entities, and thoughts,[18] as having an objective but not spatiotemporally bound existence (Platonism), but he also believed that arithmetical concepts could be defined by means of logical concepts, and arithmetical theorems derived—ultimately—from logical axioms (logicism). Moreover, Frege also developed a theory of reference, according to which statements (i.e., assertive sentences)—whether mathematical or not—when standing alone or in extensional contexts, refer to a truth value, namely, to the true or to the false. For Frege, the true and the false are not only the referents of all statements, but also in some sense the foremost Platonic entities.[19] Frege does not acknowledge the existence of states of affairs, and for him all true statements have the same reference, namely, the true. Thus, for Frege, the statements (1) 'Paris is the capital of France', (2) '2 + 2 = 4', (3) 'Every set can be well-ordered', and (4) 'The product of any family of compact topological spaces is a compact topological space', although they seem to speak about very different things, have the same reference, namely, the true. Apart from the fact that statements (2), (3), and (4) seem to be true in all possible worlds, whereas (1) does not, (3) and (4) are mathematically equivalent and are mathematically equivalent neither with (1) nor with (2), nor is (1) mathematically equivalent with (2). Frege's semantics ignores all of this and also runs counter to our intuitions that statements (1)–(4) not only express very different thoughts, but also speak about very different things. Hence, Frege's semantics seems inappropriate for mathematics and, particularly, does not adequately assess the interderivability phenomena of seemingly unrelated mathematical statements, since it does not do justice to the fact that statements (3) and (4) above are interderivable, but are interderivable neither with (1) nor with (2), nor is (1) interderivable with (2).

The first step for building a semantics of mathematics that can adequately assess the interderivability of seemingly unrelated mathematical statements consists in taking states of affairs as the referents of statements.[20] The statements '3 + 4 = 7' and '6 + 1 = 7' express different thoughts, but refer to the same state of affairs, and the statements (1) 'Every filter on a set can be extended to an ultrafilter' and (2) 'Every dual ideal on a set can be extended to a maximal dual ideal' also seem to express different thoughts—if 'filter' is not introduced into the theory as an abbreviation of 'dual ideal'—but refer to the same state of affairs, namely, to the mathematical fact that an entity with the properties of a filter can be extended to (i.e., is contained in) a filter which is maximal in the sense of not being properly contained in any other filter. Moreover, this state of affairs is clearly different from that referred to by the equations '3 + 4 = 7' and '6 + 1 = 7', although all four statements are true.

But to acknowledge states of affairs as the reference of statements is clearly not enough, since precisely in each of the two lists of mathematically equivalent statements of §1, the statements, although interderivable, speak about very different things and, thus, refer to very different states of affairs.

On the other hand, since all statements in both lists have the same truth value as the statements 'Paris is the capital of France' and 'Frege died in 1925', namely, the true, to adequately assess the interderivability of seemingly unrelated mathematical statements, a semantics appropriate for mathematics has to postulate the existence of abstract entities intermediate between states of affairs and truth values. Thus, I will speak of abstract situations of affairs, and will say that the statements in each of the two lists of mathematically equivalent statements of §1 have the same abstract situation of affairs as their reference basis, although the states of affairs referred to by them are clearly different. More simple examples of pairs of mathematical statements referring to different states of affairs but having the same abstract situation of affairs as reference basis are the somewhat trivial pair of inequalities. '5 > 3' and '3 < 5', and pairs of dual statements as, e.g., the pair (1) 'Every filter on a set can be extended to an ultrafilter' and (2) 'Every ideal on a set can be extended to a maximal ideal'. (That "abstract situations of affairs" are really "abstract" can be easily admitted if we try to apprehend them intuitively even in the latter more elementary examples considered above.)

On the other hand, the statements 'Paris is the capital of France', 'Frege died in 1925,' and '2 + 2 = 4' have very different abstract situations of affairs as their reference bases, and all these reference bases differ both from the common abstract situation of affairs of, e.g., the Axiom of Choice and Tychonoff's Theorem, and from the common abstract situation of affairs of, e.g., the Ultrafilter Theorem and the Compactness Theorem. Thus, a semantics for mathematical statements that can offer an appropriate assessment of mathematical equivalence must include the following schema (where the arrows represent functions):

Statement

↓

Thought (or Proposition) (= Sense of Statements)

↓

State of Affairs (= Reference of Statements)

↓

Abstract Situation of Affairs (= Reference Basis of Statements)

↓

Truth Value

This semantics for statements is essentially a reconstruction of Husserl's, as is the distinction—and the terminology—between state of affairs (*Sachverhalt*) and situation of affairs (*Sachlage*).[21] This distinction remains in Husserl's writings somewhat sketchy, and since I am using it exclusively for mathematical contexts, for which the distinction is much clearer and can be made much more precise than for nonmathematical ones, I have added the adjective 'abstract' to 'situation of affairs' not so much to emphasize its abstract character—which is present even in nonmathematical contexts[22]—as to underscore that the notion of abstract situation of affairs that I have introduced here for a purpose not explicitly envisaged by Husserl, should be regarded more as a sort of explanans of Husserl's notion of situation of affairs than as exactly the same notion. On the other hand, since Husserl also developed an epistemology of mathematics, combining it with the neo-Husserlian semantics of mathematics just sketched, it seems possible to defend mathematical Platonism along neo-Husserlian lines.

Finally, it seems interesting to examine the semantics of mathematical statements propounded in this paper from the point of view of the information conveyed. As is well known, Frege begins and ends his famous "*Über Sinn und Bedeutung*" with a discussion of identity statements.[23] In particular, he is interested in explaining how it is that an identity statement of the form '$a = b$', when true—no matter whether synthetically or analytically—can have a greater cognitive value and, thus, be more informative, than an identity statement of the form '$a = a$'. The identity statements '$117 = 117$' and '$117 = 9 \times 13$' are both true identity statements and, according to Frege, analytically true, but the second is much more informative than the first. To explain this situation, Frege introduces the notion of sense. '117' and '9×13' have the same reference, namely, the number 117, but their senses are different and, thus, the senses of the identity statements '$117 = 117$' and '$117 = 9 \times 13$' are different.[24] Thus, if in a mathematical (or nonmathematical) statement, standing alone or in an extensional context, we substitute a proper name (in its broad Fregean meaning) for another proper name with a different sense but with the same reference, the truth value of the statement remains the same but its cognitive value can change.

However, a mathematical statement conveys information not only at the level of senses, but also at the level of states of affairs. Mathematical (and nonmathematical) statements refer to states of affairs, and when one learns to what state of affairs a mathematical statement does refer, one obtains some information. When a mathematical statement asserts that a mathematical entity, e.g., an ultraproduct or a Hausdorff space, has a definite property, it conveys (a nontrivial) information, based on an understanding of the referents of the constituents of the statement, namely, ultraproducts, Hausdorff spaces, etc.

Moreover, at the level of abstract situations of affairs there seems to lie a deeper, more intangible and probably less universal[25] level of information which has a strong metamathematical flavor.[26] When someone grasps a statement that speaks about the interderivability of two seemingly unrelated mathematical statements, a certain information is conveyed to him, an information at a level which builds a sort of 'deep structure' of mathematics.

NOTES

1. From now on I will most frequently simply write 'interderivable' and 'mathematically equivalent' instead of 'interderivable under ZF' or 'mathematically equivalent under ZF'. Throughout this paper all four expressions are taken as synonyms.

2. Both for the equivalents of the Ultrafilter Theorem and for those of the Axiom of Choice, see G. H. Moore, *Zermelo's Axiom of Choice* (New York: Springer, 1982). See especially p. 328 for equivalents of the Ultrafilter Theorem and pp. 330–33 for equivalents of the Axiom of Choice. See also H. Rubin and J. E. Rubin's *Equivalents of the Axiom of Choice* (Amsterdam: North-Holland, 1963) or *Equivalents of the Axiom of Choice*, II (Amsterdam: North-Holland, 1985).

3. P. Benacerraf, "What Numbers Could Not Be," *Philosophy of Mathematics*, 2nd ed., ed. P. Benacerraf and H. Putnam (Cambridge: Cambridge University Press, 1983), pp. 272–94.

4. R. Eberle, *Nominalistic Systems* (Dordrecht: Reidel, 1970), p. 6.

5. M. Beeson, *Foundations of Constructive Mathematics* (Berlin: Springer, 1985), p. 47. Beeson's constructivism follows that of Errett Bishop's *Foundations of Constructive Analysis* (New York: McGraw-Hill, 1967). See Chapter III of Beeson's book for other sorts of constructivism, to which one can add, e.g., I. Kant's remarks on mathematical method in *Kritik der reinen Vernunft* (1781, 2nd ed. 1787), especially Part II: *Transzendentale Methodenlehre*, and M. Dummett's recent 'linguistic intuitionism' in *Elements of Intuitionism* (Oxford: Oxford University Press, 1977), and in "The Philosophical Basis of Intuitionistic Logic," *Philosophy of Mathematics*, ed. P. Benacerraf and H. Putnam , pp. 97–129.

6. See, e.g., E. W. Beth, *The Foundations of Mathematics* (Amsterdam: North-Holland, 1965), chapter 15, §142, especially pp. 437–39.

7. See G. Frege, *Die Grundlagen der Arithmetik*, 1884, *Centenarausgabe*, edited and with an introduction by C. Thiel (Hamburg: Meiner: 1986), especially §§7–10.

8. See E. Husserl, *Logische Untersuchungen*, Husserliana, vol. XVIII (The Hague: M. Nijhoff, 1975), especially chapter 5.

9. See, e.g., W. Quine, "Two Dogmas of Empiricism," *From a Logical Point of View*, 2nd ed. (Cambridge, MA: Harvard University Press, 1961 [1953]), pp. 20–46, especially pp. 42–46. See also his *Philosophy of Logic* (Englewood Cliffs: Prentice Hall, 1970), especially chapter 7.

10. See, e.g., W. Quine, *Word and Object* (Cambridge, MA: MIT Press, 1960).

11. Ibid., especially chapter 2.

12. See, e.g., "Two Dogmas of Empiricism," p. 43.

13. See Philip Kitcher, T*he Nature of Mathematical Knowledge* (Oxford: Oxford University Press, 1983).

14. Ibid., p. 109.
15. Ibid., p. 107.
16. Ibid., pp. 108–9.
17. See Husserl's *Logische Untersuchungen*, Husserliana, vol. XIX/2 (The Hague: M. Nijhoff, 1984), Sixth Investigation, chapter 6. See also my paper "Husserl's Epistemology of Mathematics and the Foundation of Platonism in Mathematics," in *Husserl Studies* 4 (1987): pp. 81–102, chapter 12 of the present book. A possible exception is Kurt Gödel. See his "What is Cantor's Continuum Problem?" reprinted in *Philosophy of Mathematics*, ed. P. Benacerraf and H. Putnam, pp. 470–85.
18. See, e.g., Frege's "Der Gedanke," *Kleine Schriften*, ed. I. Angelelli (Darmstadt: Wissenschaftliche Buchgesellschaft, 1967), pp. 342–62.
19. See Frege's *Grundgesetze der Arithmetik*, vol. I (Hildesheim: Olms, 1966 [1893]), § 10.
20. Compare the rest of this section with my paper "On Frege's Two Notions of Sense," *History and Philosophy of Logic* 7 (1986): 31–41, chapter 4 of the present book. See also my "Remarks on Sense and Reference in Frege and Husserl," *Kant-Studien* 73 (1982): 425–39, chapter 2 of the present book.
21. Husserl, *Logische Untersuchungen*, vol. 2, Sixth Investigation, §48 and especially *Erfahrung und Urteil*, 5th ed. rev. (Hamburg: Meiner, 1976 [1939]), §§58–65.
22. See, e.g., Husserl's *Logische Untersuchungen*, vol. 2, Sixth Investigation, §48. See also chapters 2 and 4 of this book.
23. G. Frege, "Über Sinn und Bedeutung," *Kleine Schriften*, pp. 143–62. See also my "Identity Statements in the Semantics of Sense and Reference," *Logique et Analyse* 25 (1982): 399–411, chapter 3 of the present book.
24. It is not clear what the sense of the expression '117' actually is. We can assume, for simplicity's sake, that it is the same as that of the expression '116 + 1'. In any case, it seems to be clearly different from that of the expression '9 × 13'.
25. Interderivability phenomena like those considered in this paper seem to be rather isolated phenomena. That for any mathematical statement, there are other mathematical statements that refer to different states of affairs but are mathematically equivalent to it, seems improbable and, in any case, would have to be proved.
26. It should be clear from the very beginning of this paper that the interderivability results have a metamathematical character. I have not emphasized this point to avoid somewhat esoteric terminology that could originate unnecessary confusion.

14

Guillermo E. Rosado Haddock

ON HUSSERL'S DISTINCTION
BETWEEN STATE OF AFFAIRS
(*SACHVERHALT*) AND SITUATION OF
AFFAIRS (*SACHLAGE*)

In his influential paper "Mathematical Truth"[1] Paul Benacerraf states two requirements for any account of mathematical truth to be worth considering, namely: (i) that the semantic treatment of mathematical statements does not differ essentially from the semantic treatment of nonmathematical statements, and (ii) that the account of mathematical truth harmonize with what he calls a reasonable epistemology. According to him, combinatorial accounts of mathematical truth, which tend to identify mathematical truth with derivability in a formal system, violate the first requirement, whereas Platonist philosophies of mathematics (like Gödel's) violate the second requirement. Such a violation of the second requirement, however, depends on Benacerraf's understanding of 'reasonable epistemology'. It should be clear that if one identifies 'reasonable epistemology' with empiricist theory of knowledge (causal or not), Platonist philosophies of mathematics are not easy to reconcile with reasonable epistemologies. But such an identification need not be taken for granted.

1. The Problem

In this paper, however, I am not interested in discussing the merits of Benac-erraf's sketch of a causal theory of mathematical truth, but would like to state a third requirement (i.e., a third necessary condition) for a semantics plus epistemology of mathematics, which in my opinion is not satisfied by any causal account. (iii) A semantics plus epistemology of mathematics must give a satisfactory account of the equivalence—in the sense of interderivabil-ity—of apparently unrelated mathematical statements—like the Axiom of Choice and its many mathematical equivalents.

The best known representatives of Platonism in the philosophy of mathematics, i.e., Cantor, Frege, and Gödel, did not develop enough—so far as I know—an epistemology of mathematics. Husserl, however, whose philosophy of mathematics (as developed in *Logische Untersuchungen*[2] and *Formale und transzendentale Logik*)[3] can also be considered as a sort of Platonism, tried to develop in his Sixth Logical Investigation and in *Erfahrung und Urteil*[4] such an epistemology of mathematics. It is my opinion that Husserl's sketchy epistemology of mathematics plus his somewhat scattered remarks of a semantical nature can be elaborated further to produce a semantics plus epistemology that satisfies all three requirements stated above. In this paper, however, I will limit my consideration to some of Husserl's semantical insights and will try to show rather sketchily how some of these insights can be fruitfully applied in a semantics of mathematics that satisfies the first and third of the above requirements (i.e., those which are more properly of a semantic nature, since the second is rather a requirement on epistemologies of mathematics).

2. Husserl's Semantics in *Logische Untersuchungen*

Around 1891 Husserl[5] and Frege[6] developed independently of each other very similar semantic insights. In Frege's more popular terminology, they distinguished from the physical aspect of a linguistic expression the sense and the reference of the expression, and applied this distinction to the different sorts of linguistic expressions that they recognized. Thus, in the case of Frege we have the following schema:

Sentence	Proper Name	Conceptual Word
↓	↓	↓
Sense of the Sentence (Thought)	Sense of the Proper Name	Sense of the Conceptual Word
↓	↓	↓
Reference of the Sentence (Truth Value)	Reference of the Proper Name (Object)	Reference of the Conceptual Word (Concept)
		↓
		Object that falls under the Concept

The corresponding Husserlian schema, taken from Husserl's discussions in the First, Fourth and Sixth Logical Investigations differs from Frege's in the following two points, only the last of which is of particular importance for the ensuing discussion: (i) what corresponds only approximately to Frege's conceptual words, namely, universal names, have the corresponding concept as sense, and as reference the objects that fall under the concept; and (ii) the reference of statements are states of affairs, not truth values.[7] Thus, the relation of a statement to its truth value is in Husserl more mediated than in Frege. We will see below that it is even more mediated, and anticipating somehow our further discussion, we can reconstruct schematically Husserl's semantics as follows:

Sentence	Proper Name	Universal Name
↓	↓	↓
Sense of the Sentence (Proposition or Thought)	Sense of the Proper Name	Sense of the Universal Name (Concept)
↓	↓	↓
Reference of the Sentence (State of Affairs)	Reference of the Proper Name (Object)	Reference of the Universal Name (Object that falls under the Concept)
↓		
Reference Base of the Sentence (Situation of Affairs)		
↓		
Truth Value		

3. States of Affairs and Situations of Affairs

For Husserl, as for Frege in "*Über Sinn und Bedeutung*," in each of the following pairs of expressions the expressions have different sense but the same reference:

(i) {the teacher of Alexander the Great, the most famous disciple of Plato},

(ii) {5 + 3, 7 + 1},

(iii) {the preservation theorem for unions of chains of models, the Chang-Los-Suszko Theorem}.

Thus, for Husserl, as for Frege[8] in the above mentioned article, the statements '5 + 3 > 6' and '7 + 1 > 6' have different sense but the same reference, only that this reference is not a truth value but a state of affairs. Similarly, the statements 'The teacher of Alexander the Great was born in Stagira' and 'The most famous disciple of Plato was born in Stagira' have different senses, i.e., express different thoughts, or, in Husserl's more frequent terminology, different propositions, but have the same state of affairs as reference. A similar situation occurs with the statements 'The preservation theorem for unions of chains of models is harder to prove than Henkin's Theorem' and 'The Chang-Los-Suszko Theorem is harder to prove than Henkin's Theorem'.

But states of affairs are for Husserl categorial objectualities (or objectualities of the understanding). They are not sensibly perceived—as e.g., sets are not sensibly perceived—but are constituted on the basis of sensible perception (although the relation to the sensibly given can vary in at least two ways). The objectualities of the understanding are always founded objectualities and the categorial acts are always founded on sensible acts. (The somewhat special case of categorial acts in which mathematical objectualities are constituted, need not detour us here, but it should be emphasized that even mathematical objectualities are, according to Husserl, in some sense sensibly founded). Even in the case of the apparently more 'empirically bounded' propositions, not all its components have on the side of possible fulfillments a sensible correlate. Those components of propositions that Husserl calls 'syntactical forms' do not have any sensible correlate, but help to constitute, on the side of possible fulfillments, new objectualities, objectualities of the understanding, and in particular, states of affairs.

Although he does not use his later technical term 'situation of affairs' (*Sachlage*), in §48 of the Sixth Logical Investigation,[9] Husserl distinguishes between states of affairs as categorial objectualities and a pre-categorial relation that serves as its basis. In cases like '*A* is a part of *B*' and '*B* contains *A* as a part', '*A* lies at the right of *B*' and '*B* lies at the left of *A*', and '*A* is larger than *B*' and '*B* is smaller than *A*' Husserl says somewhat unclearly that we have the same relation but differently conceived. To these two different conceptions correspond two different possible (categorial) relations that are different objectualities categorially constituted on the basis of the same pre-categorial relation.

In *Erfahrung und Urteil* Husserl states more clearly the distinction between state of affairs (*Sachverhalt*) and situation of affairs (*Sachlage*). In §59 of this book[10] Husserl says that states of affairs are categorial objectualities based on objectualities which can be receptively apprehended. To a state of affairs there corresponds in the receptivity a situation of affairs, i.e., something identical that can be made explicit in two or more ways. Moreover, equivalent predicative judgments (i.e., equivalent propositions) correspond to the same situation of affairs intuitively given. Any situation of affairs serves as foundation of two or more states of affairs. Thus, the situation of affairs of A being of greater size than B gives rise to the states of affairs that $A > B$ and that $B < A$. Although situations of affairs are founded objectualities —they are receptively apprehended complexions of objects—they are not objectualities of the understanding. They are passively constituted foundations of states of affairs.

4. A Sketch of Some Formal Considerations

In this paper, however, we are not so much interested in the passive nature of situations of affairs in contradistinction to states of affairs, as in the fact that to one and the same situation of affairs there can correspond two or more states of affairs in the same way that to one and the same state of affairs there correspond two or more propositions (thoughts), and, particularly, that equivalent propositions correspond to the same situation of affairs. Thus, in such a case, we can say that the states of affairs that correspond to equivalent propositions, if not identical, are at least equivalent.

Moreover, if we consider a natural language, and therein all possible statements, we can obtain a partition of all propositions (i.e., thoughts) expressed in the statements modulo identity of state of affairs, i.e., assigning to the same equivalence class all propositions that refer to the same state of affairs. But we can also consider all possible states of affairs referred to by statements of the language and partition the set of all those states of affairs modulo identity of situation of affairs, i.e., assigning to the same equivalence class all states of affairs that have the same situation of affairs as basis. Thus, according to the first partition, $5 + 3 > 6$ and $7 + 1 > 6$ belong to the same equivalence class, but $6 < 5 + 3$ belongs to a different equivalence class, whereas, according to the second partition, the three inequalities belong to the same equivalence class. There are also two other related partitions of less interest, namely: (i) the somewhat trivial partition of all possible statements of a language modulo identity of the proposition (i.e., thought) expressed— in the case that there are perfect synonyms in the language—and (ii) the partition of situations of affairs modulo identity of truth value. In the first case, two statements belong to the same equivalence class if they differ at most in one or more places, in which they contain nonidentical but synonymous corresponding expressions. In the last case considered, two statements that have the same truth value—no matter if they express different propositions, refer to different states of affairs, or even have different situations of affairs as reference base—belong to the same equivalence class. We can call this partition the Fregean partition, and the partitions modulo identity of state of affairs and modulo identity of situation of affairs the Husserlian partitions. Thus, according to the Fregean partition not only 5 + 3 > 6 and 6 < 5 + 3 belong to the same equivalence class, but also the statements 'Aristotle was born in Stagira' and 'Paris is the capital of France' belong to that equivalence class. (All these considerations on partitions can be easily generalized to two or more languages. In the case of the partition modulo identity of thought (i.e., of the proposition expressed) such a generalization would guarantee that the partition is not completely trivial in the sense that each equivalence class contains exactly one member).

Corresponding to the four different partitions mentioned above, we can consider transformations of statements that preserve thoughts, transformations of statements that do not necessarily preserve thoughts, but preserve states of affairs, transformations of statements that do not necessarily preserve either thoughts or states of affairs but preserve situations of affairs, and finally, transformations of statements that only preserve truth values. The transformations of each of the four sorts build groups of transformations, i.e., they all contain the identity transformation, if they contain a transformation, then they also contain its inverse transformation, and they all obey the associative law for transformations. Moreover, the group of transformations that only preserve truth values contains properly the other three groups as subgroups, whereas, the group of transformations that preserve not only truth values but also situations of affairs contains properly the two smaller groups as subgroups, and the group of transformations that preserve not only truth values and situations of affairs but also states of affairs contains properly the group of transformations that only preserve thoughts as a subgroup. All this hints at an algebraic study of the semantic notions considered, but such is a task for another time, for another place, and possibly for another person.

5. Abstract Situations of Affairs

As we know, both for Husserl and for Frege in "*Über Sinn und Bedeutung*," the expressions '7 + 1' and '5 + 3' have different senses but refer to the same entity, namely, to the number 8. Thus, the equations '5 + 3 = 8' and '7 + 1 = 8' have different senses but the same reference. For Frege this reference is a truth value, namely, the true, whereas for Husserl it is a state of affairs. That a semantics such as Frege's is completely inadequate for mathematics (and even for everyday language) can be easily seen if we consider that the most varied mathematical and nonmathematical statements have in common the truth value the true. Thus, e.g., the statements (i) 'The Ultrafilter Theorem was first proved by Tarski', (ii) 'The Ultrafilter Theorem is implied· by the Axiom of Choice', (iii) 'There exists a real positive square root of 2', (iv) 'Aristotle was the teacher of Alexander the Great', and (v) 'Paris is the capital of France' have the truth value the true. But it is clear that the relation existing between '5 + 3 = 8' and '7 + 1 = 8' is much stronger than the relation existing between any of them and any of the statements (i) through (v), or between any two, of these last five statements. The statements '5 + 3 = 8' and '7 + 1 = 8' can be obtained from each other by a simple transformation that substitutes an expression by another expression having a different sense but the same reference, whereas, neither of them can be so obtained from any of the statements (i) through (v), nor can any of these last five statements be so obtained from any of the other six statements that we are considering. Thus, a semantics for mathematics (and also a semantics for everyday language) must take into account not only the sense and the truth value of statements, but also its corresponding state of affairs.

Such distinctions, however, are not enough. In mathematics there exist what we may call duality phenomena, i.e., pairs of equivalent results that correspond to so-called dual mathematical notions. E.g., in each of the following pairs, the notions are so-called dual notions: {ideal, filter}, {open set, closed set}, {lower bound, upper bound}, {intersection, union}. The members of these (and of other such) pairs behave very similarly in many (although not necessarily in all) mathematical contexts, i.e., if a result is obtained for one of these notions, there exists in such mathematical contexts an equivalent dual result for the dual notion.

Let us consider, e.g., the notions of 'ideal' and 'filter'. A filter is sometimes called a dual ideal. An ideal that is not properly contained in any other ideal is a maximal ideal or a prime ideal. A filter that is not properly contained in any other filter is called an ultrafilter. Thus, an ultrafilter is a maximal dual ideal. Since the notion of ultrafilter is not usually defined in terms of ideals, i.e., as a maximal dual ideal, the expressions 'ultrafilter' and 'maximal dual ideal', although naming the same entity, have different senses.[11] Hence, statements like 'The Ultrafilter Theorem was first proved by Tarski' and 'The Maximal Dual Ideal Theorem was first proved by Tarski' refer to the same state of affairs but have different sense. One is obtained from the other by substituting an expression by another having different sense but the same reference.

But consider now the following statements: (I) (The Ultrafilter Theorem) 'Every filter can be extended to an ultrafilter', (II) (The Prime Ideal Theorem) 'Every ideal can be extended to a maximal ideal'. These two statements are clearly related, since they are corresponding equivalent results for dual notions, but they cannot be obtained from each other by a simple substitution of an expression by another expression having different sense but the same reference. A similar situation occurs, e.g., in general topology with the (equivalent) definitions of the notion of topological space in terms of open sets and in terms of closed sets, in which the roles of the dual notions of union and intersection are interchanged. Another similar situation occurs with the following model-theoretic trivialities: (i) 'Existential theories are closed under extension', (ii) 'Universal theories are closed under substructures'. In all of these cases we have a sort of equivalence, but one that cannot be adequately explained as a mere substitution of expressions by expressions having different sense but the same reference, i.e., the same state of affairs. On the other hand, such equivalences cannot be adequately explained as an identity of truth value. The equivalence between The Ultrafilter Theorem and The Prime Ideal Theorem is not a mere equivalence modulo identity of truth value, since modulo identity of truth value they are equivalent also to '5 + 3 = 8', to 'Existential theories are closed under extensions' and to 'Paris is the capital of France', since all of them have the truth value the true.

Thus, we are led to consider a new notion for the sort of equivalence existing between, e.g., The Ultrafilter Theorem and The Prime Ideal Theorem. We will call this notion, because of its Husserlian inspiration, the abstract situation of affairs. Hence, two statements are to have the same abstract situation of affairs if they are equivalent in the sense of interderivable on the basis of the whole of mathematics as based in Zermelo-Fraenkel set theory or any other axiomatization of similar deductive capacity.[12]

With the help of this notion of an abstract situation of affairs we can analyze not only phenomena of duality, but also the very important phenomena of seemingly unrelated mathematical results which have been shown to be mathematically equivalent. The most popular example of these phenomena of mathematical equivalence is the Axiom of Choice and its equivalents. The Axiom of Choice is equivalent (in the sense of interderivable) to the most varied mathematical statements in different areas of mathematics, e.g., to Zermelo's Well-Ordering Principle, to the Trichotomy of Cardinals and to Zorn's Lemma in set theory, to the Upward and Downward Löwenheim-Skolem-Tarski theorems in logic, to the Tychonoff's Compactness Theorem in general topology and to many other mathematical results about the most diverse states of affairs.[13] It is clear that the relation of equivalence existing between, e.g., the Axiom of Choice and the Upward Löwenheim-Skolem-Tarski Theorem cannot be explained either as an identity of state of affairs— as in the case of the equations '5 + 3 = 8' and '7 + 1 = 8'—since they do not refer to the same state of affairs, nor as an identity of truth value, since 'Paris is the capital of France' and '5 + 3 = 8' have the same truth value as the Axiom of Choice and the Upward Löwenheim-Skolem-Tarski Theorem but are not interderivable with them. On the other hand, it is also clear that our notion of an abstract situation of affairs can be adequately used to explain the relation existing between the Axiom of Choice and its mathematical equivalents (or, e.g., between the Prime Ideal Theorem and its mathematical equivalents).

Finally, when a mathematical statement is derivable from another mathematical statement as, e.g., the Prime Ideal Theorem from the Axiom of Choice we can say that the abstract situation of affairs of the derivable statement is contained in the abstract situation of affairs of the deriving statement, and that the (mathematical) content of an abstract situation of affairs is the set of all abstract situations of affairs contained in it.

NOTES

1. P. Benacerraf, "Mathematical Truth," pp. 403–20.
2. See Husserl, *Logische Untersuchungen*, vol. 1, chapter 11.
3. See Husserl, *Formale und transzendentale Logik*, chapters 1–3.
4. See Husserl, *Erfahrung und Urteil*, part 2, chapter 2 and part 3, chapters 2 and 3.
5. See e.g., Husserl's "Besprechung von E. Schröder, *Vorlesungen über die Algebra der Logik, I*," reprinted in his *Aufsätze und Rezensionen*, pp. 3–43, and his 1890, "Zur Logik der Zeichen" in the appendix to his *Philosophie der Arithmetik, mit ergänzenden Texten*, Husserliana, vol. XII, pp. 340–73. Both papers are available in English in Husserl's *Early Writings in the Philosophy of Logic and Mathematics*.
6. See, e.g., Frege's "Über Sinn und Bedeutung," reprinted in *Kleine Schriften*, pp. 143–62, his "Ausführungen über Sinn und Bedeutung" in his *Nachgelassene Schriften*, pp. 128–36, and his *Wissenschaftlicher Briefwechsel*, pp. 94–98.
7. See, e.g., Husserl's *Logische Untersuchungen*, vol. 2, Fourth Investigation, §11 and Sixth Investigation, §48.
8. As I have tried to show elsewhere (see chapter 4 of this book), Frege had two different notions of sense.
9. Husserl, *Logische Untersuchungen*, vol. 2, Sixth Investigation, §48.
10. Husserl, *Erfahrung und Urteil*, §59, pp. 285–88.
11. A proper filter F over a set I is defined as a family of subsets of I (i.e., F $\subseteq$ P(I)—the power set of I), such that (i) $\emptyset \notin$ F, (ii) if D$'$ ϵ F and D$'$ ϵ F, then D $\cap$ D$'$ ϵ F, and (iii) if D ϵ F and D $\subseteq$ D$'$, then D$'$ ϵ F. An ultrafilter over I is then defined as a filter not properly contained in any other (proper) filter over I.
12. In this informal treatment, we can ignore any differences in deductive capacity between rival set theories.
13. For a detailed study of the Axiom of Choice and its equivalents, see, e.g., G. H. Moore's *Zermelo's Axiom of Choice*.

REFERENCES

Barwise, J. and J. Perry. "Semantic Innocence and Uncompromising Situations." *Midwest Studies in Philosophy 6, The Foundations of Analytic Philosophy*. Ed. P. A. French et al. Minneapolis: University of Minnesota Press, 1981, pp. 387–403.

Benacerraf, P. "Mathematical Truth." *Philosophy of Mathematics*. 2nd ed. Ed. P. Benacerraf and H. Putnam. Cambridge: Cambridge University Press, 1984, pp. 403–20.

Frege, G. "Ausführungen über Sinn und Bedeutung." *Nachgelassene Schriften*. 2nd ed. Ed. H. Hermes et al. Hamburg: Meiner, 1983 (1969), pp. 128–36.

————. "*Über Sinn und Bedeutung.*" *Kleine Schriften*. Ed. I. Angelelli. Darmstadt: Wissenschaftliche Buchgesellschaft, 1967 (1892), pp. 143–62.

————. *Wissenschaftlicher Briefwechsel*. Ed. G. Gabriel et al. Hamburg: Meiner, 1976.

Husserl, E. *Aufsätze und Rezensionen (1890–1910)*. Husserliana, vol. XXII. The Hague: M. Nijhoff, 1979.

————. "Besprechung von E. Schröder *Vorlesungen über die Algebra der Logik I.*" In Husserl's *Aufsätze und Rezensionen*, 3–43. Translated in his *Early Writings*.

————. *Early Writings in the Philosophy of Logic and Mathematics*. Dordrecht: Kluwer, 1994.

————. *Erfahrung und Urteil*. 5th ed. Hamburg: Meiner, 1976 (1939).

————. *Formale und transzendentale Logik*. Husserliana, vol. XVII. The Hague: M. Nijhoff, 1974 (1929).

————. *Logische Untersuchungen*. Husserliana, vols. XVIII, XIX. The Hague: M. Nijhoff, 1975, 1984 (1900/01, 1913).

————. *Philosophie der Arithmetik*. Husserliana, vol. XII. The Hague: M. Nijhoff, The Hague, 1970.

————. "Zur Logik der Zeichen." In the Husserliana edition of Husserl's *Philosophie der Arithmetik*, 340–73. Translated in his *Early Writings*.

Moore, G. H. *Zermelo's Axiom of Choice*. New York: Springer, 1982.

Rosado Haddock, G.E. "Edmund Husserls Philosophie der Logik und Mathematik im Lichte der gegenwärtigen Logik und Grundlagenforschung." Ph.D. Dissertation, Rheinische Friedrich-Wilhelms-Universität, Bonn, 1973.

————. *Exposición Crítica de la Filosofía de Gottlob Frege*. Santo Domingo (Privately published and circulated), 1985.

————. "Husserl's Epistemology of Mathematics and the Foundation of Platonism in Mathematics." *Husserl Studies* 4 (1987) 81–102, chapter 12 of the present book.

————. "Identity Statements in the Semantics of Sense and Reference." *Logique et Analyse* 100 (1982): 399–411, chapter 3 of the present book.

————. "Interderivability of Seemingly Unrelated Mathematical Statements and the Philosophy of Mathematics." *Diálogos* (1992) 121–34, chapter 13 of the present book.

————. "On Frege's Two Notions of Sense." *History and Philosophy of Logic 7* (1986) 31–41, chapter 4 of the present book.

————. "Remarks on Sense and Reference in Frege and Husserl." *Kant-Studien 73* 425–39, chapter 2 of the present book.

Schmit, R. *Husserls Philosophie der Mathematik*. Bonn: Bouvier, 1981.

Schuhmann, K. *Husserl-Chronik*. The Hague: M. Nijhoff, 1977.

Weidemann, H. "Aussagesatz und Sachverhalt: ein Versuch zur Neubestimmung ihres Verhältnisses." *Grazer Philosophische Studien* 18 (1982): 75–99.

15

Guillermo E. Rosado Haddock

ON ANTIPLATONISM AND ITS DOGMAS

1. Introduction

Recent discussion in the philosophy of mathematics, especially in the Anglo-American world, occurs within the framework of some empiricist dogmas accepted as self-evident truths by philosophers presumably propounding very different views. Thus, not only Quine and Putnam, Benacerraf and Kitcher, but even philosophers like the so-called Platonist, Penelope Maddy, and the so-called nominalist, Hartry Field, and many others accept in one way or another the common core of "evident truths" that only serve the purpose of reassuring them of the "obviousness" of their common prejudice, namely: the rejection of the existence of mathematical entities as conceived by the Platonist. Their nowadays common coin argumentations, which I am going to consider in this paper, are devised in such a way that they all presuppose that there are no abstract mathematical entities.

One of the cornerstones of this common framework is the belief that the only, or at least the best, argument on behalf of Platonism in mathematics—thus, on behalf of the belief in the existence of mathematical entities, like numbers or sets, foreign to causal interaction and immune to the vicissitudes of the physico-real world—is the so-called indispensability argument wielded by Quine and Putnam, according to which the successful application of mathematics in physical science guarantees the truth of mathematical statements and (in Quine's case but not necessarily in Putnam's) the existence of mathematical entities. This argument was put forward by Quine in "On What There Is"[1] and more emphatically by Putnam in "What is Mathematical Truth?,"[2] and is more or less accepted by most philosophers in the same tradition.

Another cornerstone of the common framework is Paul Benacerraf's argumentation in "What Numbers Could Not Be,"[3] which supposedly establishes, firstly, that numbers are not sets, and, secondly, that numbers are not objects. Strictly, we have here two different arguments with two different purposes. I shall call them "Benacerraf's first and second ontological arguments" (in that order). A third and extremely important component of the common framework is Benacerraf's epistemological argument against Platonism in mathematics, put forward in his paper "Mathematical Truth."[4] Briefly, that argument tells us that since we do not have epistemological access of a causal nature to any abstract mathematical entity, it is at least unnecessary to invoke those entities to explain our mathematical knowledge. A fourth—probably, less generally accepted—component of the common framework is Putnam's argument against realism in mathematics wielded in his paper "Models and Reality."[5] I will call that argument "Putnam's Skolemization Argument," since it intends to extract epistemological consequences from the so-called "Skolem Paradox." In what follows we shall see that all four components of the common framework have a common "foundation" in the above-mentioned prejudice against the existence of abstract mathematical entities and our possible access to them.

2. Putnam's Skolemization Argument

Let us consider first the most important features of Putnam's argumentation in "Models and Reality." At the beginning of the paper Putnam tells us that there are three different stands with respect to the problem of reference and truth, especially, reference and truth in mathematics, namely: (1) the Platonist, or extremely realist stand, which he repeatedly stigmatizes as postulating the existence of "nonnatural" or "mysterious" mental powers that give us access in an "irreductible" and "unexplained" way to mathematical entities and truths; (2) the verificationist, which substitutes the notion of verification for the classical notion of truth—Putnam says "verification or proof," although it is a consequence of Gödel's first incompleteness theorem that the notion of proof is, in general, no adequate surrogate for the notion of truth in classical mathematics; and (3) the moderate realist stand, which he characterizes as aiming to preserve the centrality of the classical notions of truth and reference without presupposing "nonnatural" mental powers. Although Putnam does not say it explicitly, when he speaks about moderate realism, he seems to have in mind the conception that he had propounded in "What is Mathematical Truth?" according to which[6] to be a realist means to maintain that (1) the statements of the theory under consideration (in this case, the whole of mathematics) are true or are false, and that (2) it is something in the physico-real world that makes them true. Thus, such a kind of realism does not need to commit itself to the existence of abstract mathematical entities, but only to the objectivity of mathematics. Indeed, Putnam's realism in

"What is Mathematical Truth?" is related to the thesis that the indispensability argument is the only (or at least, the best) argument on behalf of realism in mathematics, since he maintains[7] that "the criterion of truth in mathematics, as in physics, is the practical success of our ideas." This sort of pragmatism concerning the notion of truth induces Putnam to follow in that paper in the footsteps of Quine and consider mathematical knowledge as corrigible, thus, as neither absolute nor immune to revision by experience. (Although it is not directly relevant for our present purposes, it should be mentioned that Quine's holism—which should not be confused with the much sounder Duhem thesis—is another dogma of Anglo-American analytic philosophy, and, ironically, one that was introduced in response to the two dogmas of pre-Quinean empiricism discussed by Quine in his duly famous paper.[8] Contrary to the Quinean dogma, revisions in the physical and biological sciences never revise the mathematics—in the sense of considering the theorems of the mathematical apparatus used by the scientist as false—nor revise other scientific theories that are considered completely isolated from the one under consideration. Such a revision would be considered unsound scientific practice.)[9] Continuing with Putnam, he is going to argue that the so-called Skolem Paradox represents a very strong case against such a moderate realism, but leaves untouched both Platonism and verificationism. Let me now present the paradox together with Putnam's assessment of it, which is an uncritical adoption of Skolem's rendering.

Let us consider a system of axioms for set theory, e.g., the system ZF of Zermelo-Fraenkel set theory. One of the theorems of ZF asserts that there are uncountably many sets. The Skolem Paradox occurs when we attempt to formalize ZF in a language—like those of first order—for which the Löwenheim-Skolem Theorem is valid. Such a theorem says that any theory expressed in a (finite or countable) first-order language—or in any language for which the theorem is valid—which has a model, also has a countable model. Thus, if we formalize ZF in a first order language—as is usually done since Skolem—ZF will have a model with a countable universe, although one of its theorems establishes that there exist uncountably many sets.[10]

Skolem's solution, adopted by Putnam and by most contemporary logicians and philosophers of mathematics, is essentially the following. The existence (or nonexistence) of a set is not an absolute feature of sets, but depends on the language under consideration. Hence, the existence (or nonexistence) of a set of ordered pairs, as required to establish a bijection between two given sets, is relative to the language under consideration. When we say that a model for a formalization of ZF in a first order language is uncountable, we are not considering all the possible bijections between the set of natural numbers and the universe of the model, but only those that exist inside the model. But the possibility is not excluded that outside the model there exists a set of ordered pairs as required to establish the desired correspondence.

Therefore, a model can be uncountable relative to the language under consideration, but countable as seen from outside the model.

Putnam argues[11] that the Skolem Paradox establishes that no interesting theory (in the sense of "first-order theory") can, by itself, determine its universe of objects up to isomorphism. Moreover, he adds[12] that Skolem's argument can be extended to show that if the theoretical constraints do not determine the universe of discourse, then possible additional restrictions of an operational nature cannot determine it either. Putnam even maintains[13] that such an argument shows that a formalization of all science, or of all our beliefs, could not eradicate undesirable countable interpretations.

I am not concerned here with an assessment of the negative impact of Skolem's paradox on moderate realism, but with the two conceptions of mathematical truth and reference that, according to Putnam, remain unaffected by Skolem's paradox, namely, Platonism and verificationism. Of special concern are the grounds that Putnam has for adopting verificationism and rejecting Platonism. The reason given by Putnam[14] to reject the Platonist stand is that it invokes "nonnatural" cognitive faculties, and this he considers "epistemologically otiose and devoid of conviction as science." It is interesting to observe that this same reason, namely, the rejection of cognitive faculties that he calls "nonnatural"—"natural" for him would probably be only sense perception with some sort of causal link between the knower and the objects of knowledge—makes him avoid a much less artificial rendering of the so-called Skolem Paradox, namely, that first-order languages— and, in general, all those languages in which the Löwenheim-Skolem Theorem is valid—are inadequate to formalize set theory, in an analogous, but not exactly identical way to that in which first-order languages are inadequate to formalize arithmetic. In second-order languages you can formalize both set theory and arithmetic much more adequately, since for such languages neither the Löwenheim-Skolem Theorem nor any of its Tarskian variants are valid, and, hence, the Skolem Paradox cannot be construed for them. (Moreover, in the case of arithmetic, second-order arithmetic is categorical, i.e., all its models are isomorphic, whereas first-order arithmetic, thanks to the compactness theorem, has nonstandard countable models, i.e., countable models not isomorphic to the standard model, as was shown precisely by Skolem.) Indeed, we should not forget that the Löwenheim-Skolem Theorem and its Tarskian variants are limitative results, which establish the noncategoricity of theories. Of course, second-order languages have other probably undesirable properties like noncompactness and semantic incompleteness. However, decidability, for example, seems to be a desirable property, and propositional languages are decidable, whereas (full) first-order languages are not. But that has not hindered logicians from preferring first-order languages to propositional ones because of their expressive power. This same ground could be given for preferring second-order languages to first-order ones. However, no matter how this rivalry between first- and second-order languages (and possibly others) is finally decided, the fact is that Skolem's paradox has a parochial nature, and this fact weakens considerably Putnam's argumentation in "Models and Reality." Thus, even if his argumentation against moderate realism on the basis of the so-called Skolem Paradox were correct, his preference for verificationism over Platonism, and his rejection of a much more natural rendering of the paradox than the one he adopts from Skolem are based only on the prejudice against the existence of abstract mathematical entities and our epistemological access to them.

3. The Quine-Putnam Indispensability Argument

Let us consider now the belief that the only (or, at least, the best) argument on behalf of realism in mathematics is the success in the application of mathematics to physical science. According to this view, as Putnam tells us in "What is Mathematical Truth?,"[15] realism in the philosophy of mathematics is based both in mathematical experience and in physical experience, and "the rendering under which mathematics is true has to be compatible with the application of mathematics outside of mathematics."[16] Moreover, Putnam even maintains[17] that, in view of the integration of mathematics with physics, it is not possible to be a realist with respect to physics but a nominalist with respect to mathematics.

First of all, it sounds somewhat queer that mathematics, the most exact of all sciences, with the possible exception of logic—if it is sound to distinguish between them—has to justify the truth of its theorems and the existence of the entities about which she presumably speaks, by referring to the success of its applications to physical science. As observed by James R. Brown in his paper "Π in the Sky,"[18] the truth of elementary arithmetical statements as, e.g., "There is an immediate successor of 3 in the natural number series," or "2 is less than 3," is much more evident than the truth of any statement in the physical sciences. Something analogous occurs with elementary statements about sets, e.g., 'If the set M contains x and y as its sole members, and the set C contains y, v, and w as its sole members, then the intersection of M and C contains y as its sole member, and the union of M and C contains x, y, v, and w as its only members'. Moreover, such statements are not only much more evident than the laws of physics, but also seem to be true in all possible worlds. Indeed, an existential mathematical statement as, e.g., 'There is a prime number greater than 100', whose truth seems to convey the existence of the objects spoken about, seems to be true in every possible world and, thus, such objects seem to exist in every possible world. On the other hand, statements in the physical sciences seem to be only contingently true, since it is not really difficult to imagine possible worlds governed by different physical laws.

It should be mentioned here that the Quine-Putnam tradition seems to have an inadequate understanding of the role of mathematics in physical science. As has been correctly argued by James R. Brown in the above mentioned paper,[19] Quine's claim that not only physical science, but also mathematics, is tested by experience via the so-called observational sentences, seems not to be warranted by the history of science. In that history, Brown argues,[20] when there is an unexpected empirical result, scientists have never concluded that it is the mathematics used (or part of it) that has been falsified and requires modification. They have concluded correctly that it is the physical theory (or part of it) that has been falsified. Brown claims against Quine[21] that the role of mathematics in physical science does not consist in constituting additional hypotheses, but in offering models (in the sense in which the word 'model' is used in the natural sciences). In the face of adverse empirical results the model is substituted by a more adequate one, without claiming that the mathematical theory has been falsified. This theory has simply been shown to be inadequate as a model for those features of the physical world that the scientist wanted to explain. Thus, e.g., it is not the case that Euclidean geometry has been falsified by general relativity. It has simply been shown that Euclidean geometry is not an adequate model for the description of the spatio-temporal structure of our physical universe. (We are not claiming here—and Brown most surely was not—that mathematics is unrevisable. The antinomies of naive set theory clearly yielded a revision in mathematics. But revisions in mathematics are internally motivated, not the result of any empirical testings. Of course, there are also shifts of interest in mathematics, and these can be partly motivated by external sources. But shift of interest is not revision. A mathematical theory can lose its interest, e.g., if from the theoretical standpoint there is not much still to be discovered, or its results are surpassed by a more general theory. It could also be the case that the physical theories to which it is applied are falsified or lose interest, in which case there is an additional, but not decisive reason, for the abandonment of research in the mathematical theory.)

Let us return to the Quine-Putnam argument, and compare it with the following similar fictional situation. Someone wants to argue that words have meaning and that statements express thoughts. However, instead of arguing directly, he claims that if it were not the case that words have meaning and that statements express thoughts, it would not be possible to explain how it is that there is literature and that it can be read and "understood" by different people. Now, languages have existed, in their oral manifestations, long before literature, as we conceive it, appeared on the face of the earth. It seems completely unreasonable to think that before the invention of literature words did not have meaning and statements did not express thoughts, or, at least, that there was no way to establish that words have meaning and that statements express thoughts. Indeed, it is not hard to imagine a possible world in which people communicate orally as well or as badly as we do in our world, although in that possible world there is no literature. What is not possible is a world in which there were literature but there were no languages. In a similar way, the historical origins of mathematics can be traced back many centuries before the advent of physical science, which, as is conceived nowadays, seems to trace its origins up to Galileo (and the physical theories currently considered as true are products of the last two centuries). Now, although applicable to the physical (and other) sciences, mathematical theorems seem to be true even if all actually accepted physical theories were false and, thus, the claim that only after the advent of modern physical science can we argue that mathematical theorems are true seems really amazing, to say the least. It is also extremely unreasonable to think that before the advent of modern physical science there was no way to establish the existence of mathematical entities, thus, e.g., that there exists an immediate successor of 3 in the natural number series. Moreover, it is perfectly conceivable that there exists a world in which all mathematical theorems known to present-day mathematicians are true (supposing that current mathematics is consistent), and that mathematicians know as much mathematics as they actually know, but in which none of the physical laws accepted as true nowadays were known to humanity. What is not possible is a world in which physical science were as developed as it actually is, but in which our present mathematical theories (especially those applicable to present-day physical science) were not valid, or, at least, were not considered to be valid. If we are going to argue on behalf of the existence of mathematical entities or of the truth of mathematical theorems, we have to do it, as, e.g., Brown correctly maintains,[22] from within mathematics itself.

Quine's and Putnam's belief that the indispensability argument represents the only (or, at least, the best) argument on behalf of mathematical realism and the truth of mathematical theorems, is based on an inadequate view both of science and of our cognitive capacities. His criticism of logical empiricism notwithstanding, Quine defends a sort of empiricism that is also incapable of doing justice to the eminently theoretical character of physical science. His whole conception of knowledge and experience originates in a behaviorism of doubtful scientific credentials. The whole epistemological tradition that originates with Quine, and of which Putnam and Benacerraf are prominent members, limits our cognitive capacities to sense perception, possibly garnished with a causal dressing. Its rejection of any other argument on behalf of mathematical realism is bounded to their fear of admitting what Putnam, in an axiologically loaded terminology, has called "nonnatural mental powers" and "mysterious mental powers"—but which could be more correctly called "categorial intuition" or "intellectual intuition"—or even of overtly admitting the existence of abstract mathematical entities. This is the reason why in "On What There Is"[23] Quine refers to mathematics (and even to physics) as a convenient myth, as a useful fiction of high explanatory value. Indeed, according to Quine, we cannot free ourselves of theoretical fictionalism as soon as we transcend the so-called sensory data, without taking into account that precisely these pure sensory data are the first great fiction. (As has been argued by many authors,[24] it is extremely difficult to isolate constituents of physical theories that are not theoretically contaminated.) On the other hand, Putnam made more explicit than Quine his reluctance to admit the existence of abstract mathematical entities when in "What is Mathematical Truth?"[25] he defended a sort of realism without mathematical entities. Prima facie, it looks as if the indispensability argument—contrary to other possible arguments on behalf of mathematical realism—would allow Putnam to defend a sort of mathematical realism, without having to admit the existence of numbers, sets and other "undesirable" entities. (I am not going to dwell here on the issue of the cogency of such a mathematical realism, which has since been abandoned by its proponent.) What seems much clearer, however, is that Quine's view of mathematics as a convenient myth, together with his adhesion to the indispensability argument and Putnam's view of a mathematical realism without mathematical entities paved the way for Field's philosophy of mathematics.

4. Paul Benacerraf's Ontological Arguments

In "What Numbers Could Not Be"[26] Paul Benacerraf has claimed that since there is more than one way—possibly infinitely many ways—to characterize numbers as sets, e.g., von Neumann's characterization and Zermelo's characterization, and since they possess incompatible properties, numbers cannot be sets. Benacerraf argues[27] that at most one of those characterizations, which are not even extensionally equivalent, can be true. But since there is no ground to prefer one of them over any other, numbers cannot be sets. A second argument of Benacerraf in the same paper attempts to show that numbers cannot be objects. On this point he refers both to Russell and Quine, who had conceived arithmetic as the study of recursive progressions. The only numerical properties that would seem relevant for arithmetic are those had in virtue only of being members of a recursive progression. No single numerical property that would individualize them as particular objects would be relevant for arithmetic. Hence, Benacerraf concludes that numbers are not even objects.

Benacerraf's two ontological arguments should be clearly distinguished, since they are not only different, but have different purposes. Let us consider the first one. The fact that there are various characterizations of numbers by means of different sets is not an unusual event in mathematics. As, e.g., Field has observed,[28] the real numbers are sometimes identified with Dedekind cuts and sometimes with equivalence classes of Cauchy sequences. In the same vein, ordered pairs, topological spaces and other mathematical entities can be characterized in different ways. Mathematicians, Platonists or not, recognize the existence of such a variety of characterizations without extracting from those situations any Benacerrafian argument. In particular, the fact that in different axiomatizations of set theory we can show that entities characterized in very different ways possess the properties that we usually attribute to numbers, although they do not have any other property in common, does not allow us to conclude anything about the nature of numbers. The situation described by Benacerraf is not very dissimilar to that which occurs when two radically different senses have the same referent, as is the case of the senses of the following two definite descriptions: "the French leftist who was Leon Trotsky's secretary from 1932 to 1939" and "the mathematician and historian of logic who edited *A Source Book in Mathematical Logic*." (If we consider not the definite descriptions but the corresponding conceptual expressions, the similarity seems even more plausible. The referent is in that case—following Husserl or Carnap, not Frege—the same unit set.) The fact that Jean van Heijenoort can be referred to in such extremely different ways, and that his life from more or less 1948 onwards has almost nothing to do with his former life (except for serving as a consultant for the Trotsky archives)—which makes him an excellent choice to illustrate Kripke's puzzle in "A Puzzle about Belief"—does not allow us to extract any ontological conclusions about van Heijenoort. It could well be the case that numbers were sets, or that they were any other kind of abstract entity, and that due to our cognitive limitations we would be forced to characterize them by means of sets which, apart from the properties that they have in common with numbers, do not possess any other property in common and, thus, are both inadequate but "manageable." (Moreover, it is possible that Benacerraf is going too far when he attributes to set-theoretical reductionists the identification of numbers with particular sets. Indeed, in his paper reproduced in Benacerraf's and Putnam's anthology Carnap underscores[29] that the logicist merely produces, by means of explicit definitions in a system of logic, constructions of logical objects that, in virtue of those definitions, have the properties that numbers have. Perhaps, if asked by Benacerraf to comment on the situation under discussion, both Zermelo and von Neumann would have answered: "This is the way in which numbers are represented in my system and that is the way in which numbers are represented in his system. I am not attempting to tell you what numbers really are." Even in Frege's *Die Grundlagen der Arithmetik* there is a passage[30] in which he admits that his identification of the number 0 with the set which is the extension of the concept "different from itself" involves some arbitrariness, since prima facie the extension of any other concept under which no object falls could have been identified with the number 0. On the basis of that passage someone could try to render Frege's endeavor not as claiming to have shown that the number 0 is the extension of the concept "different from itself," but as claiming that by means of that and the subsequent definitions, one can construe a system of logical objects for which one can prove, from logical axioms only, all the properties usually assigned to numbers. It should be clear that I am not arguing here for such an interpretation of Frege, but merely indicating that it would have textual evidence from that passage.)

Now, even if Benacerraf had succeeded in showing that numbers are not sets, that would not allow him to conclude that numbers are not objects. It could well occur that numbers were not sets, and even that there were no mathematical entity to which all others were reducible, but that there were various fundamental mathematical entities, e.g., numbers, sets, relations, functions, etc., and that, because of their level of abstraction, it were possible to characterize each of them in one or more ways in terms of each of the others. Thus, e.g., relations could be characterized, following Frege, as functions of the same number of arguments whose value is a truth value, and, on the other hand, functions of n arguments could be characterized as relations of $n+1$ arguments uniquely determined in their last arguments. Moreover, relations could also be characterized as sets of ordered pairs, and, on the other hand, sets could be characterized as relations of a single argument. Nonfundamental mathematical entities could then be characterized either as specializations of the fundamental entities, or as combinations of them, or as combinations of their specializations, or as equivalence classes of some already recognized entities.

With respect to Benacerraf's claim that numbers are not objects, since arithmetic is the theory of recursive progressions, and, thus, only structural properties are relevant for arithmetic, i.e., properties that do not individualize them, the following comments seem appropriate. First of all, it is not sufficiently clear from Benacerraf's paper in which sense is the system of natural numbers indistinguishable from other systems of (as Benacerraf would say) "supposed" objects that constitute recursive progressions. He could be thinking of some kind of formalization of Dedekind-Peano arithmetic, or of something like the hierarchy of (species of) structures that constitutes mathematics according to the school of Nicolas Bourbaki. Let us suppose first that Benacerraf is considering some kind of formalization of Dedekind-Peano arithmetic in a system of logic. Then you have to distinguish at least two cases, namely: (1) the formalization is made in first-order logic, or (2) the formalization is made in second-order logic. In this last case we obtain a categorical theory, i.e., a theory all of whose models are isomorphic, but the theory would have—at least according to the Quinean tradition—a strong ontological commitment not to the taste of Benacerraf. On the other hand, if the formalization is made in first-order logic, the theory is not categorical —not even $\aleph_0$-categorical (because of a theorem of Skolem)—and, thus, it would have models nonisomorphic to the standard one—even some of cardinality $\aleph_0$. In that case, to determine the standard model completely and exclude such undesirable models, one has precisely to take into account those properties of the system of numbers that would distinguish it from the other models.

On the other hand, if Benacerraf is thinking that the system of natural numbers is a particular case of a structure of recursive progression in the same way in which any group is a particular case of the group structure as determined by the group axioms, then he is not justified in denying the existence of numbers. Of course, a group G and a group G* share the same group structure. But that does not mean that they do not exist—in the sense in which Platonists conceive the existence of mathematical entities. Indeed, the fact (if general relativity is true) that the space-time in which we live has the structure of a particular Riemannian manifold of four dimensions with variable curvature does not allow us to conclude either that physics is the study of Riemannian manifolds of four dimensions with variable curvature, or that the space-time in which we live and the space-time points of which it is constituted do not exist. There are various thousands of copies of Benacerraf's and Putnam's *Philosophy of Mathematics*, and all of them have the same "structure," since they share all the "relevant" properties (for potential readers of the book). By a reasoning similar to that applied to numbers, Benacerraf should conclude that none of the copies of his and Putnam's book exist. And if a mischievous Platonist genetic engineer succeeded in producing an exact human copy of Paul Benacerraf, by a reasoning similar to that applied to numbers, Benacerraf should conclude—his Cartesian cogito notwithstanding—that he really does not exist. Of course, Benacerraf will not extract in such cases the conclusion analogous to the one he extracted in the case of numbers. But this difference brings to the fore the hidden ground behind Benacerraf's argumentation, namely, his prejudice against the existence of mathematical entities. Thus, Benacerraf has not established that numbers are not objects: he has presupposed that they are not.

5. Benacerraf's Epistemological Argument

In his excessively influential paper "Mathematical Truth"[31] Benacerraf has claimed that treatments of the nature of mathematical truth have been motivated by two different kinds of concerns that are not easy to reconcile, namely: (1) the concern of having an homogeneous semantic theory, in which the semantics of mathematical statements parallels that of non-mathematical statements, and (2) the concern of the compatibility of the treatment of mathematical truth with a reasonable epistemology. Benacerraf considers that there is only one adequate semantic treatment of mathematical statements that is similar to that of non-mathematical statements, namely, the one offered by Tarskian semantics. This claim, however, does not seem completely correct. With respect to natural languages, Kripke has shown that Tarski's semantics is not completely adequate—as Tarski himself very well knew, but some so-called Tarskians had forgotten—and he has proposed a more adequate one that seems to have been almost immediately superseded by the Gupta-Herzberger-Belnap revision theory of truth. (The relation between this last theory and Tarski's theory of truth applied to natural languages seems to be similar to that between relativity theory and Newtonian mechanics, namely, the Tarskian theory is false in the light of the Gupta-Herzberger-Belnap theory, but in very special limiting cases they coincide.) On the other hand, although Tarskian semantics seems, in general, to be adequate for formalized languages, I am convinced that as soon as a substantial portion of mathematics is formalized, Tarskian semantics will require some modification to do justice to the fact that in mathematics there are many statements that are mathematically equivalent, although they seem to be theoretically unrelated and even belong to somewhat distant areas of mathematics.

My interest here, however, is in the other sort of concern mentioned by Benacerraf and presented by him as a sort of requisite, namely: that the treatment of mathematical truth be compatible with a reasonable epistemology. In other words, Benacerraf claims[32] that an acceptable semantics should be compatible with a reasonable epistemology. But reasonable epistemology is for Benacerraf[33] one that admits essentially sense perception as the sole form of knowledge, but that, according to him,[34] differs from former empiricism because it has integrated a causal component. Thus, for someone to know that some particular statement is true there has to exist a causal relation between that person and the referents of the names, predicates and even quantifiers[35] occurring in the statement. Moreover, Benacerraf also believes in a causal theory of reference, as that propounded by Kripke and Putnam. Specifically, Benacerraf claims[36] that in the case of medium-sized objects there should exist a direct causal reference to the facts known and to the objects that constitute them, whereas other sorts of knowledge, among which he includes our knowledge of laws and general theories, should be explained as based in some way or other on our knowledge of medium-sized objects. It should be clear from such a characterization of a "reasonable epistemology" that it practically excludes per definitionem the possibility of having knowledge of abstract mathematical entities. But since Benacerraf requires[37] of the truth of mathematical statements—which, at least, in the case of existential statements seem to involve the existence of the entities spoken about—that they do not make it impossible that at least some of the mathematical truths be known according to the canons of his "reasonable epistemology," and since abstract mathematical entities are not causally related to us, his requirement also practically excludes per definitionem their existence.

Now, it seems very strange that the existence of mathematical entities and the truth of mathematical theorems (e.g., existential ones) be in jeopardy on the basis of an epistemological view, since in the history of philosophy not a single epistemological theory has succeeded in establishing itself with the firmness of at least the theories in the less developed areas of the natural sciences. More strange, however, is the fact that, as John P. Burgess has argued in his "Epistemology and Nominalism,"[38] one of the authors cited by Benacerraf as a propounder of the causal theory of knowledge, namely, Alvin I. Goldman, has later questioned his own causal theory of knowledge, and—what is much more important—contrary to what Benacerraf does in "Mathematical Truth" had limited his causal theory to contingent knowledge, in contrast to necessary knowledge and, thus, as Burgess comments, seems to have excluded the application of the causal theory to mathematics. Benacerraf's convenient misinterpretation of Goldman is exclusively motivated by his desire to exclude abstract mathematical entities from any possible epistemological access. (Indeed, even Field—of all people—in "Fictionalism, Epistemology and Modality"[39] recognizes that nowadays nobody believes in a causal theory of knowledge like that on which Benacerraf bases his epistemological critique of mathematical Platonism. Nonetheless, as we shall see below, Field commits himself in his writings to some sort of modified causal theory of knowledge, when he admits the Quine-Putnam indispensability argument as the only possible argument on behalf of realism in mathematics.)

But the causal theory of knowledge is an inadequate epistemology even for the physical sciences. First of all, as was observed by James R. Brown in "Π in the Sky,"[40] such a view has difficulties with generalizations. Indeed, it seems very difficult to reconcile an epistemology based on sense perception, together with a causal component and possibly some sort of induction, with the laws of high generality and the sophisticated theories of physics. The causal dressing does not add anything to the solution of the difficulties of a similar character that have haunted other variants of empiricism and, especially, logical empiricism, in their attempts to explain the nature of physical theories. Empiricism, with or without a causal dressing, is a nonstarter in the philosophy of physical science. On the other hand, as has been underscored by various authors, such a view is especially inadequate in microphysics. As Michael D. Resnik comments in "Beliefs About Mathematical Objects,"[41] physicists sometimes theorize about physical particles before having the least empirical evidence about them.[42]

On the other hand, even in macrophysics scientists often postulate the existence of physico-geometrical entities, as the so-called singularities of space-time in general relativity and cosmology—e.g., the center of black holes—which presumably have causal effects but with which the possibility of being directly causally connected seems excluded, and even indirect causal connections seem theoretically loaded. Now, if the causal epistemology presupposed by Benacerraf in his rejection of any possible access to abstract mathematical entities seems to be inadequate to explain our knowledge of physics, its inadequacy to explain mathematical knowledge should be even more evident, since mathematical knowledge is both more abstract and firmer than physical knowledge. Moreover, even if there were some causal connection present in mathematical knowledge, it would have a contingent nature and, thus, would not be decisive for a discipline that seems to produce necessary knowledge. Once again, only the prejudice against the existence of mathematical entities and our possible cognitive access to them constitutes the real ground for Benacerraf's argument.

I have shown that the four cornerstones of current discussion in the philosophy of mathematics are really empiricist dogmas of a new sort, based essentially on the prejudice against full-blown ontological and epistemological Platonism. Although I have considered them separately, those dogmas are in some sense intertwined and tend to reinforce each other. I will now briefly examine the impact of the above displayed argumentation on the most daring of current philosophies of mathematics, namely, Field's program.[43]

6. The Consequences for Field's Program

Hartry Field's philosophy of mathematics is, in some sense, the culmination of that philosophical current originating in Quine, Putnam, and Benacerraf that we have been considering, since it boldly extracts the ultimate consequences from its framework of shared beliefs. If the only argument on behalf of realism in mathematics and on behalf of the truth of mathematical statements is the indispensability argument, if mathematics is—as Quine said in "On What There Is"[44]—a convenient fiction, if any epistemological access to abstract mathematical entities is excluded, since they do not relate causally with us, and if we cannot characterize numbers uniquely and do not even know if they are objects at all, then why not explicitly conclude that such entities do not exist, and that existential mathematical statements are strictly false, even though they are useful devices in the derivation of physical statements, but without any ontological import?

Prima facie, however, Field's views seem to abandon the Quine-Putnam tradition, and in some details he even clashes with it. Nonetheless, most of his views are really radicalizations of ideas to be found in Quine, Putnam, and Benacerraf. As is well known, Field claims that mathematics does not need to be a corpus of truths to be successfully applied to the physical sciences. Moreover, he argues that existential mathematical statements are all false (and the universal ones just vacuously true), since there do not exist the mathematical entities whose existence would make such statements true. According to Field, the property that mathematical statements need to have in order to be successfully applied to the physical sciences is not truth, but conservativeness, which is a strong form of consistency. (Mere consistency would be insufficient, since a theory can be consistent and still imply false consequences about the physical world.) [45] A mathematical theory is conservative if it is consistent with any internally consistent theory about the physical world. In other words, a conservative mathematical theory, when added to the corpus of physical statements does not imply any physical statement not implied by the corpus of physical statements—even though, in practice, obtaining that statement would be much more complicated without the help from the mathematical theory. (It should be mentioned here that Field conceives the role of mathematics in physical science in a very similar fashion to Quine's and, thus, Brown's critique of Quine mentioned above, would also apply to Field.)

Field's view of mathematics has probably been the most discussed and criticized in recent years by people working in the philosophy of mathematics in the Anglo-American world, and many of these critiques, e.g., those by Hale, Burgess, Brown, and Irvine, seem adequate. Thus, e.g., Burgess seems to be correct when he claims that Field's program—which is really just a project—is not only of very improbable success, but even if successful in its nominalist reconstruction of the physical sciences, such a reconstruction would be of little interest to physicists. Indeed, as Burgess puts it in "Epistemology and Nominalism,"[46] ontological economy does not seem to have played any important role in the history of science. On the other hand, as many have argued[47] and, moreover, is sufficiently clear, Field's views conflict with the manifest evidence of many elementary mathematical theorems, e.g., arithmetical theorems that seem to be not only true and even necessarily so, but also trivially so. One would have to offer extremely good reasons to at least consider the possibility that they were not necessarily true and that their manifest obviousness is merely an illusion. However, ontological economy and the rejection of abstract mathematical entities seem to be very poor reasons to embrace such an anti-intuitive enterprise with such a low probability of success as Field's. Nonetheless, I am not interested here in criticizing Field's views directly, but will try to argue that they presuppose the truth of both Quine's and Putnam's claim concerning the indispensability argument and of Benacerraf's ontological and epistemological arguments. Hence, with the collapse of the foundations, the whole structure will fall down, including Field's views.

First of all, it should be clear from Field's writings that the only argument on behalf of mathematical realism that he takes seriously is the Quine-Putnam indispensability argument. On the other hand, as can be seen from his discussion of Benacerraf's ontological argumentation in "Fictionalism, Epistemology and Mathematics,"[48] Field accepts Benacerraf's ontological arguments and extracts from them the strong conclusion that, since numbers are not objects, they do not exist at all. With regard to Benacerraf's epistemological argument, it should be said that, notwithstanding the above-mentioned passage of "Realism and Anti-Realism about Mathematics," in which Field says that nobody believes anymore in the causal theory of knowledge propounded by Benacerraf, it should also be clear that he presupposes some sort of causal theory of knowledge. Thus, in "Fictionalism, Epistemology and Mathematics"[49] he argues that there exists a remarkable difference between the roles played by mathematical and physical entities in our explanations of the physical world. The role played by physical entities in those explanations is usually causal, namely, as causal agents that produce the phenomena that are to be explained. But since mathematical entities are supposed to be acausal, their role in such explanations has to be different. Now, Field confesses[50] that explanations which involve reference to entities not causally connected with the phenomena to be explained seem to him rather strange. Here we have a possibly somewhat more cautious version than Benacerraf's of the causal theory of knowledge.

Now, in the foregoing sections of this essay I have shown that the Quine-Putnam thesis with respect to the indispensability argument, as well as both Benacerraf's epistemological argument and his ontological arguments (and also Putnam's skolemization argument) rest on the unwarranted assumption that we cannot have epistemological access to abstract mathematical entities, and even on the stronger assumption that mathematical entities do not exist. But since Field's program presupposes both the validity of Benacerraf's arguments and the Quine-Putnam claim that the indispensability argument is the only argument on behalf of mathematical realism that is worth considering, the pillars of Field's views turn out to be extremely shaky, and this makes such a desperate philosophy of mathematics devoid of any interest or attractiveness (except perhaps as a technical exercise in logical ingenuity).

However, not everything said by Field about mathematics is devoid of interest (as is to be expected of such an ingenious author). There are some interesting points made by him that should be mentioned here. Thus, e.g., I think that Field is correct when he asserts that the so-called Frege-Wright argument, based on the fact that in the languages usually known to logicians numerical expressions are singular terms, is insufficient to establish mathematical realism. Nonetheless, the analogy used by Field between the presumed mathematical fiction and literary fiction ignores completely the manifest necessity with which some elementary mathematical statements impose themselves upon us, which makes us suspect that the so-called mathematical myth would be not merely a convenient one—as Quine and Field would like us to believe—but a necessary one. On the other hand, Field seems to have better grasped the difference between mathematics and physics than Quine and Putnam when he claims that mathematics is conservative but physics is not, although his grounds for such a claim are, of course, different from those that I would give. In my case it is simply that, as Bob Hale puts it in his paper "Nominalism,"[51] if you consider that mathematical theorems are necessarily true, then you are committed to the conservativeness of mathematics. Field's claim that conservativeness and truth are mutually independent also seems convincing, since, as he argues, physics is not conservative but is probably true—if present physics is not true, one can expect that at some time in the future the physics considered true at that moment will really be true—and, on the other hand, conservativeness does not imply truth either, since ZF + AC and ZF + ¬AC seem both to be conservative but only one could be true. Finally, it should be pointed out that by claiming that mathematics is conservative but physics is not, and, thus, by being a nominalist (or fictionalist) with regard to mathematical existence but a realist with regard to that of physical entities, Field separates himself from the tradition on which he dwelled, since, contrary to that tradition, which considers mathematics (and even logic) as continuous with physics and, thus, as an empirical science, he clearly states in "Realism and Anti-Realism about Mathematics"[52] that he cannot accept that mathematics is continuous with physics. Once more I coincide with Field, but for different reasons, namely, since for me mathematical theorems seem to be necessary whereas physical statements are not. It is interesting to observe, on this issue, that if Field's views are consistent, he turns out to be a counterexample to Putnam's claim in "What is Mathematical Truth?"[53] that it is not possible to be a realist with regard to physical entities but a nominalist with regard to mathematical ones.

7. Opening the Doors

Let us conclude this paper with some comments on our knowledge of mathematical entities. I will follow closely the little known epistemology of mathematics developed by Edmund Husserl in his masterpiece *Logische Untersuchungen*,[54] specifically in the Sixth Investigation.

We usually say that an empirical statement like 'The ball is red' or 'Joe is taller than Charles' is true or false, depending on how things are in the world, thus, depending on which states of affairs hold. Statements, pace Frege, refer to certain states of affairs, and are true if those states of affairs hold, and false if they do not hold, e.g., if the ball under consideration is not red but green. Sense perception presents us with the existing states of affairs, and in this respect confirms or disconfirms the proposition expressed by the statement. But in sense perception only physical objects and sensible properties are given. Only the "material" constituents of statements, namely, terms and predicates, have "correlata" in sense perception (or even in imagination). Particles like the connectives 'and', 'or', 'if . . . , then', the quantifiers, and even expressions like 'is larger than' have no correlata in sense perception. Nonetheless, we usually say that we can empirically confirm or disconfirm statements which contain such particles, and we recognize the difference in truth conditions between the statements 'Mary and Julia are in the park' and 'Mary or Julia is in the park'. Constituents of statements that do not have any correlata in sense perception can be called 'formal constituents of statements', and the act of knowledge that fulfills such a constituent of statements is not a simple sense perception, but a categorial perception. Categorial perception builds on sense perception, but does not reduce to sense perception, and makes us acquainted with categorial objects. Categorial perception does not modify the underlying sense perception, since that would be a distortion that would produce another sensible object, but rather leaves untouched the sense perceptions on which it is founded and in which the sensible objects are given to us on which the new objectualities are built.

Husserl considers sets and states of affairs as examples of categorial objectualities. Hence, e.g., the state of affairs that Joe is taller than Charles is a categorial objectuality that builds on and structures itself on the sensible objects Joe and Charles given in sense perception. The state of affairs that Joe is taller than Charles is a categorial objectuality and, indeed, a different one from the state of affairs that Charles is shorter than Joe. (To both of them underlies the proto-relation, called by Husserl "situation of affairs," that Joe has a bigger size than Charles. I will not comment here on Husserl's notion of a situation of affairs, since it is not required to understand what follows.)[55] Analogously, given various objects in sense perception, in categorial perception we are given the set of those objects. Such a set is not given to us in sense perception—we do not see it with our own eyes nor touch it—but rather builds on what is given by sense perception and is—to use Husserl's terminology—"constituted" in categorial perception. The set is a new objectuality founded on sensible objects, and given to us in a categorial perception founded on sense perception. (Instead of 'perception', from now on I will frequently say 'intuition'—both sensible and categorial—and, thus, following Husserl, include also the imagination, since for most of our present purposes an imaginative act can play the same role as a perceptual act.)

We have seen that categorial perception is founded on sense perception but does not reduce to it, and that categorial objectualities are founded on sensible objects but do not reduce to them. Now, once categorial objectualities of this first level—like sets or relations—are given to us, new categorial intuitions can be built on the corresponding categorial intuitions of the first level, and in such categorial intuitions of the second level new categorial objectualities of second level are constituted—e.g., relations between sets, say bijections between sets, and also sets of relations, sets of sets (as, e.g., the power set of a given set), and so forth. In this way, repeating this process indefinitely, a hierarchy of categorial intuitions is obtained and a corresponding hierarchy of categorial objectualities is given to us, so that in categorial intuitions of the *n*th level categorial objectualities of the *n*th level are constituted.

From what has been said up to now, it does not seem clear how it is that the objects of pure mathematics are independent of experience, since the categorial objectualities given in the different levels of the hierarchy of categorial intuitions seem to be founded on sensible objects. Indeed, there are categorial objectualities that seem to possess a sensible component, and Husserl calls them "mixed categorial objectualities." An additional component plays a decisive role in the constitution of mathematical objectualities, namely, categorial (or formal) abstraction, which should not be confused with generalization. One can say somewhat schematically that for Husserl mathematical intuition is categorial intuition plus categorial abstraction. Categorial abstraction, as categorial intuition, is neither a mysterious nor a nonnatural process, but something perfectly usual in mathematical knowledge. If we have a relation, like that of being taller than, between Joe and Charles, we can substitute the terms by indeterminates, thus, by variables, say 'x' and 'y', and the relation of being taller than by a transitive, asymmetric, and irreflexive relation. Analogously, given a concrete set M, we can substitute its elements by indeterminates, and if we also substitute by indeterminates the concrete objects that belong to another concrete set C, and, moreover, we abstract, in general, from the peculiarities of the two sets, we can then consider bijections between one of those sets and a (not necessarily proper) subset of the other, and then consider their respective cardinalities.

It is in the manner described, namely, by means of categorial intuition purified by categorial abstraction that the basic mathematical objectualities are constituted. Other mathematical entities can then be obtained either by combining different objectualities to form more complex objectualities—as Husserl and Bourbaki have seen[56]—or by means of the formation of equivalence classes based on a congruence relation—as discussed by Frege[57]— or by other similar formal means. There is nothing nonnatural or mysterious in this cognitive process, since categorial intuition builds on sensible intuition, and its objectualities build on and structure themselves on the objects of sense perception—which, after all, is even insufficient to give us a physical world coherently structured, as studied by the physical and biological sciences. On the other hand, categorial abstraction is a perfectly common procedure in mathematics, which is responsible for the level of formal generality attained by mathematics, as exemplified by universal algebra, general topology, category theory, and other areas of contemporary mathematics. Without fear of paradox, it can be claimed that, although categorial intuition builds on sensible intuition, there is no trace of sensible foundation in mathematical intuition.[58]

APPENDIX 1

In the course of my refutation of the skolemization argument—and else-where—I have argued that second-order logic is more adequate than first-order logic to render mathematical theories not only, in general, because of its superior expressive power, but, specifically, because it is immune to the cardinality-indeterminacy related to the Löwenheim-Skolem-Tarski theorems. However, it seems at first sight that second-order logic is vulnerable to another, probably worse, form of indeterminacy. Second-order logic admits some nonstandard semantics besides the standard one. In particular, there is the Henkin nonstandard semantics with its nonfull models rivaling the standard semantics, all of whose models are full models, i.e., they have as many relations and operations as possible. As is well known, for Henkin's and other similar nonstandard semantic renderings of second-order logic one can obtain a semantic completeness result with its not less famous corollaries, namely, compactness and the Löwenheim-Skolem-Tarski theorems, whereas for the standard semantic rendering such theorems are false. In what follows, I will argue that such an indeterminacy in the semantic rendering of second-order logic is in an important sense an illusion and, moreover, that it is possible to obtain corresponding deviant semantic renderings for first-order logic, namely, renderings under which first-order logic would be decidable.

Firstly, there is a theorem of Per Lindström that establishes that there is no proper extension of first-order logic for which both the Compactness Theorem and the Löwenheim-Skolem Theorem are true. Second-order logic is clearly a proper extension of first-order logic in any reasonable sense of the word 'extension' and, thus, at least one of these two theorems should be false for second-order logic. This is precisely what occurs with second-order logic endowed with its standard semantics. Under the standard rendering, both the Compactness Theorem and the Löwenheim-Skolem Theorem are false. Since, as we mentioned above, those theorems can be obtained for the Henkin and other nonstandard renderings, it is a corollary of Lindström's Theorem that such nonstandard semantics are inadequate for second-order logic. In fact, what such semantic renderings really do is to interpret second-order logic in first-order logic, for which, as is well known, the Compactness Theorem and the Löwenheim-Skolem Theorem (and also the Semantic Completeness Theorem) are true.

Moreover, one can obtain for first-order logic similar deviant semantic renderings, which essentially interpret first-order logic in propositional logic and allow us to derive a decidability result for "theoremhood" in such a "first-order" logic without contradicting Church's undecidability result. It is, however, a corollary of Church's theorem that such nonstandard semantics are inadequate for first-order logic (in the same fashion as Henkin's and other nonstandard semantics are inadequate for second-order logic).

To make my ideas somewhat precise, let us consider a first order language with the usual logical and auxiliary signs—we need not specify them except for the sign '$\neg$' for negation and the now usual sign for the universal quantifier '$\forall$'—denumerably many individual variables $x_1, x_2, \ldots$, denumerably many monadic predicates $A_1, A_2, \ldots$, and for $n \geq 2$, denumerably many n-adic predicates $R_1^n, R_2^n, \ldots$. We can now proceed in any of two different ways to obtain a monadic rendering of the n-adic predicates. We can either establish a one to one correspondence between the n-adic predicate letters and the monadic predicate letters in a fashion similar to Cantor's one to one correspondence between the natural numbers and the rationals, or we can simply reinterpret the n-adic predicates as "new" monadic predicates, deleting all but the first individual variable of any n-tuple of variables to which it was supposed to apply, or even deleting all individual variables and the corresponding auxiliary signs. In the first and second case we would have a rendering of first-order logic in monadic first-order logic, whereas in the last case the rendering would be in propositional logic. Thus, the formula $R_i^n (x_1, x_2, \ldots, x_n)$ of first-order logic would be assigned the formula $A_k (x_1)$ under the first rendering, where A_k is the monadic predicate assigned to R_i^n under the one-to-one correspondence, the formula $R_i^n (x_1)$ under the second rendering, where R_i^n is seen just as a new monadic predicate, and the letter R_i^n, which is now simply a propositional variable, under the third rendering. In any of the three renderings, logical signs are rendered by themselves, as are also rendered the auxiliary signs (if not deleted, as occurs with many of them under the last rendering), and vacuous quantifiers are deleted. Thus, e.g., under the first rendering the closed sentence $(\forall x_1) \ldots (\forall x_n) \neg R_i^n (x_1, \ldots, x_n)$ is assigned to $(\forall x_1) \neg A_k (x_1)$, where A_k is the monadic predicate paired with R_i^n under the one-to-one correspondence, whereas it is assigned to $(\forall x_1) \neg R_i^n (x_1)$ under the second rendering and to $\neg R_i^n$ under the third rendering. One can easily establish that all three renderings are consistent in the sense that if T is a first-order theory and T* its corresponding theory under any of the three renderings, and there is a formula ψ^* in $L(T^*)$ such that $\vdash \psi^*$ and $\vdash \neg\psi^*$ in T*, then $\vdash \psi$ and $\vdash \neg\psi$ in T, where ψ is the formula in the original first-order language that corresponds to ψ^*. Hence, T* is consistent if T is. Moreover, one can easily show that the interpretation of a first-order formula is a theorem of monadic first-order logic under any of the first two renderings or of propositional logic under the third rendering if and only if the original formula is a theorem of full first-order logic.[59] But both monadic first-order logic and propositional logic are decidable. Thus, under any of these renderings first-order logic would be decidable. This is a very similar situation to that which occurs when you give second-order logic a nonstandard Henkin interpretation.[60] To interpret an n-adic predicate by a monadic predicate is consistent, but you are not giving the n-adic predicate its full or adequate interpretation. As indicated above, Church's theorem can play for first-order logic a similar task as Lindström's theorem for second-order logic, namely, that of excluding such consistent but clearly deviant interpretations. Hence, after all, in the important sense under discussion, second-order logic is not essentially more indeterminate than first-order logic.

APPENDIX 2

In this paper I have argued only indirectly against Field's philosophy of mathematics, namely, by showing that it is based on theses developed by Quine, Putnam, and Benacerraf that, although widely accepted, are really empiricist dogmas of post-Quinean Anglo-American analytic philosophy. On the other hand, in my paper "Interderivability of Seemingly Unrelated Mathematical Statements and the Philosophy of Mathematics" (chapter 13 of the present book) I have argued that more traditional philosophies of mathematics, like constructivism, formalism, pre-Fieldian nominalism, empiricism and Fregean Platonism, have serious difficulties in assessing the (meta-)mathematical fact that there exist mathematical statements, e.g., the Axiom of Choice and Tychonoff's Theorem,[61] which seem to be completely unrelated with regard to their content, but nonetheless are interderivable. We spared Field's philosophy of mathematics of a similar critique. However, it should be sufficiently clear that for a conception of mathematics which denies the existence of mathematical entities and, thus, for which all existential mathematical statements are false and all universal ones vacuously true, the existence of seemingly unrelated mathematical statements must look somewhat puzzling. For, first of all, it is very doubtful that, e.g., the Axiom of Choice and all its many equivalents can be rendered in the same logical form, since some of them, e.g., the Trichotomy of Cardinals, are clearly universal statements (and, thus, vacuously true, according to Field), whereas others seem to admit an existential rendering (and, according to Field, would have to be false). Hence, in that case, if Field is right, we would have interderivable statements with different truth values. (But even if all those interderivable statements admitted the same formal rendering, e.g., as universal statements, the fact is that not all universal statements in the language of mathematics are interderivable with them. Thus, even under such an extremely improbable assumption as that all those statements have the same logical form, we would not be able to give an adequate assessment of the interderivability of seemingly unrelated mathematical statements.)

A related but more simple and straightforward argument against Field's philosophy of mathematics is the following. Consider any theory T in a first-order language L. As is well known, a theory T is model complete exactly when for any two models M and M* of T, if M is a substructure of M*, then M is an elementary substructure of M*.[62] A criterion for the model completeness of a theory T in L is that for any existential sentence φ in L there exists a universal sentence ψ in L such that the interderivability of φ and ψ is a theorem of T. But, according to Field, φ is false, since it is existential, and ψ vacuously true, since it is universal. Hence, if T is consistent and Field's conception were correct, φ and ψ could not be interderivable in T and, thus, T could not be model complete. Thus, if Field were right, there could not be any model complete theory. However, there are model complete theories. Therefore, Field's conception is false. Hence, Field's philosophy of mathematics, as it stands, is not only totally unwarranted, as was argued in the main text, and intuitively false, but also demonstrably false.

NOTES

1. W. Quine, "On What There Is," *From a Logical Point of View*, pp. 1–19.
2. H. Putnam, "What is Mathematical Truth?," *Philosophical Papers*, vol. I, pp. 60–78.
3. P. Benacerraf, "What Numbers Could Not Be," *Philosophy of Mathematics*, ed. Benacerraf and Putnam, pp. 272–94.
4. P. Benacerraf, "Mathematical Truth," *Philosophy of Mathematics*, ed. Benacerraf and Putnam, pp. 408–20.
5. H. Putnam, "Models and Reality," *Philosophy of Mathematics*, ed. Benacerraf and Putnam, pp. 421–44.
6. See Putnam, "What is Mathematical Truth?," *Philosophical Papers*, vol. I, pp. 67–68.
7. Ibid., pp. 60–61.
8. See W. Quine, "Two Dogmas of Empiricism," *From a Logical Point of View*, pp. 20–46.
9. For a comparison of Quine's holism with Duhem's thesis, see D. Gillies, *Philosophy of Science in the Twentieth Century*.
10. For an authoritative exposition of Skolem's paradox—although with the same usual interpretation—see H. D. Ebbinghaus, J. Flum, and W. Thomas, *Mathematical Logic*, p. 108 and especially pp. 112–13.
11. Putnam, "Models and Reality," *Philosophy of Mathematics*, ed. Benacerraf and Putnam, pp. 442–43.
12. Ibid., p. 423.
13. Ibid.
14. Ibid., p. 430.
15. Putnam, "What is Mathematical Truth?," *Philosophical Papers*, vol. 1, p. 73.
16. Ibid., p. 74.
17. Ibid.
18. J. .R. Brown, "Π in the Sky," *Physicalism in Mathematics*, ed. A. D. Irvine, p. 98.
19. Ibid., pp. 101–02.

20. Ibid., p. 102.
21. Ibid.
22. Ibid., p. 99.
23. Quine, "On What There Is," *From a Logical Point of View*, p. 18.
24. See, e.g., M. Bunge, *Philosophy of Physics*, pp. 3 and 226. See also D. Shapere, *Reason and the Search for Knowledge*, especially chapter 16. For some recent discussion on the nature of physical theories, see, e.g., *Reduction in Science*, ed. W. Balzer, D. A. Pearce, and H. J. Schmidt.
25. Putnam, "What is Mathematical Truth?," *Philosophical Papers* vol. 1, pp. 69–74.
26. See Benacerraf, "What Numbers Could Not Be," *Philosophy of Mathematics*, ed. Benacerraf and Putnam.
27. Ibid., pp. 284–85.
28. See H. Field, "Fictionalism, Epistemology and Mathematics," *Realism, Mathematics and Modality*, pp. 1–52, especially pp. 20–21. See also S. MacLane, *Mathematics: Form and Function*, p. 106.
29. See R. Carnap's "The Logicist Foundations of Mathematics," *Philosophy of Mathematics*, ed. Benacerraf and Putnam, pp. 41–52, especially p. 43.
30. G. Frege, *Die Grundlagen der Arithmetik*, p. 88.
31. See Benacerraf, "Mathematical Truth," *Philosophy of Mathematics*, ed. Benacerraf and Putnam, pp. 408–20.
32. Ibid., p. 409.
33. Ibid.
34. Ibid., p. 413.
35. Ibid.
36. Ibid.
37. Ibid., p. 409.
38. J. Burgess, "Epistemology and Nominalism," *Physicalism in Mathematics*, ed. A. D. Irvine, pp. 1–15. See p. 6.
39. Field, "Fictionalism, Epistemology and Modality," *Realism, Mathematics and Modality*, p. 25.
40. Brown, "Π in the Sky," *Physicalism in Mathematics*, ed. Irvine, p. 100.
41. See M. Resnik, "Belief About Mathematical Objects" in *Physicalism in Mathematics*, ed. Irvine, pp. 41–71, especially pp. 45–46.
42. On this issue, see also J. R. Brown's detailed argumentation both in "Π in the Sky," *Physicalism in Mathematics*, ed. Irvine, pp. 111–18 and in chapter 5 of his *The Laboratory of the Mind*. However, the situation on which Brown bases his argumentation does not seem as clear as Brown would like.
43. I will not attempt to deal here with all aspects of Field's views, but only with those relevant to the foregoing discussion, which are without doubt the most central. Thus, e.g., nothing is said about Field's treatment of modalities in some of his recent papers and very little about his discussion with Hale and Wright.
44. Quine, "On What There Is," *From a Logical Point of View*, pp. 17–18.
45. See Field's "Realism and Anti-Realism about Mathematics," *Realism, Mathematics and Modality*, pp. 53–78, especially p. 55.
46. Burgess, "Epistemology and Nominalism," *Physicalism in Mathematics*, ed. Irvine, pp. 11–12.
47. E.g., Irvine in the introduction to his *Physicalism in Mathematics*, p. xiii.
48. Field, "Fictionalism, Epistemology and Mathematics," *Realism, Mathematics and Modality*, pp. 20–22.

49. Ibid., p. 19.
50. Ibid., pp. 18–19.
51. B. Hale, "Nominalism," *Physicalism in Mathematics*, ed. Irvine, pp. 121–44. See p. 123.
52. Field, "Realism and Anti-Realism About Mathematics," *Realism, Mathematics and Modality*, pp. 59–61.
53. Putnam, "What is Mathematical Truth?," *Philosophical Papers*, vol. 1, p. 74.
54. E. Husserl, *Logische Untersuchungen*.
55. For a discussion of this Husserlian notion, see my "Remarks on Sense and Reference in Frege and Husserl," chapter 2 of the present book, "On Frege's Two Notions of Sense," chapter 4 of the present book and "On Husserl's Distinction between State of Affairs (*Sachverhalt*) and Situation of Affairs (*Sachlage*)," chapter 14 of the present book.
56. See Husserl's *Logische Untersuchungen*, vol. 1, chapter 11, and also his *Formale und transzendentale Logik*, especially part 1. See also my dissertation "Edmund Husserls Philosophie der Logik und Mathematik im Lichte der gegenwärtigen Logik und Grundlagenforschung." For Bourbaki, see his "The Architecture of Mathematics."
57. See Frege, *Die Grundlagen der Arithmetik*, especially §§62–69.
58. For a much more detailed treatment of Husserl's epistemology of mathematics see chapter 12 of the present book.
59. This argumentation is similar to that used in, e.g., E. Mendelson's classic *Introduction to Mathematical Logic* to establish the consistency of first-order logic relative to the consistency of propositional logic.
60. Someone could argue that this is really a syntactic interpretation of a theory in another theory, whereas the Henkin interpretation of second-order logic is a semantic one. However, one could transform the three syntactic interpretations of first-order logic given above in semantic interpretations precisely in a fashion similar to that of Henkin's construction of the canonical model in his proof of the semantic completeness of first-order logic, in which the semantic interpretation is essentially a mirror image of the new theory (endowed with all the desired properties of syntactic completeness, etc.) obtained on the basis of the original theory.
61. Tychonoff's Theorem affirms that the product of a family of compact topological spaces is a compact topological space.
62. Speaking somewhat loosely, we can say that a model M of a theory T in a language L is a substructure of another model M* of T if and only if the universe of M is a subset of the universe of M* and for any atomic formula of L (in n variables) and for any n-tuple of members of the universe of M, the n-tuple satisfies the atomic formula in M if and only if it satisfies the formula in M*. It is easy to prove that the same is valid for all quantifier-free formulas in L. M is an elementary substructure of M* if and only if the same is valid for all formulas of L (including those that contain quantifiers).

REFERENCES

Balzer, W., D. A. Pearce, and H. J. Schmidt, eds. *Reduction in Science*. Dordrecht: Kluwer, 1984.

Benacerraf, P., and H. Putnam, eds. *Philosophy of Mathematics.* 2nd ed. Cambridge: University Press, 1983.

Bourbaki, N. "The Architecture of Mathematics." *American Mathematical Monthly* 57 (1950) 221–32.

Brown, J. *The Laboratory of the Mind.* London: Routledge and Kegan Paul, 1991.

Bunge, M. *Philosophy of Physics.* Dordrecht: Reidel, 1973.

Dummett, M. *Frege, Philosophy of Mathematics.* London: Duckworth.

Ebbinghaus, H., J. Flum, and W. Thomas. *Mathematical Logic.* Berlin: Springer, 1984.

Field, H. *Science without Numbers.* Princeton: Princeton University Press, 1980.

———. *Realism, Mathematics and Modality.* Oxford: Blackwell, 1989.

Frege, G. *Die Grundlagen der Arithmetik.* Hamburg: Centenarausgabe, Meiner, 1986 (1884).

———. *Kleine Schriften.* 2nd ed. Ed. I. Angelelli. Hildesheim: Olms, 1990.

Gillies, D. *Philosophy of Science in the Twentieth Century.* London: Routledge and Kegan Paul, 1993.

Goldman, A. I. "A Causal Theory of Knowing." *Journal of Philosophy* 64 (1967): 357–72.

Gupta, A. "Truth and Paradox." *Journal of Philosophical Logic* 11 (1982): 1-60.

Hale, B. *Abstract Objects.* Oxford: Blackwell, 1987.

Herzberger, H. "Notes on Naive Semantics." *Journal of Philosophical Logic* 11 (1982): 61–102.

Husserl, E. *Formale und transzendentale Logik.* Husserliana, vol. XVII. The Hague: M. Nijhoff, 1974 (1929).

———. *Logische Untersuchungen.* Husserliana, vols. XVIII and XIX. The Hague: M. Nijhoff, 1975, 1984 (1900/01, 1913).

Irvine, A. D., ed. *Physicalism in Mathematics.* Dordrecht: Kluwer, 1990.

Kitcher, P. *The Nature of Mathematical Knowledge.* Oxford: Oxford University Press, 1983.

Kripke, S. "An Outline of a Theory of Truth." *Journal of Philosophy* 72 (1975): 690–716.

———. 1979. "A Puzzle About Belief." In *Meaning and Use.* Ed. A. Margalit. Dordrecht: Reidel, 1979, 239–83.

Mac Lane, S. *Mathematics: Form and Function.* Berlin: Springer, 1986.

Maddy, P. *Realism in Mathematics.* Oxford: Clarendon, 1990.

Mendelson, E.. *Introduction to Mathematical Logic.* Princeton: Princeton University Press, 1964.

Putnam, H. *Mathematics, Matter and Method, Philosophical Papers.* Vol. I. Cambridge: Cambridge University Press, 1975.

Quine, W. *From a Logical Point of View.* Cambridge, MA: Harvard University Press, 1953.

Rosado Haddock, G. E. 1973. "Edmund Husserls Philosophie der Logik und Mathematik im Lichte der gegenwärtigen Logik und Grundlagenforschung." Ph.D. dissertation, Rheinische Friedrich-Wilhelms-Universität, Bonn, 1973.

———. "Husserl's Epistemology of Mathematics and the Foundation of Platonism in Mathematics." *Husserl Studies* 4 (1987): 81–102, chapter 12 of the present book.

———. "Interderivability of Seemingly Unrelated Mathematical Statements and the Philosophy of Mathematics." *Diálogos* 59 (1992): 121–34, chapter 13 of the present book.

———. "On Frege's Two Notions of Sense." *History and Philosophy of Logic* 7, no. 1 (1986): 31–41, chapter 4 of the present book.

———. "On Husserl's Distinction between State of Affairs (*Sachverhalt*) and Situa-

tion of Affairs (*Sachlage*)." In *Phenomenology and the Formal Sciences.* Ed. T. Seebohm et al. Dordrecht: Kluwer, 1991, 31–48, chapter 14 of the present book.

————. "Remarks on Sense and Reference in Frege and Husserl." *Kant-Studien* 74,
no. 4 (1982): 425–39, chapter 2 of the present book.

Shapere, D. *Reason and the Search for Knowledge.* Dordrecht: Reidel, 1984.

Shapiro, S. *Foundations without Foundationalism.* Oxford: Clarendon, 1991.

Tarski, A. "The Concept of Truth in Formalized Languages." Translated in A. Tarski,
Logic, Semantics, Metamathematics. 2nd ed. Indianapolis: Hackett, 1983 (1956).

BIBLIOGRAPHY

Angelelli, Ignacio. "Friends and Opponents of the Substitutivity of Identicals in the History of Logic." In *Studies on Frege*. Vol. 2. Ed. M. Schirn, 141–66.

———. "Die Zweideutigkeit von Freges Sinn und Bedeutung." *Allgemeine Zeitschrift für Philosophie* 3 (1978): 62–66.

Aristotle. *Works*. Cambridge, MA: Harvard University Press, 1983.

Bachelard, Suzanne. *A Study of Husserl's Formal and Transcendental Logic*. Evanston, IL: Northwestern University Press, 1968.

Balzer, W., D. A. Pearce, and H. J. Schmidt, eds. *Reduction in Science*. Dordrecht: Kluwer, 1984.

Barwise, Jon, and John Etchemendy. *The Liar: An Essay on Truth and Circularity*. New York: Oxford University Press, 1987.

Barwise, Jon, and John Perry. "Semantic Innocence and Uncompromising Situations." *Midwest Studies in Philosophy*. Vol. 6, *The Foundations of Analytic Philosophy*. Ed. P. A. French et al. Minneapolis, MN: University of Minnesota Press, 1981, 387–403.

Becker, Oskar. "The Philosophy of E. Husserl." In Elveton, ed., 40–72, translation of "Die Philosophie Edmund Husserls," *Kant-Studien* 35 (1930): 119–50.

Beeson, Michael J. *Foundations of Constructive Mathematics*. Berlin: Springer, 1985.

Bell, David. *Husserl*. London: Routledge and Kegan Paul, 1990.

Benacerraf, Paul. "Mathematical Truth." In Benacerraf and Putnam, eds., 403–20.

———. "What Numbers Could Not Be." In Benacerraf and Putnam, eds., 272–94.

Benacerraf, Paul, and H. Putnam, eds. *Philosophy of Mathematics*. 2nd ed. rev. Cambridge, UK: Cambridge University Press, 1983 (1964).

Beth, Evert Willem. *The Foundations of Mathematics*. 2nd ed. rev. Amsterdam: North Holland, 1965 (1959).

Bishop, Erret. *Foundations of Constructive Analysis*. New York: McGraw-Hill, 1967.

Bolzano, Bernard. *Paradoxes of the Infinite*. London: Routledge and Kegan Paul, 1950. Translation of his *Paradoxien des Unendlichen* (1831).

———. *Theory of Science*. Berkeley, CA: University of California Press, 1972. Partial translation of his *Wissenschaftslehre* (1837) by R. George.

———. *Theory of Science*. Dordrecht: Reidel, 1973. Partial translation of his *Wissenschaftslehre* (1837) by B. Terrell.

Bourbaki, Nicholas. "The Architecture of Mathematics." *American Mathematical Monthly* 57 (1950): 221–32.

Brown, James R. *The Laboratory of the Mind*. London: Routledge and Kegan Paul, 1991.

Bunge, Mario. *Philosophy of Physics*. Dordrecht: Reidel, 1973.

Burge, Tyler. "Frege on Truth." In Haaparanta and Hintikka, eds., 97–154.

Cantor, Georg. *Contributions to the Founding of the Theory of Transfinite Numbers*. New York: Dover, 1955 (1915). Translation of "Beiträge zur Begründung der transfiniten Mengenlehre." *Mathematische Annalen* 46 (1895): 481–512.

————. *Georg Cantor Briefe.* Ed. H. Meschkowski and W. Nilson. New York: Springer, 1991.

————. *Gesammelte Abhandlungen.* Ed. E. Zermelo. Berlin: Springer, 1932.

————. *Gesammelte Abhandlungen zur Lehre vom Transfiniten.* Halle: Pfeffer, 1890. In his *Gesammelte Abhandlungen*, 165–246. See Cantor's "Mitteilungen zur Lehre vom Transfiniten."

————. *Grundlagen einer allgemeinen Mannigfaltigkeitslehre. Ein mathematisch-philosophischer Versuch in der Lehre des Unendlichen.* Leipzig: Teubner, 1883. Also published in his *Gesammelte Abhandlungen*, 165–246.

————. "Mitteilungen zur Lehre vom Transfiniten." *Zeitschrift für Philosophie und philosophische Kritik* 91 (1887): 81–125; 92 (1888): 240–65. In Cantor's *Gesammelte Abhandlungen*, 378–439. Also published as *Gesammelte Abhandlungen zur Lehre vom Transfiniten.*

————. "Principien einer Theorie der Ordnungstypen"(dated November 6, 1884). First published by I. Grattan-Guinness in *Acta Mathematica* 124 (1970): 65–107.

————. "Rezension von Freges *Grundlagen der Arithmetik.*" *Deutsche Literaturzeitung* 6 (1885): 728–29. Also in Cantor's *Gesammelte Abhandlungen*, 440–41.

Carnap, Rudolf. *The Logical Structure of the World.* Berkeley, CA: University of California Press, 1969. Translation of *Der Logische Aufbau der Welt.* 4th ed. Hamburg: Meiner, 1974 (1928).

————. *Logical Syntax of Language.* Expanded version. London: Routledge and Kegan Paul, 1937. Translation of *Die Logische Syntax der Sprache* (1934).

————. *Meaning and Necessity.* Chicago: University of Chicago Press, 1947.

Caton, Charles. "The Idea of Sameness Challenges Reflection." In *Studies on Frege.* Vol. 2. Ed. M. Schirn, 167–80.

Cavaillès, Jean. *Méthode axiomatique et formalisme.* Paris: Hermann, 1981 (1937).

————. *Philosophie mathématique.* Paris: Hermann, 1962.

————. *Sur la logique et la théorie de la science.* Paris: P.U.F., 1947.

Charraud, Nathalie. *Infini et inconscient, essai sur Georg Cantor.* Paris: Anthropos, 1994.

Church, Alonzo. *Introduction to Mathematical Logic.* Princeton, NJ: Princeton University Press, 1956 (1944).

————. "Review of M. Farber, *The Foundations of Phenomenology.*" *Journal of Symbolic Logic* (9): 1944, 63–65.

Couturat, Louis. *De l'infini mathématique.* Paris: Blanchard, 1973 (1896).

da Silva, Jairo J. "Gödel and Transcendental Phenomenology." Forthcoming.

————. "Husserl's Phenomenology and Weyl's Predicativism." *Synthese* 110 (1997): 277–96.

————. "Husserl's Philosophy of Mathematics." *Manuscrito* 16, no. 2 (1993): 121–48.

————. "Husserl's Two Notions of Completeness." Forthcoming in *Synthese.*

Dauben, Joseph. *Georg Cantor: His Mathematics and Philosophy of the Infinite.* Princeton, NJ: Princeton University Press, 1979.

Davidson, Donald, and G. Harman, eds. *Semantics of Natural Language.* Dordrecht: Reidel, 1972.

Dedekind, Richard. *Essays on the Theory of Numbers.* New York: Dover, 1963.

————. "The Nature and Meaning of Numbers." In *Essays on the Theory of Numbers.* New York: Dover, 1963, 32–115.

Demopoulos, William. "Frege, Hilbert, and the Conceptual Structure of Model Theory." *History and Philosophy of Logic* 15 (1994): 211–25.

Demopoulos, William, ed. *Frege's Philosophy of Mathematics*. Cambridge, MA: Harvard University Press, 1995.

Desanti, Jean-Toussaint. *Les idéalités mathématiques*. Paris: Seuil, 1968.

Dreben, Burton, and A. Kanamori. "Hilbert and Set Theory." *Synthese* 110 (1997): 77–136.

Dreyfus, Herbert. "Sense and Essence: Frege and Husserl." *International Philosophical Quarterly* 10 (1970).

Dugac, Pierre. *Richard Dedekind et les fondements des mathématiques*. Paris: Vrin, 1976.

Dummett, Michael. *Elements of Intuitionism*. Oxford: Oxford University Press, 1977.

———. *Frege, Philosophy of Language*. 2nd ed. rev. London: Duckworth (also Cambridge, MA: Harvard University Press), 1981 (1973).

———. *Frege, Philosophy of Mathematics*. London: Duckworth (also Cambridge, MA: Harvard University Press,), 1991.

———. *The Interpretation of Frege's Philosophy*. London: Duckworth (also Cambridge, MA: Harvard University Press), 1981.

———. "The Philosophical Basis of Intuitionistic Logic." In Benacerraf and Putnam, eds., 97–129.

Ebbinghaus, H. D., J. Flum, and W. Thomas. *Mathematical Logic*. Berlin: Springer, 1984.

Eberle, Rolf. *Nominalistic Systems*. Dordrecht: Reidel, 1970.

Eccarius, W. "Georg Cantor und Kurd Lasswitz: Briefe zur Philosophie des Unendlichen." *NTM Schriften, Gesch. Naturwiss. Technik, Med.* 22 (1985): 7–28.

Ehrlich, Philip. "From Completeness to Archimedean Completeness." *Synthese* 110 (1997): 57–76.

Eley, Lothar. "Einleitung des Herausgebers." In Husserl's *Philosophie der Arithmetik, mit ergänzenden Texte*, Husserliana, vol. XII, xiii–xxix.

Elliston, Frederick A., and Peter McCormick, eds. *Husserl: Expositions and Appraisals*. Notre Dame, IN: Notre Dame University Press, 1981.

Elveton, R. O., ed. *The Phenomenology of Husserl: Selected Critical Readings*. Chicago: Quadrangle Books, 1970.

Farber, Marvin. *The Foundations of Phenomenology*. Cambridge, MA: Harvard University Press, 1943.

Field, Hartry. *Realism, Mathematics and Modality*. Oxford: Blackwell, 1989.

———. *Science without Numbers*. Princeton: Princeton University Press, 1980.

Findlay, J.N. "Translator's Introduction." In Husserl's *Logical Investigations*, 1–40.

Føllesdal, Dagfinn. "Gödel and Husserl." In Hintikka, ed., 427–46.

———. "Husserl and Frege: A Contribution to Elucidating the Origins of Phenomenological Philosophy." In Haaparanta, ed., 3–47. Translation of his 1958 Norwegian Master's Thesis.

———."Husserl's Notion of Noema." *Journal of Philosophy* 66 (1969): 680–87.

———."An Introduction to Phenomenology for Analytic Philosophers." In R.E. Olson and A.M. Paul, eds., 417–29.

———. "Introductory Note" to Gödel's *Collected Works*, vol. 3, 364–73.

Fraenkel, A. A. "Georg Cantor." *Jahresbericht der deutschen Mathematiker Vereinigung* 39 (1930): 189–266. Abridged in Cantor's *Gesammelte Abhandlungen*, 452–83.

Frege, Gottlob. *The Basic Laws of Arithmetic*. Berkeley, CA: University of California Press, 1964. Translation of *Grundgesetze I* (1893).

———. "*Begriffsschrift*, a formula language, modeled upon that of arithmetic for pure thought." In van Heijenoort, 5-82. Translation of *Begriffsschrift*, Halle: Nebert, 1879; reprinted Hildesheim: Olms, 1964.

———. *Collected Papers on Mathematics, Logic and Philosophy.* Oxford: Blackwell, 1984. Translation of Frege's *Kleine Schriften.*

———. "Comments on Sense and Meaning 1892–1895." In his *Posthumous Writings,* 118–25. Translation of "Ausführungen über Sinn und Bedeutung 1892–1895," in his *Nachgelassene Schriften,* 128–36.

———. "Draft Towards a Review of Cantor's *Gesammelte Abhandlungen zur Lehre vom Transfiniten.*" In his *Posthumous Writings,* 68–71.

———. *Foundations of Arithmetic.* 2nd rev. ed. Oxford: Blackwell, 1986. Translation of his *Die Grundlagen der Arithmetik.*

———. "Function and Concept." In his *Translations from the Philosophical Writings,* 21–41. Translation of "Funktion und Begriff," in *Kleine Schriften,* 125–42.

———. "The Thought." In his *Collected Papers,* 351–72. Translation of "Der Gedanke," in his *Kleine Schriften,* 342–62.

———. "Gottlob Frege: Briefe an Ludwig Wittgenstein." *Grazer philosophische Studien* 33/34 (1986): 5–33.

———. *Gottlob Freges Briefwechsel mit D. Hilbert, E. Husserl und B. Russell.* Ed. Gottfried Gabriel. Hamburg: Meiner, 1980.

———. *Die Grundlagen der Arithmetik.* Hamburg: Centenarausgabe, 1986. Originally published Breslau: Koebner, 1884.

———. *Grundgesetze der Arithmetik I.* Jena: Pohle, 1893.

———. *Grundgesetze der Arithmetik I, II.* Hildesheim: Olms: 1966 (1893, 1903).

———. "Introduction to Logic." In his *Posthumous Writings,* 203–50. Translation of his "Einleitung in die Logik."

———. *Kleine Schriften.* 2nd ed. Ed. I. Angelelli. Hildesheim: Olms, 1990 (1967).

———. *Nachgelassene Schriften.* Ed. H. Hermes et al. Hamburg: Meiner, 1983 (1969).

———. "On Concept and Object." In his *Translations from the Philosophical Writings,* 42–55. Translation of *Über Begriff und Gegenstand,* in his *Kleine Schriften,* 167–78.

———. "On Formal Theories of Arithmetic." In his *Collected Papers,* pp. 112–21.

———. "On Sense and Meaning." In his *Translations from the Philosophical Writings,* 56–78. Translation of "Über Sinn und Bedeutung," in his *Kleine Schriften,* 143–62.

———. *Philosophical and Mathematical Correspondence.* Oxford: Blackwell, 1980. Translation of his *Wissenschaftlicher Briefwechsel.*

———. *Posthumous Writings.* Oxford: Blackwell, 1979. Translation of his *Nachgelassene Schriften.*

———. "Reply to Cantor's Review of *Grundlagen der Arithmetik.*" In his *Collected Papers,* 122. Translation of "Erwiderung auf Cantors Rezension der *Grundlagen der Arithmetik,*" *Deutsche Literaturzeitung* 6, no. 28 (1885): sp. 1030.

———. "Review of Dr. E. Husserls *Philosophy of Arithmetic.*" *Mind* 81, no. 323 (July 1894): 321–37 and in Frege's *Collected Papers,* 195–209. Translation of his "Rezension von E. Husserl: *Philosophie der Arithmetik,*" *Zeitschrift für Philosophie und philosophische Kritik* 103 (1894): 313–32, in his *Kleine Schriften,* 179–92.

———. "Review of Georg Cantor, *Zur Lehre vom Transfiniten. Gesammelte Abhandlungen aus der Zeitschrift für Philosophie und philosophische Kritik.*" In his *Collected Papers,* 178–81, in the original German in his *Kleine Schriften,* 163–66.

———. *Translations from the Philosophical Writings.* 3rd ed. Ed. P. Geach and M. Black. Oxford: Blackwell, 1980 (1952).

———. *Wissenschaftlicher Briefwechsel.* Ed. G. Gabriel et al. Hamburg: Meiner, 1976.

Garciadiego, Alejandro. *Bertrand Russell and the Origins of the Set-theoretic 'Paradoxes'.* Basel: Birkhäuser, 1992.

Gerlach, H., and Sepp, H., eds. *Husserl in Halle.* Bern: Peter Lang, 1994.

Gillies, D. *Philosophy of Science in the Twentieth Century.* London: Routledge and Kegan Paul, 1993.

Gilson, Lucienne. *Méthode et métaphysique selon Franz Brentano.* Paris: Vrin, 1955.

———. *La Psychologie descriptive selon Franz Brentano.* Paris: Vrin, 1955.

Gödel, Kurt. *Collected Works.* 3 vols. New York: Oxford University Press, 1990–95.

———. "The Modern Development of Mathematics in the Light of Philosophy." In his *Collected Works,* vol. 3, 374–87.

———. "On Formally Undecidable Propositions of *Principia Mathematica* and related systems I." In his *Collected Works,* vol. 1, 144–95 and in van Heijenoort, ed., 595–617. Translation of his "Über formal unentscheidbare Sätze der *Principia Mathematica* und verwandter Systeme I" (1931).

———. "Russell's Mathematical Logic." In his *Collected Works,* vol. 2, 119–41 and Benacerraf and Putnam, eds., 447–69.

———. "What is Cantor's Continuum Problem?" In his *Collected Works,* vol. 2, 147–87, and Benacerraf and Putnam, eds., 470–85.

Goldman, A. I. "A Causal Theory of Knowing." *Journal of Philosophy* 64 (1967): 357–72.

Grattan-Guinness, Ivor. "Bertrand Russell on His Paradox and the Multiplicative Axiom: An Unpublished Letter to Philip Jourdain." *Journal of Philosophical Logic* 1 (1972): 103–10.

———. *Dear Russell-Dear Jourdain: A Commentary on Russell's Logic Based on His Correspondence with Philip Jourdain.* London: Duckworth, 1977.

———. "Georg Cantor's Influence on Bertrand Russell." *History and Philosophy of Logic* 1 (1980): 61–93.

———. "How Russell Discovered His Paradox." *Historia Mathematica* 5 (1978): 127–37.

———. "Numbers, Magnitudes, Ratios and Proportions in Euclid's *Elements*: How Did He Handle Them?" *Historia Mathematica* 23 (1996): 355–75.

———. "Preliminary Notes on the Historical Significance of Quantification and The Axiom of Choice in Mathematical Analysis." *Historia Mathematica* 2 (1975): 475–88.

———. "Psychology in the Foundations of Logic and Mathematics: the Cases of Boole, Cantor and Brouwer." *History and Philosophy of Logic* 3 (1982): 33–53.

———. "Towards a Biography of Georg Cantor." *Annals of Science* 27, no. 4 (1971): 345–91.

Gupta, Anil. "Truth and Paradox." *Journal of Philosophical Logic* 11 (1982): 1–60.

Haaparanta, Leila, ed. *Mind, Meaning and Mathematics, Essays on the Philosophical Views of Husserl and Frege.* Dordrecht: Kluwer, 1994.

Haaparanta, Leila, and Jaakko Hintikka, eds. *Frege Synthesized.* Dordrecht: Reidel, 1986.

Hale, Bob. *Abstract Objects.* Oxford: Blackwell, 1987.

———. "Nominalism." In Irvine, ed. 121–44.

Hallett, Michael. *Cantorian Set Theory and Limitation of Size.* Oxford: Clarendon, 1984.

Hatcher, W.S. *The Logical Foundations of Mathematics.* Oxford: Pergamon Press, 1982. A revised version of his *Foundations of Mathematics,* W. B. Saunders & Co., 1968.

Herzberger, H. G. "Notes on Naive Semantics." *Journal of Philosophical Logic* 11 (1982): 61–102.

Hilbert, David. "Über den Zahlbegriff." *Jahresbericht der Deutschen Mathematiker Vereinigung* 8 (1900): 180–84.

———. "On the Infinite." In van Heijenoort, ed., 369–92.

Hill, Claire Ortiz. *Word and Object in Husserl, Frege and Russell, the Roots of Twentieth Century Philosophy.* Athens, OH: Ohio University Press, 1991.

———. "Frege's Letters." In Hintikka, ed., 97–118.

———. "From Empirical Psychology to Phenomenology: Husserl on the Brentano Puzzle." In Poli, ed., 151–67.

———. "La logique des expressions intentionnelles." Mémoire de Maîtrise, Université de Paris-Sorbonne, April 1, 1979.

———. *Rethinking Identity and Metaphysics.* New Haven, CT: Yale University Press, 1997.

Hintikka, Jaakko. "Is There Completeness in Mathematics after Gödel?" *Philosophical Topics* 17, no. 2 (Fall 1989): 71–90.

———. "On the Development of the Model-Theoretic Viewpoint in Logical Theory." *Synthese* 77 (1988): 1–36.

———. "The Idea of Phenomenology in Wittgenstein and Husserl." *Austrian Philosophy Past and Present.* Ed. K. Lehrer and J. C. Marek. Kluwer: Dordrecht, 1997, 101–23.

———. *The Intentions of Intentionality and Other New Models for Modalities.* Dordrecht: Reidel, 1975.

———. *Knowledge and Belief.* Ithaca, NY: Cornell University Press, 1962.

———. *Models for Modalities.* Dordrecht: Reidel, 1969.

Hintikka, Jaakko, ed. *From Dedekind to Gödel, Essays on the Development of the Foundations of Mathematics.* Dordrecht: Kluwer, 1995.

Hintikka, Jaakko, and D. Davidson, eds. *Words and Objections, Essays on the Work of W.O. Quine.* Boston: Reidel, 1969.

Hintikka, Jaakko, and G. Sandu. "The Skeleton in Frege's Cupboard: The Standard vs. Non-Standard Distinction." *Journal of Philosophy* 89 (1992): 290–315.

Husserl, Edmund. *Aufsätze und Rezensionen 1890–1910.* Husserliana, vol. XXII. The Hague: M. Nijhoff, 1979.

———. "Besprechung von E; Schröder, *Vorlesungen über die Algebra der Logik I.*" In Husserl's *Aufsätze und Rezensionen.* Husserliana, vol. XXII, 3–43. Translated in his *Early Writings*, 52–91.

———. *Briefwechsel, Die Brentanoschule.* Vol. I. Dordrecht: Kluwer, 1994.

———. *The Crisis of European Sciences and Transcendental Phenomenology: An Introduction to Phenomenological Philosophy.* Evanston, IL: Northwestern University Press, 1970. Translation of his *Krisis der europaïschen Wissenschaften.*

———. *Early Writings in the Philosophy of Logic and Mathematics.* Dordrecht: Kluwer, 1994.

———. *Einleitung in die Logik und Erkenntnistheorie, Vorlesungen 1906/07.* Husserliana, vol. XXIV. Dordrecht: Kluwer, 1984.

———. *Erfahrung und Urteil.* 5th ed. Hamburg: Meiner, 1976 (original ed. 1939; first Meiner ed. with preface by Lothar Eley 1972).

———. *Experience and Judgement.* London: Routledge and Kegan Paul, 1973. Translation of his *Erfahrung und Urteil.*

———. *Formal and Transcendental Logic.* The Hague: M. Nijhoff, 1969. Translation of his *Formale und transzendentale Logik*, 2nd ed. Husserliana, vol. XVII. The Hague: M. Nijhoff, 1974 (1929).

———. *Ideas, General Introduction to Pure Phenomenology.* New York: Colliers,

1962. Translation of his *Ideen zu einer reinen Phänomenologie und einer phänom-enologischen Philosophie I*. Rev. ed. Husserliana, vol. III. The Hague: M. Nijhoff, 1976 (1913).

———. *Introduction to the Logical Investigations: A Draft of a Preface to the Logical Investigations (1913)*. The Hague: M. Nijhoff, 1975.

———. *Die Krisis der europäischen Wissenschaften und die transzendentale Phänome-nologie*. Husserliana, vol. VI. The Hague: M. Nijhoff, 1954 (1936).

———. *Logical Investigations*. New York: Humanities Press, 1970. Translation of the 2nd edition of his *Logische Untersuchungen*.

———. *Logik und allgemeine Erkenntnistheorie, Vorlesungen 1917/18, mit ergänzen-den Texten aus der ersten Fassung 1910/11*. Husserliana, vol. XXX. Dordrecht: Kluwer, 1996.

———. *Logische Untersuchungen*. Halle: Niemeyer, 1900/01 (2nd ed. rev., 1913). Also published as Husserliana, vols. XVIII, XIX/I–II. The Hague: M. Nijhoff, 1975, 1984.

———. "On the Concept of Number." In McCormick and Elliston, eds., 92–120. Translation of his "Über den Begriff der Zahl: Psychologische Analysen" avail-able in his *Philosophie der Arithmetik, mit ergänzenden Texten*, Husserliana, vol. XII, 289–339.

———. "On the Logic of Signs (Semiotic)." In his *Early Writings*, 20–51. Transla-tion of his "Zur Logik der Zeichen."

———. "Personal Notes." In his *Early Writings*, 490–500. Translation of his "Per-sönliche Aufzeichnungen," *Philosophy and Phenomenological Research* 16 (1956): 293–302.

———. *Philosophie der Arithmetik*. Halle: Pfeffer, 1891.

———. *Philosophie der Arithmetik, mit ergänzenden Texten (1890–1901)*. Husser-liana, vol. XII. The Hague: M. Nijhoff, 1970.

———. "Recollections of Franz Brentano." In McCormick and Elliston, eds., 342–49. Also in Mc Alister's *The Philosophy of Franz Brentano*, 47–55. Transla-tion of "Errinerungen an Franz Brentano," in O. Kraus, ed.

———. "Review of Melchior Palaygi's *Der Streit der Psychologisten und Formalisten in der modernen Logik*." In his *Early Writings*, 197–206 and *The Personalist* 53 (Winter 1972). Translation of his "Rezension von Palaygi," *Zeitschrift für Psy-chologie und Physiologie der Sinnesorgane* 31 (1903).

———. *Studien zur Arithmetik und Geometrie, Texte aus dem Nachlass (1886–1901)*. Husserliana, vol. XXI. The Hague: M. Nijhoff, 1983.

———. "Über den Begriff der Zahl: Psychologische Analysen." Halle: Heyneman-sche Buchdrückerei, 1887. In his *Philosophie der Arithmetik, mit ergänzenden Texten*, Husserliana, vol. XII, 289–339.

———. *Vorlesungen über Bedeutungslehre, Sommersemester 1908*. Husserliana, vol. XXVI. Dordrecht: M. Nijhoff, 1987.

———. "Zur Logik der Zeichen." In his *Philosophie der Arithmetik, mit ergänzenden Texte*, Husserliana, vol. XII, 340–73. Translated in his *Early Writings*, 20–51.

Husserl, Malvine. "Skizze eines Lebensbildes von E. Husserl." *Husserl Studies* 5 (1988): 105–25.

Illemann, Werner. *Husserls vorphänomenologische Philosophie*. Leipzig: Hirzel, 1932.

Irvine, A. D., ed. *Physicalism in Mathematics*. Dordrecht: Kluwer, 1990.

Ishiguro, Hidé. *Leibniz's Philosophy of Logic and Language*. Cambridge: Cambridge University Press, 1990.

Jourdain, Philip. "The Development of the Theories of Mathematical Logic and the Principles of Mathematics." *Quarterly of Pure and Applied Mathematics* 48 (1912): 219–315.

————. "The Development of the Theory of Transfinite Numbers." *Archiv der Mathematik und Physik* 14 (1908/09): 289-311; 16 (1910): 21–43; 22 (1913/14), 1–21. In his *Selected Essays on the History of Set Theory and Logics*, 33–99.

————. *Selected Essays on the History of Set Theory and Logics*. Ed. I. Grattan-Guinness. Bologna: CLUEB, 1991.

Kaplan, David, and Richard Montague. "A Paradox Regained." *Notre Dame Journal of Formal Logic* 1, no. 3 (July 1960): 79–90.

Kerry, Benno. "Über Georg Cantors Mannichfaltigkeitsuntersuchungen." *Vierteljahrsschrift für wissentschaftliche Philosophie* 9 (1885): 191–232.

Kersey, Ethel. "The Noema, Husserlian and Beyond: An Annotated Bibliography of English Language Sources." *Philosophy Research Archives* 9, Microfiche supplement, 62–90, 1984 (1983).

Khatchadourian, Haig. "Kripke and Frege on Identity Statements." In *Studies on Frege*, vol. 2, Schirn, ed., 269–98.

Kienzle, Bertram. "Notiz zu Freges Theorien der Identität." In *Studies on Frege*, vol. 2, Schirn, ed., 217–19.

Kilminster, C.W. *Russell*. Brighton: The Harvester Press, 1984.

Kitcher, Philip. "Frege, Dedekind and the Philosophy of Mathematics." In Haaparanta and Hintikka, eds., 299–343.

————. *The Nature of Mathematical Knowledge*. Oxford: Oxford University Press, 1983.

Kline, Morris. *Mathematical Thought from Ancient to Modern Times*. Vol. 3. New York: Oxford University Press, 1972.

Kraus, Oskar, ed. *Franz Brentano: Zur Kenntnis seines Lebens und seiner Lehre*. Munich: Beck'sche, 1919.

Kreiser, Lothar. "W. Wundts Auffassung der Mathematik-Briefe von G. Cantor an W. Wundt." *Wissenschaftliche Zeitschrift, Karl-Marx Universität, Ges. u. Sprachwissenschaft* 28 (1979): 197–206.

Kripke, Saul. "Identity and Necessity." In *Identity and Individuation*, M. K. Munitz, ed., 135–64.

————. "Naming and Necessity." In Davidson and Harman, eds., 253–355 and 763–69.

————. *Naming and Necessity*. Oxford: Blackwell, 1980 (1972).

————. "An Outline of a Theory of Truth." *Journal of Philosophy* 72 (1975): 690–716. Also in Martin, ed., *Recent Essays on Truth and the Liar Paradox*, 53–82.

————. "A Puzzle About Belief." In Margalit, ed., 239–83.

Kusch, Martin. *Language as Calculus vs. Language as Universal Medium*. Dordrecht: Kluwer, 1989.

Landgrebe, Ludwig, and Jan Patocka. *Edmund Husserl: zum Gedächtnis*. New York: Garland, 1980.

Lipps, Hans. *Die Verbindlichkeit der Sprache*. Frankfurt: Klostermann, 1958.

Lohmar, Hans. "Husserls Phänomenologie als Philosophie der Mathematik." Ph.D. dissertation, Cologne, 1987.

Lotze, Hermann. *Logic*. New York: Garland, 1980. Translation of his *Logik*, 1888.

————. *Metaphysik*. Leipzig: Hirzel, 1879.

MacLane, Saunders. *Mathematics: Form and Function*. Berlin: Springer, 1986.

Maddy, Penelope. *Realism in Mathematics*. Oxford: Oxford University Press, 1990.

Mahnke, Dietrich. "From Hilbert to Husserl: First Introduction to Phenomenology, especially that of formal mathematics." *Studies in the History and Philosophy of Science* 8 (1966): 71–84. Translation of "Von Hilbert zu Husserl: Erste Einführung in die Phänomenologie," *Unterrichtsblätter für Mathematik und Naturwissenschaft* (29): 1923, 34–37.

Majer, Ulrich. "Husserl and Hilbert on Completeness: A Neglected Chapter in Early Twentieth Century Foundations of Mathematics." *Synthese* 110 (1997): 37–56.

Mancosu, Paolo. *From Brouwer to Hilbert: The Debate on the Foundations of Mathematics in the 1920s.* New York: Oxford University Press, 1998.

Marcus, Ruth Barcan. "Extensionality." *Mind* 69 (1960): 55–62.

———. *Modalities.* New York: Oxford University Press, 1993.

Margalit, A., ed. *Meaning and Use.* Dordrecht: Reidel, 1979.

Martin, Robert, ed. *The Paradox of the Liar.* New Haven: Yale University Press, 1970.

———, ed. *Recent Essays on Truth and the Liar Paradox.* Oxford: Clarendon, 1984.

Maxsein, Agnes. "Die Entwicklung des Begriffs 'A priori' von Bolzano über Lotze zu Husserl." Ph.D. dissertation, Giessen, 1933.

McAlister, Linda. *The Philosophy of Franz Brentano.* London: Duckworth, 1976.

McCormick, Peter, and Elliston, Frederick A., eds. *Husserl: Shorter Works.* Notre Dame: Notre Dame University Press, 1981.

Mendelson, Elliott. *Introduction to Mathematical Logic.* Princeton, NJ: Van Nostrand, 1964.

Meschkowski, Herbert. "Aus den Briefbüchern Georg Cantors." *Archive for the History of the Exact Sciences* 2, no 6 (1965): 503–19.

———. *Probleme des Unendlichen. Werk und Leben Georg Cantors.* Braunschweig: Vieweg, 1967.

Miller, J. Philip. *Numbers in Presence and Absence.* The Hague: M. Nijhoff, 1982.

Mohanty, J. N. *Husserl and Frege.* Bloomington, IN: Indiana University Press, 1982.

———. "Husserl and Frege: a New Look at their Relationship." *Research in Phenomenology* 4 (1974): 51–62.

———. "Husserl's Formalism." In Seebohm, et al., eds., 93–105.

Montague, Richard. "Syntactical Treatments of Modality, with Corollaries on Reflection Principles and Finite Axiomatizability." *Proceedings of a Colloquium on Modal and Many-Valued Logics, Acta Philosophica Fennica* 16, Helsinki, 23–26 August, 1963, 153–67.

Moran, Dermot. *Introduction to Phenomenology.* London: Routledge, 2000.

Moore, Gregory H. *Zermelo's Axiom of Choice.* New York: Springer, 1982.

Munitz, Milton K., ed. *Identity and Individuation.* New York: New York University Press, 1971.

Neemann, U. "Husserl und Bolzano." *Allgemeine Zeitschrift für Philosophie* 2 (1977): 52–66.

Olson, R. E., and A.M. Paul, eds. *Contemporary Philosophy in Scandinavia.* Baltimore: Johns Hopkins Press, 1972.

Osborn, Andrew. *The Philosophy of E. Husserl in its Development to his First Conception of Phenomenology in the Logical Investigations.* New York: International Press, 1934. Reprinted in 1949.

Panza, Mario, and J.M. Salanskis. *L'objectivité mathématique, platonismes et structures formelles.* Paris: Masson, 1995.

Peckhaus, Volker. *Hilbertsprogramm und kritische Philosophie, das Göttinger Modell interdisziplinärer Zusammenarbeit zwischen Mathematik und Philosophie.* Göttingen: Vandenhoeck & Ruprecht, 1990.

Picker, Bernold. "Die Bedeutung der Mathematik für die Philosophie Edmund Husserls." *Philosophia Naturalis* 7 (1962): 266–355, his 1955 Münster dissertation.

Plato. *Dialogues.* Cambridge, MA: Harvard University Press, 1926–1946.

Plessner, Helmuth. *Husserl in Göttingen*. New York: Garland, 1980.

Poli, Roberto. "Husserl's Conception of Formal Ontology." *History and Philosophy of Logic* 14 (1993): 1–14.

Poli, Roberto, ed. *The Brentano Puzzle*. Ashgate, UK: Aldershot, 1998.

Purkert, Walter, and Hans Ilgauds. *Georg Cantor 1845–1918*. Basel: Birkhäuser, 1991.

Putnam, Hilary. *Mathematics, Matter and Method, Philosophical Papers*. Vol. 1. Cambridge: Cambridge University Press, 1975.

Quine, Willard. *From a Logical Point of View*. 2nd ed. Cambridge, MA: Harvard University Press, 1961 (1953).

———. "Logic and the Reification of Universals." In his *From a Logical Point of View*, 102–29.

———. *Mathematical Logic*. New York: Norton, 1940.

———. *Methods of Logic*. 2nd ed. London: Routledge and Kegan Paul, 1962 (1952).

———. "New Foundations for Mathematical Logic." In his *From a Logical Point of View*, 80–101.

———. *Philosophy of Logic*. Englewood Cliffs, NJ: Prentice Hall, 1970.

———. "The Problem of Interpreting Modal Logic." *Journal of Symbolic Logic* 12, no. 2 (June 1947): 43–48.

———. "Promoting Extensionality." *Synthese* 98 (1994): 143–51.

———. *Set Theory and Its Logic*. Cambridge, MA: Harvard University Press, 1969.

———. *Ways of Paradox*. Cambridge, MA: Harvard University Press, 1976.

———. *Word and Object*. Cambridge, MA: M.I.T. Press, 1960.

Reid, Constance. *Hilbert*. New York: Springer, 1970.

———. *Courant in Göttingen and New York*. New York: Springer, 1979.

Reinach, Adolf. "Zur Theorie des negativen Urteils." Translated in Smith, ed., 315–77.

Resnik, Michael. *Frege and the Philosophy of Mathematics*. Ithaca, NY: Cornell University, 1980.

Rodriguez-Consuerga, Francisco. *The Mathematical Philosophy of Bertrand Russell: Origins and Development*. Basel: Birkhäuser, 1991.

Rosado Haddock, Guillermo E. "Edmund Husserls Philosophie der Logik und Mathematik im Lichte der gegenwärtigen Logik und Grundlagenforschung." Ph.D. dissertation, Rheinische Friedrich-Wilhelms-Universität, Bonn, 1973.

———. "Essay Review of M. Schirn (ed.), *Frege: Importance and Legacy*." *History and Philosophy of Logic* 19, no. 4 (1998): 249–66.

———. *Exposición Crítica de la Filosofía de Gottlob Frege*. Santo Domingo: 1985.

———. "Husserl's Relevance for the Philosophy and Foundations of Mathematics." *Axiomathes* VIII, nos. 1–3 (1997): 125–42.

———. "Necessità a posteriori e contingenza a priori in Kripke: alcune note critiche." *Nominazione* 2 (June 1981): 205–19.

———. "On Necessity and Existence." *Diálogos* 68 (1996): 57–62.

———. "On the Semantics of Mathematical Statements." *Manuscrito* 11, no. 1 (1996): 149–75.

———. "Review of Schirn M. (ed.) *Studies on Frege*." *Diálogos* 38 (November 1981): 157–83.

Rubin, H., and Rubin, J. E. *Equivalents of the Axiom of Choice*. Amsterdam: North Holland, 1963.

———. *Equivalents of the Axiom of Choice II*. Amsterdam: North Holland, 1985.

Russell, Bertrand. *Essays in Analysis*. Ed. D. Lackey. London: Allen and Unwin, 1973.

———. *Introduction to Mathematical Philosophy.* London: Allen and Unwin, 1919.
———. "Knowledge by Acquaintance and Knowledge by Description." In his *Mysticism and Logic,* 200–21.
———. *Logic and Knowledge.* London: Allen and Unwin, 1956.
———. "Mathematics and Metaphysicians." In his *Mysticism and Logic,* 75–95.
———. "Meinong's Theory of Complexes and Assumptions." *Mind* 13 (1904). Reprinted in his *Essays on Analysis,* 21–76.
———. "My Mental Development." In Schilpp, ed., 3–20.
———. *My Philosophical Development.* London: Allen and Unwin, 1975 (1959).
———. *Mysticism and Logic.* London: Allen and Unwin, 1986 (1917).
———. "On Some Difficulties in the Theory of Transfinite Numbers and Order Types." *Proceedings of the London Mathematical Society* 4, series 2, part 1, 29–53. Reprinted in his *Essays in Analysis.*
———. "Review of A. Meinong's *Über die Bedeutung des Weberschen Gesetzes.*" *Mind* 8 (1899): 251–56.
———. "Review of A. Meinong's *Über die Erfahrungsgrundlagen unseres Wissens.*" *Mind* 15 (1906): 412–15.
———. "Review of A. Meinong's *Über die Stellung der Gegenstandstheorie im System.*" *Mind* 16 (1907): 436–39. Reprinted in *Essays on Analysis,* 89–93.
———. "Review of A. *Meinong's Untersuchungen zur Gegenstandstheorie und Psychologie.*" *Mind* 14 (1905): 530–38. Reprinted in *Essays on Analysis,* 77–88.
———. *Principles of Mathematics.* New York: Norton, 1903.
———. *Problems of Philosophy.* Oxford: Oxford University Press, 1967 (1912).
Russell, Bertrand, and Alfred N. Whitehead. *Principia Mathematica to *56.* 2nd ed. Cambridge: Cambridge University Press, 1964 (1927, 1910).
Scanlon, John. "'*Tertium Non Datur.*' Husserl's Conception of a Definite Multiplicity." In Seebohm, et al., eds., 139–47.
Schilpp, Paul, ed. *The Philosophy of Bertrand Russell.* Library of Living Philosophers, vol. V. La Salle, IL: Open Court, and Northwestern: Evanston, IL, 1944.
Schirn, Matthias. "Axiom V and Hume's Principle in Frege's Foundational Project." *Diálogos* 66 (1995): 7–20.
Schirn, Matthias, ed. "Identität und Identitätsaussage bei Frege." In *Studies on Frege,* vol. 2, M. Schirn, ed., 181–215.
———, ed. *Frege: Importance and Legacy.* Berlin: de Gruyter, 1996.
———, ed. *The Philosophy of Mathematics Today.* New York: Oxford University Press, 1998.
———, ed. *Studies on Frege.* 3 vols. Stuttgart-Bad Cannstatt: Frommann-Holzboog, 1976.
Schmit, Roger. *Husserls Philosophie der Mathematik: platonische und konstruktivische Momente in Husserls Mathematik Begriff.* Bonn: Bouvier, 1981.
Schuhmann, Karl. *Husserl-Chronik.* The Hague: M. Nijhoff, 1977.
———. "Husserls doppelter Vorstellungsbegriff: die Texte von 1893." *Brentano Studien* 3 (1990/91): 119–36.
Sebestik, Jan. *Logique et mathématique chez Bernard Bolzano.* Paris: Vrin, 1992.
Seebohm, Thomas, et al., eds. *Phenomenology and the Formal Sciences.* Dordrecht: Kluwer, 1991.
Shapere, Dudley. *Reason and the Search for Knowledge.* Dordrecht: Reidel, 1984.
Shapiro, Stewart. *Foundations without Foundationalism.* Oxford: Oxford University Press, 1991.
Shwayder, David. "On the Determination of Reference by Sense." In *Studies on Frege,* vol. 3, M. Schirn, ed., 85–95.

Simons, Peter. "Three Essays in Formal Ontology." In Smith, ed., 111-260.
———. "Meaning and Language." In Smith and Smith, eds., 106–37.
Sluga, Hans. *Gottlob Frege.* London: Routledge and Kegan Paul, 1980.
———. "Semantic Content and Cognitive Sense." In Haaparanta and Hintikka, eds., 47–64.
Smith, Barry. *Austrian Philosophy: The Legacy of Franz Brentano.* La Salle, IL: Open Court, 1994.
Smith, Barry, and Kevin Mulligan. "Pieces of a Theory." In Smith, ed., 15–109.
Smith, Barry, ed. *Parts and Moments.* Munich: Philosophia, 1982.
Smith, Barry, and David W. Smith, eds. *The Cambridge Companion to Husserl.* Cambridge: Cambridge University Press, 1995.
Smith, David W., and Ronald McIntyre. *Husserl on Intentionality.* Dordrecht: Reidel, 1982.
Solomon, Robert. "Husserl's Concept of the Noema." In Elliston and McCormick, eds., 168–81.
———. "Sense and Reference: Frege and Husserl." *International Philosophical Quarterly* 10 (1970): 387–401.
Spiegelberg, Herbert. *The Context of the Phenomenological Movement.* The Hague: M. Nijhoff, 1981.
———."Franz Brentano (1838–1917): Forerunner of the Phenomenological Movement." In his *The Phenomenological Movement,* 27–48.
———. *The Phenomenological Movement.* 3rd ed. The Hague: M. Nijhoff, 1982 (1960, 1965).
Suszko, R. "The Reification of Situations." In Wolenski, ed., 247–70.
Tarski, Alfred. "The Concept of Truth in Formalized Languages." In his *Logic, Semantics and Metamathematics,* 152–278. Translation of his "Der Wahrheitsbegriff in den Sprachen der deduktiven Disziplinen."
———. *Logic, Semantics and Metamathematics.* 2nd ed. Indianapolis, IN: Hackett, 1983 (1956).
Thiel, Christian. "'Einleitung' to the Centenarausgabe of Frege's *Die Grundlagen der Arithmetik.*" Hamburg: Meiner, 1986.
———. "Gottlob Frege: Die Abstraktion." In *Studies on Frege,* vol. 1, M. Schirn, ed., 243–64.
———. *Sense and Reference in Frege's Logic.* Dordrecht: Reidel, 1968. Translation of his *Sinn und Bedeutung in der Logik Gottlob Freges,* Meisenheim am Glan: Anton Hain, 1965.
Tieszen, Richard. "Gödel's Path from the Incompleteness Theorems (1931) to Phenomenology (1961)." *The Bulletin of Symbolic Logic* 4, no. 2 (June 1998): 181–203.
———. "Gödel's Philosophical Remarks on Logic and Mathematics." *Mind* 107, no. 425 (January 1998): 219–32.
———. "Kurt Gödel and Phenomenology." *Philosophy of Science* 59 (1992): 176–94.
———. *Mathematical Intuition.* Dordrecht: Kluwer, 1989.
Tiles, Mary. *The Philosophy of Set Theory, An Historical Introduction to Cantor's Paradise.* Oxford: Blackwell, 1989.
Tragesser, Robert. "How Mathematical Foundation All But Came About: A Report on Studies Toward a Phenomenological Critique of Gödel's Views on Mathematical Intuition." In Seebohm, et al., eds., 195–213.
van Heijenoort, Jan, ed. *From Frege to Gödel: A Source Book in Mathematical Logic, 1879–1931.* Cambridge, MA: Harvard University Press, 1967.
Wang, Hao. "The Axiomatization of Arithmetic." *Journal of Symbolic Logic* 22, no. 2

(June 1957): 145–58.

———. *Beyond Analytic Philosophy*. Cambridge, MA: M.I.T. Press, 1986.

———. *From Mathematics to Philosophy*. London: Routledge and Kegan Paul, 1974.

———. *A Logical Journey, From Gödel to Philosophy*. Cambridge, MA: M.I.T. Press, 1996.

———. *Reflections on Kurt Gödel*. Cambridge, MA: M.I.T. Press, 1987.

Webb, Judson. "Hilbert's Formalism and Arithmetization of Mathematics." *Synthese* 110 (1997): 1–14.

Weidemann, Hermann. "Aussagesatz und Sachverhalt ein Versuch zur Neubestimmung ihres Verhältnisses." *Grazer Philosophische Studien* 18 (1982): 75–99.

Weinberg, Joan. "Abstraction in the Formation of Concepts." *Dictionary of the History of Ideas*. Ed. P. Wiener, 1–9.

Weiner, Joan. "Putting Frege in Perspective." In Haaparanta and Hintikka, eds., 9–27.

Whitehead, Alfred North. *An Introduction to Mathematics*. Oxford: Oxford University Press, 1958 (1911).

Wiener, P., ed. *Dictionary of the History of Ideas*. New York: Charles Scribner's Sons, 1968.

Wiggins, David. *Sameness and Substance*. Oxford: Blackwell, 1980.

Willard, Dallas. "Husserl on a Logic That Failed." *Philosophical Review* 89, no. 1 (1980): 46–64.

———. *Logic and the Objectivity of Knowledge*. Athens, OH: University of Ohio Press, 1984.

Wolenski, Jan, ed. *Philosophical Logic in Poland*. Dordrecht: Kluwer, 1994.

Wright, Crispin. *Frege's Conception of Numbers as Objects*. Aberdeen: Aberdeen University Press, 1983.

Zeller, Eduard. *Platonische Studien*. Tübingen: Osiander, 1839.

———. *Die Philosophie der Griechen (Sokrates und die Sokratiker. Plato und die alte Akademie)*. 3rd ed. Leipzig: Fues's Verlag, 1875.

———. *Die Philosophie der Griechen (Aristoteles und die alten Peripatetiker)*. 3rd ed. Leipzig: Fues's Verlag, 1879.

Index